Food Systems Sustainability and Environmental Policies in Modern Economies

Abiodun Elijah Obayelu
Federal University of Agriculture - Abeokuta (FUNAAB), Nigeria

A volume in the Advances in Environmental Engineering and Green Technologies (AEEGT) Book Series

IGI Global
DISSEMINATOR OF KNOWLEDGE

Published in the United States of America by
IGI Global
Engineering Science Reference (an imprint of IGI Global)
701 E. Chocolate Avenue
Hershey PA, USA 17033
Tel: 717-533-8845
Fax: 717-533-8661
E-mail: cust@igi-global.com
Web site: http://www.igi-global.com

Library of Congress Cataloging-in-Publication Data

Names: Obayelu, Abiodun Elijah, 1972- editor.
Title: Food systems sustainability and environmental policies in modern economies / Abiodun Elijah Obayelu, editor.
Description: Hershey, PA : Engineering Science Reference, 2017.
Identifiers: LCCN 2017019881| ISBN 9781522536314 (hardcover) | ISBN 9781522536321 (ebook)
Subjects: LCSH: Food supply. | Food security. | Environmental policy.
Classification: LCC HD9000.5 .F5996 2017 | DDC 338.1/9--dc23 LC record available at https://lccn.loc.gov/2017019881

This book is published in the IGI Global book series Advances in Environmental Engineering and Green Technologies (AEEGT) (ISSN: 2326-9162; eISSN: 2326-9170)

British Cataloguing in Publication Data
A Cataloguing in Publication record for this book is available from the British Library.

All work contributed to this book is new, previously-unpublished material.
The views expressed in this book are those of the authors, but not necessarily of the publisher.

For electronic access to this publication, please contact: eresources@igi-global.com.

Advances in Environmental Engineering and Green Technologies (AEEGT) Book Series

ISSN:2326-9162
EISSN:2326-9170

Editor-in-Chief: Sang-Bing Tsai, University of Electronic Science and Technology of China Zhongshan Institute, China & Ming-Lang Tseng, Lunghwa University of Science and Technology, Taiwan & Yuchi Wang, University of Electronic Science and Technology of China Zhongshan Institute, China

MISSION

Growing awareness and an increased focus on environmental issues such as climate change, energy use, and loss of non-renewable resources have brought about a greater need for research that provides potential solutions to these problems. Research in environmental science and engineering continues to play a vital role in uncovering new opportunities for a "green" future.

The **Advances in Environmental Engineering and Green Technologies (AEEGT)** book series is a mouthpiece for research in all aspects of environmental science, earth science, and green initiatives. This series supports the ongoing research in this field through publishing books that discuss topics within environmental engineering or that deal with the interdisciplinary field of green technologies.

COVERAGE

- Policies Involving Green Technologies and Environmental Engineering
- Green Technology
- Electric Vehicles
- Contaminated Site Remediation
- Pollution Management
- Sustainable Communities
- Biofilters and Biofiltration
- Green Transportation
- Industrial Waste Management and Minimization
- Air Quality

IGI Global is currently accepting manuscripts for publication within this series. To submit a proposal for a volume in this series, please contact our Acquisition Editors at Acquisitions@igi-global.com or visit: http://www.igi-global.com/publish/.

The Advances in Environmental Engineering and Green Technologies (AEEGT) Book Series (ISSN 2326-9162) is published by IGI Global, 701 E. Chocolate Avenue, Hershey, PA 17033-1240, USA, www.igi-global.com. This series is composed of titles available for purchase individually; each title is edited to be contextually exclusive from any other title within the series. For pricing and ordering information please visit http://www.igi-global.com/book-series/advances-environmental-engineering-green-technologies/73679. Postmaster: Send all address changes to above address.

Titles in this Series

For a list of additional titles in this series, please visit:
https://www.igi-global.com/book-series/advances-environmental-engineering-green-technologies/73679

Innovative Strategies and Frameworks in Climate Change Adaptation Emerging Research ...
Alexander G. Flor (University of the Philippines Open University, Philippines) and Benjamina Gonzalez Flor (University of the Philippines, Phiippines)
Engineering Science Reference • ©2018 • 165pp • H/C (ISBN: 9781522527671) • US $165.00

Utilizing Innovative Technologies to Address the Public Health Impact of Climate Change ...
Debra Weiss-Randall (Florida Atlantic University, USA)
Engineering Science Reference • ©2018 • 295pp • H/C (ISBN: 9781522534143) • US $185.00

Economical and Technical Considerations for Solar Tracking Methodologies and ...
S. Soulayman (Higher Institute for Applied Sciences and Technology, Syria)
Engineering Science Reference • ©2018 • 647pp • H/C (ISBN: 9781522529507) • US $245.00

Hydrology and Best Practices for Managing Water Resources in Arid and Semi-Arid Lands
Christopher Misati Ondieki (Kenyatta University, Kenya) and Johnson U. Kitheka (South Eastern Kenya University, Kenya)
Engineering Science Reference • ©2018 • 266pp • H/C (ISBN: 9781522527190) • US $205.00

Computational Techniques for Modeling Atmospheric Processes
Vitaliy Prusov (University of Kyiv, Ukraine) and Anatoliy Doroshenko (National Academy of Sciences, Ukraine)
Information Science Reference • ©2018 • 460pp • H/C (ISBN: 9781522526360) • US $205.00

Advanced Nanomaterials for Water Engineering, Treatment, and Hydraulics
Tawfik A. Saleh (King Fahd University of Petroleum and Minerals, Saudi Arabia)
Engineering Science Reference • ©2017 • 384pp • H/C (ISBN: 9781522521365) • US $200.00

Handbook of Research on Inventive Bioremediation Techniques
Jatindra Nath Bhakta (University of Kalyani, India)
Engineering Science Reference • ©2017 • 624pp • H/C (ISBN: 9781522523253) • US $260.00

For an entire list of titles in this series, please visit:
https://www.igi-global.com/book-series/advances-environmental-engineering-green-technologies/73679

IGI Global
DISSEMINATOR OF KNOWLEDGE

701 East Chocolate Avenue, Hershey, PA 17033, USA
Tel: 717-533-8845 x100 • Fax: 717-533-8661
E-Mail: cust@igi-global.com • www.igi-global.com

Editorial Advisory Board

List of Reviewers

Edgar Muhoyi, *University of Zimbabwe, Zimbabwe*
I. J. Ogwu, *University of Abuja, Nigeria*
Olugbire Oluseyi Olutoyin, *Forestry Research Institute of Nigeria, Nigeria*
Kabir Kayode Salman, *University of Ibadan, Nigeria*
F. A. Sowemimo, *University of Ibadan, Nigeria*
Adigun Toyin, *Landmark University, Nigeria*

Table of Contents

Detailed Table of Contents

Food is indispensable to life. It plays an important role in the economy but what is not well known is the impact of production and consumption that food has on the environment. The nexus of food systems and the environment are complex and driven by many economic, socio-cultural, and environmental factors considered to be important in the contemporary global arena. As the world population grows, there is an increased demand on the already stretched food system and fragile environment. Processes along the food chain from agricultural production to food consumption produce outputs other than consumable food that are returned to the natural environment such as pollution or waste. This chapter sheds light on the links in food systems and environment in developing countries. A major finding is that the existing food systems that were supposed to produce adequate food for all are placing major stress on environmental assets including soil, water, fisheries, and biodiversity. For food systems to be sustainable, all hands must be on deck.

Recent studies related the link between food consumption and impacts on environment and health. These may present variations according to the dietary patterns of different populations. This chapter assesses the impacts of six dietary patterns while emphasizing protein overconsumption and sustainability of food systems in a world where one billion people are hungry and several more suffer from conditions related to obesity. The chapter shows the nutritional disparity existent in different dietary patterns and potential to make changes. Changes in dietary patterns are an opportunity to contribute for environmental and health benefits. The analysis was based on a set of environmental indicators such as greenhouse gas (GHG) emissions and land use demand, while providing a nutritional balance. The methodology comprehended a life cycle assessment in order to quantify the GHG emissions and the land use demand for food production. Finally, a review is made to focus on the benefits of shifting from current diet patterns to more sustainable ones, such as the Mediterranean.

This section connects the sustainability of agricultural systems with climate change and other environmental issues.

Small-scale irrigation schemes (SSIS) in developing countries have been crucial, but the evidence about their performance has not been sufficiently analyzed. This chapter documents such evidence by reviewing and classifying the performance indicators. It also assesses literature on whether there are discernible trends in the efficiency of SSIS, identifies and classifies SSIS constraints, and characterizes various channels through which SSIS might affect poverty. Objectives are achieved via a systematic review of literature from 1990 to 2017. Results indicate a lack of standardization of irrigation performance indicators, and there is evidence that irrigation has boosted agricultural performance. Even though SSIS were associated with higher productivity than rain-fed agriculture, they performed below their full potential due to undervaluation of irrigation water by irrigation authorities, farmer characteristics, costs, institutional setups, the policy environment, and design, cultural, community, and environmental issues. SSIS are important tools for poverty reduction, and relevant policy implications are outlined.

This chapter explains the implication of South Africa's transport fuel 2% blending. Using dry grain sorghum as feedstock with guaranteed food security has lower emission of 24.93kg/ha with emerging farmers who constituted 30% of the suppliers with a 3-year payback period. Using irrigated sorghum with food security as a priority has a relatively lower emission level of 11.47kg/ha from emerging farmers with a 9-year payback period. Using sugar beet has lower emission level of 0.12kg/ha with emerging farmers and a 3-year payback period. Soil organic content has significant influence on emissions from land use practices. Commercial sugar beet ethanol production caused high emission (4.84kg/ha) but has a short payback period of only 2 years which enhanced household food consumption by 12.5% and 31.50% under food security not a priority and food security as a priority, respectively. In all, grain sorghum food and beverage gross domestic product (GDP) increased, respectively, by 8%, 0.19%, and 0.23% under food security as not a priority, and increased by 20.83%, 0.44%, and 0.61% in opposite scenario, respectively.

Section 3
Sustainable Consumption and Health: Contemporary Issues and Policies

This section discusses the contemporary issues and policies on sustainable consumption and health.

In recent decades, we have seen the rise of the sustainable food consumption field and its push for disciplinary space in development studies. This chapter turns to the original impetus of sustainable food consumption and the question of how

The success of sustainable agriculture depends not just on the motivations, skills, and knowledge of individual farmers, but on action taken by the agricultural extension agency in harmonising the activities of other agencies involved in agricultural development. This chapter looks into the coordinating roles of agricultural extension with other agencies involved in agricultural development towards sustainable food systems in Nigeria. The agricultural extension agency needs to coordinate other agencies in rural development to ensure unity of purpose and avoid duplication of efforts. Agricultural extension is considered the best institution to coordinate other agencies involved in agricultural development towards sustainable food system in Nigeria because of the edge agricultural extension has at grassroots level in light of community organizing and empowerment role. Government in all tiers should support the agricultural extension agency by ensuring that all development partners liaise with the agency before carrying out their work to enhance sustainable rural development.

The combined effects of declining soil fertility, continuous mono-cropping, poor crop residues management, and limited resources are considered the major constraints to increased crop productivity in Sub-Saharan Africa. It is for this reason that most African governments in Sub-Saharan Africa have been implementing farm input support programmes to boost smallholder production. While substantial amounts of resources are committed to support such programmes, evidence suggests that the increased use of modern inputs such as inorganic fertilizers on the main staple food crops appear to be only marginally profitable or even unprofitable. There is a renewed realization that the use of fertilizer input alone to raise farm productivity is likely to be impeded, if sufficient attention is not given to complementary interventions such as integrated soil fertility management technologies and extension services. This chapter provides evidence from several African countries on the role of complementary interventions in enhancing profitability, effectiveness, and efficiency with which farm inputs such as inorganic fertilizer and improved seed are applied.

It highlights their rights, obligations, and responsibilities, as well as voluntary guidelines and implications for a rights-based approach to food. The chapter also reviews policies that have been developed toward ensuring a right to food in Nigeria.

Section 6
Farmland Conservation and Environmental Stewardship

This section deals with conservation of farmland and changes in the environment in the affecting food systems.

The production of cereals, tubers, and vegetables largely depends on the application of organic and inorganic fertilizers to offset their nutrients requirement. In this chapter, the authors identify different soil fertility management practices the maize farmers are using and examine the economic benefits of such practices in maize production. To complete the study, 237 maize farmers across Kogi and Kwara States, Nigeria were investigated. Descriptive statistics, gross margin, and multinomial logit tools were used to analyze the data. The results show that majority of the maize farmers (41.40%) use only inorganic material. Labour employed in the application of fertility materials, distance to the source of fertility materials, the quantity of seed planted, educational attainment, and gender of the maize farmer were the determinants of the use of fertility management practices relative to integrated soil fertility management (ISFM). The analyses of the results show that the use of ISFM for maize production is the most profitable method with a profitability ratio of 2.29.

The chapter tested the hypotheses that the National Fadama Development Project had no significant effect on the income of Fadama farmers and that there is no difference in the profit of sugar cane and rice, the two major crops cultivated in the project. Primary data were collected in 2007 through the use of questionnaire randomly administered to 150 farmers in Niger state who are Fadama beneficiaries. Data collected were analyzed using descriptive statistics, t-test, and gross margin analysis. The results of the study showed that sugar cane gave the highest gross

Foreword

It is my unreserved pleasure to write this forward to the book of readings, *Food Systems Sustainability and Environmental Policies in Modern Economies*, as a valuable and instructive text for all actors in consumer economics and indeed issue around environment as applied to food production and sustainability all over the world. My salutation goes to the insightful editor of this book for thinking out the concept and engineering of this text. I am convinced, after going through the write-ups that this piece will deliver the needed information and fill the niche needs of researchers, graduate and undergraduate students interested in this very important field germane to the human capital existence and development.

This book is divided into six sections comprising twelve chapters which can be summarised into basic information on food sustainability options (Chapters 1, 2 and 5), food production and agricultural systems (Chapters 3, 9, and 11), managing food resources (Chapters 1 and 6), food for health (Chapters 6 and 10), alternative uses of foods for biofuels and implication on food security (Chapters 4), as well as management of agricultural value chain development (Chapters 3, 7, 8, 9, and 12). Of special interest are Chapters 1 (which draws the ramifying implications of the food systems, sustainability, environmental factors governing food systems along the value chain, among others), Chapter 2 (which deals with the effect of changing dietary pattern and the environment vis-à-vis the consumers' accessibility and the divide between food inaccessibility and accessibility as well as the greenhouse effects) and Chapter 4 (which deals with biofuels production from staple foods on farmers' income and food security as a prospect towards green economy in Africa)

I have implicit confidence in the quality of erudite authors patronised in this text as well as their literary acumen add a lot of value to the book and will stimulate readership of the diverse treatise which draws examples from Africa's situation, thus making the contents applicable for African readership in the first instance and the world at large.

I therefore, congratulate the editor of this book for yet another precocious effort in the annals of food systems development, sustainability and environmental effects, towards the development of the broad field of consumer economics and a veritable and invaluable addition to knowledge as well as a worthwhile and educative expose for seekers of knowledge (researchers, graduate and undergraduate students) in this field, for academic development as well as for aesthetics.

Idris A. Ayinde
Federal University of Agriculture – Abeokuta (FUNAAB), Nigeria

Preface

INTRODUCTION

Food systems play a key economic and social role in all societies and are fundamental to ensuring sustainable development. Sustainable food systems are critical to improve food security and nutrition as well as for poverty alleviation but the current food system in most parts of the world is destroying the very environment on which future food production rely. The world's food system is facing unprecedented challenges. The global population is set to reach 8.1 billion in 2025, with 95% of population growth driven by developing countries. The food system must ensure this growing population has access to the nutrition it needs to flourish, especially as climate change re-shapes agricultural production. This means tackling the twin nutritional challenges facing the world: hunger and nutrient deficiencies, along with unhealthy diets and obesity. In the presence of climate change, increasingly challenging land-use conflicts, rising health and social costs on from both individual and societal, demographic change and a growing global population, sustainability problems arising from food systems will likely become more serious in the future. The absence of sustainability measures implies a social and moral responsibility governance, policy makers, farmers, and consumers to do their part in combating hunger. Nature has gracefully provided the necessary inputs to feed us, and we have in many occasions taken these precious gifts for granted. Working toward a sustainable food supply is crucial, but it will not be achieved without a shift in our current food systems which are not sustainable. Food systems affect everything from the nutrient value of crops and food products, to the health of livestock, the environment, and even the economy. Efforts are needed to understand other mechanisms that lead to an unhealthy diet; including gaps in environmental and policy supports to enable more locally based food distribution

This book, "Food Systems' Sustainability and Environmental Policies in Modern Economies" contains materials that will enable the readers to be able identify the interactions between food systems, contemporary issues and environmental policies, as well as the opportunities for effective policy, fiscal, social and/or technical interventions.

GLOBAL CHALLENGES ON FOOD SYSTEMS AND THE NEED FOR URGENT SOLUTION

With limited global resources, and in the face of environmental changes, meeting future food security challenges will first require a shift in thinking from just 'producing food' (and other sectoral interests) to food systems. Food security no doubt in many developing and emerging countries is under increasing stress due to convergence of issues (such as: climate change, resource depletion, dysfunctional farm policies, and loss of biodiversity) which are threats to the availability of healthy food for an ever-increasing population. A food system embraces all the elements (environment, people, inputs, processes, infrastructure, institutions, markets and trade) and activities that relate to the production, processing, distribution and marketing, preparation and consumption of food and the outputs of these activities, including socio-economic and environmental outcomes. A sustainable food system (SFS) is a dynamic concept that embraces the interconnectedness of all the food-related activities and the environment within which these activities occur. Conditions that ensure sustainability in food systems vary widely across countries and regions, as well as across different stakeholders (e.g. poor or marginal food producers who may chronically or seasonally lack productive capacity to cover own food needs, or urban low-income and non-food producers). Sustainable food systems take into account environmental, health, social and economic concerns in the production and consumption of food. Sustainable food production for example, uses systems and practices that can be continued indefinitely without depleting non-renewable resources (e.g., soil, energy, biological diversity), causing ecological harm, or widening social inequities (within and across communities, countries, or generations). Sustainable food systems need to increase agricultural productivity, improve climate resilience, and reduce greenhouse gas emissions for agriculture and related land use change. Because of the complex social-ecological systems involving multiple interactions between human and natural components in food systems, research studies on this important issue are necessary in order to be able to redirect our food systems and policies towards better-adjusted goals, and improved food and nutritional security and societal welfare. Globally, food systems are responsible for 60% of global terrestrial biodiversity loss, around 24% of the global greenhouse gas emissions, 33% of degraded soils, the depletion of 61% of 'commercial' fish populations, and the overexploitation of 20% of the world's aquifers. These pressures on our natural resource base are expected to significantly increase with population, urbanization and super-marketization trends, as well as dietary shifts to more resource-intensive food. By 2050, an expected 40% of the world population will be living in severely water-stressed river basins and greenhouse gas emissions from agriculture may increase from 24% to 30%. There are also a number of alarming disparities worldwide that reveal the impacts of current food systems on our health. Nearly 800 million people are hungry; over 2 billion suffer from micronutrient deficiencies, while over 2 billion people are obese.

Building sustainable food systems has become a popular motto and a major endeavor to redirect our food systems and policies towards better-adjusted goals and improved societal welfare. Sustainable food systems need support from a vast range of stakeholders from the farmers to the final consumer, from governments to private sector to civil society. Policy measures for sustainable food systems, should increase agricultural productivity and gender sensitive agriculture production, enhance climate resilience, reduce greenhouse gas emissions from agriculture and related land use change, improve nutrition, strengthen value chains and improve market access. Sustainable food systems may therefore depends on the success of the Zero Hunger Challenges elements, such as zero food waste, 100 percent access to food and increasing smallholder's income. This book contributes to the emerging literature on food system transformation.

ORGANIZATION OF THE BOOK

The book is made up of 12 chapters distributed among six sections.

Section 1: Overview and General State of Food Systems

The first section provides an overview of food system in developing countries. In Chapter 1, Obayelu A. Elijah gives an account of the links in food system, the environment and sustainability in developing countries. He takes the stock of what we know and some of the things we do not know so far about food systems. Gomes, Pedro Pinheiro in the 2nd chapter details through grey literature the impacts of six dietary patterns on consumption and sustainability of food systems

Section 2: Sustainable Agricultural Systems, Climate Change, and Environmental Issues

This 2nd section is very important because of the effects of climate change on agricultural production. Edgar Muhoyi and Josue Mbonigaba in the 3rd chapter demonstrate how small-scale irrigation schemes (SSIS) can be used to sustain agricultural systems in the presence of climate change and environmental challenges. The 4th chapter by Labintan Constant, Winkler, Harald and Obayelu Abiodun Elijah investigates the implications of the use of biofuel production in transitive low carbon development countries with a case study of South Africa. The aforementioned according to the authors can have implications on food security and agrarian transformation in African countries.

Section 3: Sustainable Consumption and Health – Contemporary Issues and Policies

This section of the book zeros in on contemporary issues and policies relating to sustainable consumption and health. Henry E. Alapiki and Luke Amadi in Chapter 5 provide details information on sustainable food consumption in the Neo Liberal Order stating the challenges and policy implications. Josue Mbonigaba in Chapter 6 brings on board the effects of unsustainable production and consumption of food on health and policy across developed and less developed countries.

Section 4: System Innovations and the Transition to Sustainable Food Systems in Developing and Emerging Economies

This section embarks on an exploration of some key innovations in the course of transition to sustainable food systems in developing and emerging countries. Akinnagbe Oluwole Matthew in Chapter 7 discusses how harmonization of coordinating roles of agricultural extension with other agencies involved in agricultural development can help in sustainable food systems in Nigeria. Joseph S. Kanyamuka, Charles B.L Jumbe and Jacob Ricker-Gilbert write in Chapter 8 on the making agricultural input subsidies more effective and profitable in Africa. They established that despite substantial amounts of resources committed to support agricultural subsidies programmes in Africa, evidence suggests that the increased use of modern inputs such as inorganic fertilizers on the main staple food crops appear to be only marginally profitable or even unprofitable. In Chapter 9, Evans S. Osabuohien, Uchechukwu E. Okorie and Romanus A. Osabohien examined the importance of indigenous institutional arrangements in rice production and processing activities. The authors suggest the need for actively involving the rice farmers through the Rice Growers Association of Nigerian (RGAN) for an effective rice production and processing system in Nigeria.

Section 5: Right Approaches to Sustainable Food System in Developing and Emerging Countries

This section which happens to be the shortest but an important section focuses on the focuses on the right things to do for sustainable food system. Ajayi, Clementina Oluwafune, and Adenegan, in Chapter 10, examine the need for a rights-based approach to food and nutrition security in Nigeria.

Section 6: Farmland Conservation and Environmental Stewardship

The section centres on issues relating to farmland conservation. Ogwu, I. J., Omotesho, O. A. and Muhammad-Lawal, A. in Chapter 11 investigate the effects of different soil fertility management practices on maize production. The authors suggest that Integrated Soil Fertility Management (ISFM) for maize production is the most profitable method. Chapter 12 by Abu, G. A., Okpachu, A. S. and Mallam, M. examine the economic impact assessment of the national fadama development project on rural farming communities in Niger State of Nigeria. The authors concludes that Fadama projects positively impacted on beneficiaries by sustainably increasing farm income

Abiodun Elijah Obayelu
Federal University of Agriculture – Abeokuta (FUNAAB), Nigeria

Acknowledgment

The editor appreciates the Almighty God who has made my dream of this book project come true and my family members for the moral support and a platform that enabled the completion of this book.

I wish to acknowledge the concerted efforts of all the contributing authors, the reviewers, and the members of the Editorial Advisory Board (EAB). The idea of the book project stemmed from the fact that countries in the world keep producing food hence the level of hunger, food insecurity, nutritional insecurity and environmental challenges keep increasing. The questions that therefore came to my mind were: what kinds of food systems do we have? Are they sustainable? Do we have environmental policies? Are the environmental policies supportive to good food systems? Are there new solutions to sustainable food systems? I packaged these thoughts and wrote a proposal for IGI Global and the ideas were accepted, which lead to the creation of a call for proposals to start gathering chapters for the book.

Thanks to IGI Global for the intellectual stimulation and various networks that helped in widely disseminating the information, especially the African Growth and Development Policy Modelling Consortium (AGRODEP).

Abiodun Elijah Obayelu
Federal University of Agriculture – Abeokuta (FUNAAB), Nigeria

Section 1

Overview and the General State of Food Systems

This section provides an overview of food systems and the current issues in the literature.

Chapter 1
Integrating Environment, Food Systems, and Sustainability in Feeding the Growing Population in Developing Countries

Abiodun Elijah Obayelu
Federal University of Agriculture – Abeokuta (FUNAAB), Nigeria

ABSTRACT

Food is indispensable to life. It plays an important role in the economy but what is not well known is the impact of production and consumption that food has on the environment. The nexus of food systems and the environment are complex and driven by many economic, socio cultural, and environmental factors considered to be important in the contemporary global arena. As the world population grows, there is an increased demand on the already stretched food system and fragile environment. Processes along the food chain from agricultural production to food consumption produce outputs other than consumable food that are returned to the natural environment such as pollution or waste. This chapter sheds light on the links in food systems and environment in developing countries. A major finding is that the existing food systems that were supposed to produce adequate food for all are placing major stress on environmental assets including soil, water, fisheries, and biodiversity. For food systems to be sustainable, all hands must be on deck.

DOI: 10.4018/978-1-5225-3631-4.ch001

INTRODUCTION

One of the main challenges facing most developing countries is how to simultaneously provide enough food while conserving the natural resources to produce food for the present and future generations. Fact have emerged that despite production of enough food to feed the population by some countries, almost 800 million people (representing about 12.9 percent of the population in developing regions) are hungry (FAO, IFAD and WFP, 2015); 161 million under-five year olds have been estimated to be stunted (WHO, 2013); and over 2 billion suffer from micronutrient deficiencies, in particular vitamin A, iodine, iron and zinc (Ng, *et al.,* 2014; FAO, 2013). At the same time, the number of overweight/obese people has reached more than 1.4 billion adults globally (representing about 30 percent of the total adult population). Obesity-related health conditions are rising rapidly both in developing and developed countries (WHO, 2015); and around 30 percent of the food produced worldwide (about 1.3 billion tons) is lost or wasted every year (FAO, 2011). This shows that a lot of people are probably not getting enough food and most eat diets poor in quality because the type of food systems currently are not sustainable. Food waste alone represents around 3-5% of global warming impacts, more than 20% of biodiversity pressure, and 30% of all of the world's agricultural land (EU, 2014). A food system governs what we eat; and there has been increasing concern at all levels of governance and in different policy sectors, civil society, academia and business that the food system today is not sustainable and endangers both health and the future of the planet (Kickbusch, 2010). Food system is said to be sustainable if it ensures food security and nutrition for all in such a way that the economic, social and environmental bases to generate food security and nutrition of future generations are not compromised (HLPE, 2014). Food system therefore is at the heart of the 2030 Agenda for Sustainable Development. Food production places enormous demand upon the environment. Food systems vary significantly from country to country in terms of actors, technology and type of resources used (UNEP, 2016). Food systems comprise all aspects of food production (the way the food is grown or raised; harvested or slaughtered; processed, packaged, or otherwise prepared for consumer purchase), food distribution (where and how the food is sold to consumers and how the food is transported) and food consumption (Ericksen, 2008; Farmar-Bowers, 2013; Pearson *et al.,* 2014). The various outcomes of the food system contribute to food security and health on the one hand and environmental degradation on the other (Ericksen, 2008).

Most of the food we consume is no longer produced in self-sufficient, but travels (and often a long way) from producer to consumer. Most of the food consumed are processed elsewhere and arrives in packaged forms through trade or the exchange of technologies or resources.

This chapter provides an overview of the current status of food systems, their interaction with the environment, and an assessment of the sustainability of current agricultural practices and effects on food production, underlying causes of unsustainable production and consumption patterns.

SITUATION ANALYSIS OF CURRENT FOOD SYSTEMS IN DEVELOPING COUNTRIES

The form of food system is a major issue in the politics of sustainable consumption and production (SCP) because of its impact on the environment, individual and public health, social cohesion, and the economy. Current food systems are not delivering food security and healthy food for everyone, nor are they sustainably using the limited resources. Emphasis is placed on 'modern' food systems in industrialized and emerging regions while 'traditional' food systems are still more prominent in rural areas of developing countries. Modern food systems are observed to have resulted in polluted soil, air, and water; eroded soil; dependence on imported oil; and loss of biodiversity. This variety in food systems, in combination with the social and natural environment in which they operate, has important implications on the possible pathways towards sustainable food systems and on the logic of intervention. In developing regions, there is a rapidly evolving replacement of traditional food systems by modern food systems. This trend is driven by macro-trends such as urbanization, increased wealth and other socio-economic and demographic developments. Countries with the biggest nutritional challenges are India, Nigeria and South Africa with the Food Sustainability Index (FSI) of 40.51, 45.60 and 49.34 respectively (Economist Intelligence Unit, 2015). India has a high rate of nutritional deficiency, and high prevalence of under- and malnourishment, while South Africa has a high adoption of fast foods, and limited purchasing power for buying healthy foods.

One of the findings by Liverpool-Tassie *et al.,* (2016) is that urban and rural households in Nigeria have rapidly transforming diets. Consumption has diversified greatly, shifting beyond self-sufficiency to heavy reliance on food purchases and with a heavy shift into consumption of processed foods'. Food system transformation is characterized with supermarket growth in cities and imports still dominate the processed foods sector. Supermarkets can lead to higher consumption of processed foods and may be associated with higher rates of obesity among adults (Rischke *et al.,* 2015). The extent of supermarket penetration in food markets is therefore relevant for food policy.

Minten *et al.* (2013); Hassen *et al.* (2016) documented Ethiopia's food systems and how changes in the systems were driven in part by public investments to develop modern varieties of teff which is the main staple cereal. Ghana's traditional food systems are based on varieties of staple crops including cassava, maize, millet, plantain, rice, and sorghum. Ghanaian diets are just shifting across staples as people move to cities. The urban consumers care about packaging as a sign of product quality (Hollinger and Staatz, 2015).

MAJOR FINDINGS

Food Systems-Environment-Sustainability Nexus

Drawing from the United Nations Secretary General's Zero Hunger Challenge, a food system is defined as a system that embraces all the elements (environment, people, inputs, processes, infrastructure, institutions, markets and trade) and activities that relate to the production, processing, distribution and marketing, preparation and consumption of food and the outputs of these activities, including socio-economic and environmental outcomes (HLPE, 2014). (See Figure 1). A productive food system is one that produces crops, livestock, and fisheries using the fewest resources possible. The food we grow, produce, consume, trade, transport, store and sell is the essential connecting thread between people, and the environment. The food we produce and the food choices we make are creating stresses on our environment through climate, wildlife and habitats, driving unethical trade agreements, challenging animal welfare and putting pressure on water usage. Ericksen (2008) Ingram (2011) and Tendall *et al.*(2015) considered food systems as social-ecological systems that consist of mutually interacting biophysical and social components. It encompasses the activities of food production, processing, distribution, retail, and consumption and involve multidimensional (e.g., social, economic, institutional, environmental political, cultural) processes.

Food systems are inextricably linked to the environment and sustainability. Food environment' being the physical, social and economic surroundings influence what people eat and plays a major role in determining food systems in terms of food consumption pattern. Food companies, restaurants, food vendors and retailers are actively influencing this food environment to tempt people to make certain choices. Sustainability in any given food system encompasses a broad range of factors such as farming practices, local and seasonal sourcing, supply chain policies, consumer decisions, waste and packaging (Ng, 2015). There is no one model of a sustainable food system, but a set of principles that constitute sustainability. Food systems (production, processing, packaging, distribution, retailing, consumption and waste management) are webs of people and the resources and behaviors they affect. It is interconnected with the environment and sustainability. Food system does not only overlap in the area of food production, but also comprise the diverse set of institutions, technologies and practices that govern the way food is marketed, processed, transported, accessed and consumed. The food system is linked to the environment through a range of inputs at each stage and outputs that are returned to the environment as greenhouse gases, waste water, packaging and food waste, as well as land degradation and loss of biodiversity. Sustainable food system according to Pothukuchi and Jufman, (1999) is secure, reliable and resilient to change (including climate change, rising energy prices, etc) and accessible and affordable to all members of society. It is energy efficient, an economic power generator for farmers, whole communities and regions', healthy and safe and environmentally beneficial. Sustainable food system uses creative water reclamation and conservation strategies for agricultural irrigation, balances food imports with local capacity, adopts regionally-appropriate agricultural practices and crop choices, works towards organic farming, contributes to both community and ecological health. It builds soil quality and farmland through the recycling of organic waste, supports multiple forms of urban as well as rural food production, and ensures that food processing facilities are available to farmers and processors, celebrated through community events, markets, restaurants. It also preserves biodiversity in agro-ecosystems as well as in the crop selection; has a strong educational focus to create awareness of food and agricultural issues, and is fairly traded by providing a fair wage to producers and processors locally and abroad.

Sustainable food system ensures that all residents have access to healthy and affordable food options, minimizes the environmental impact of food production and transport; facilitates and encourages local food production and processing; creates local jobs that provide fair working conditions and a living wage; benefits local economies by supporting local food producers, retailers and businesses and maximizes resources through collection and reuse of organics (compost) and other food related by products (i.e. fats, oils, grease). These activities lead to a number of food and nutrition security outcomes including access to, and utilisation of food, in addition to food availability; and to other socioeconomic outcomes (e.g. employment). Food systems crucially depend on the environment: land, soil, water, terrestrial and marine biodiversity, minerals (essential nutrients for crops and animals) and fossil fuels.

A food system therefore also encompasses the interdependent sets of enterprises, institutions, activities and relationships that collectively develop and deliver material inputs to the farming sector, produce primary commodities, and subsequently handle, process, transport, market and distribute food and other agro-based products to consumers. Food systems differ from one location or country to another in terms of actors involved and characteristics of their relationships and activities (UNEP, 2016). Examples of activities that support sustainable food systems: farmers' markets, community gardens, regional food hubs, healthy corner store initiatives and/or programs to increase access to supermarkets, farm-to-school programs, food policy and/or security councils, local food procurement policies, food waste collection programs.

These processes not only produce food but also other outputs that are returned to the natural environment, such as greenhouse gases, waste water, as well as packaging and food waste. The link that exists in food systems, the environment and sustainability are therefore complex. It encompasses scientific, economic, cultural and social factors. Consumers in both developed and developing countries are more interested in knowing what lies beyond their food market. They want to know where the foods come from, and how such food was grown because of the consciousness of their health which is a function of the food they eat. Sustainability of the environment and food systems which is multi-faceted issue has therefore become an important international issue as more foods are needed to meet up with the increasing population. The current systems of increasing food production through over-utilization of resources such as overfishing in many parts of the world has exhausted most of the world's fisheries. Overgrazing is leading to increasing desertification and over-cultivation has degraded soils. Felling of forests is disrupting rainfall and water supplies thereby enhancing climate change. All these activities are altering food system in most countries, with implications for yield, quality and affordability.

The failure to observe the simple principle of sustainability poses an ever-growing threat to food supplies that is needed to feed future generations. Sustainable food production uses systems and practices that can be continued indefinitely without depleting non-renewable resources (such as soil, energy, and biological diversity), causing ecological harm, or widening social inequities (within and across communities, countries, or generations). This by implication means that a sustainable food system does not utilize chemical pesticides, synthetic fertilizers, or genetically modified seeds or participate in practices that might contaminate the air, water, or soil. Sustainable farmers grow a variety of crops, use crop rotation, and manage waste which benefits not only the ecosystem but also the farm workers, consumers, and general public.

Food Systems and Value Chains

Food systems differ greatly across and within region and nation. They are increasingly interconnected through global and regional markets. They involve billions of actors at different levels along the food value chain. Food production activities comprises such factors as land use and tenure, soil management, crop breeding and selection, crop management, livestock breeding and management and harvesting (FAO, 1997). Food distribution involves a series of post-harvest activities including the processing, transportation, storage, packaging and marketing of food as well as activities related to household purchasing power, traditions of food use (including child feeding practices), food exchanges and gift giving and public food distribution.

Figure 1. Components of a food system. Source: (HLPE, 2014)

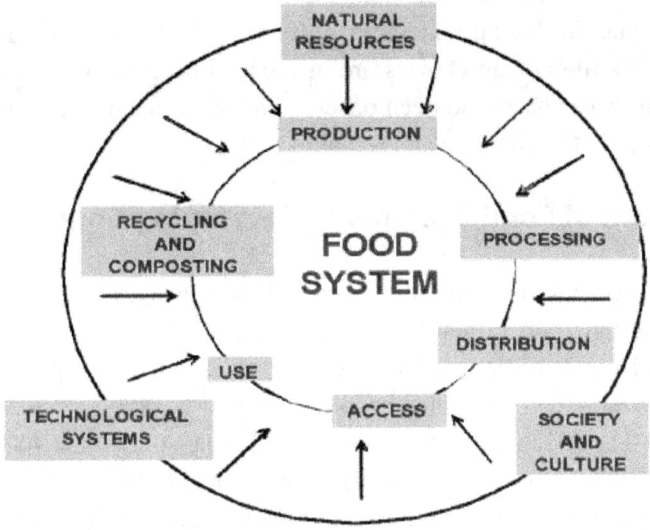

Activities related to food utilization and consumption include those involved in the preparation, processing and cooking of food at both the home and community levels, as well as household decision-making regarding food, household food distribution practices, cultural and individual food choices and access to health care, sanitation and knowledge. Diet shifts and growing demands for processed foods in developing countries is creating opportunities for adding value to primary agricultural products.

Determinants of Food System and Sustainability

The food system is highly complex and is driven by many economic, cultural and environmental factors. Better understanding of these drivers and how they interact could help to improve public policies. A food system operates within, and is influenced by, the social, political, economic and environmental context (Goodman, 1997; UNEP, 2016). Food systems gathers all the elements (environment, people, inputs, processes, infrastructures, institutions) and activities that relate to the production, processing, distribution, preparation and consumption of food and the outputs of these activities, including socio-economic and environmental outcomes (HLPE, 2014). All these factors therefore determine the nature of the food system. Achieving sustainable food systems depends on the ability to ensure zero food waste, 100 per cent access to food and increasing smallholder's income. Any policy measures for sustainable food systems must be able to link food production, distribution, consumption and the environment within which these activities occur (Figure 2).

A food system's sustainability is influenced by natural and human factors. These factors interact with each other within a food system. For example, the availability of water and land for food production is influenced by human actions such as use of fossil fuel, while human choices are influenced by environmental conditions. Food loss and waste have also been observed as determinants of sustainability of food systems in both economic and social terms.

Management of Food Systems in Developing Countries

Developing countries are going through significant changes in their food systems (Regmi and Meade, 2013; Reardon, 2015; Tschirley *et al.,* 2015a). The new food system model has implications for education and employment policies, as labor markets shift to respond to changing production systems both on- and off-farm (Tschirley et al., 2015b). For instance, the Nigerian traditional food system is characterized by low return on investments, crude and ineffective farm implements, low irrigation, and expensive inputs such as fertilizers, improved planting materials, low yielding plants and livestock. The food system is heavily reliant on fossil fuels. Large amounts of fuel are required to power farm equipment, to produce artificial fertilisers, and to transport food to supermarkets. Small producers often lack access to affordable inputs, roads and transport, electricity, irrigation, extension services, appropriate technology and market infrastructure. For example, the use of fossil fuels

Figure 2. Determinants of the Food System. Source: Nugent et al., (2011)

leads to carbon dioxide emissions (and to air pollution, depending on the burning process), while the use of minerals typically leads to nutrient emissions to ground and surface water. In Kenya, poor land preparation, inadequate soil management, and ineffective pest and disease control, leads to significant losses in food production. The inability to effectively monitor and enforce compliance and rules governing offshore territorial waters has also curtailed full exploitation of the offshore fishing potential (Alila and Atieno, 2006).

Sustainable Food Systems in the Existing Policy Frameworks and Development Agendas

Food systems are not only complex but are also characterized by many difference perceptions about the critical issues it involved, the challenges as well as way forward. Sustainable food system is complex systems that depend on a range of policy areas such as effect of climate change, water, ecosystems, the supply chain relationships, global economics, nutrition and obesity. The form of food systems carried out by people determine access to nutritious food and a way of achieving the United Nations Secretary-General Zero Hunger Challenge and Sustainable Development Goals 2, 3, 5, 6, 9, 12, 14, and 15 (Table 1)

Both developed and developing countries currently have a large number of laws, financial and other regulations that are influencing directly or indirectly on food systems.

Drawing from the transition theory (De Haan and Rotmans, 2011) three governance dynamics or 'pathways' can be distinguished that have the power to reshape current food systems, and hence their interactions with natural resources. These governance dynamics are (i) reforms by governments and international institutions; (ii) adaptations by food system actors; and (iii) alternative (niche) innovators.

CONCLUSION AND RECOMMENDATIONS

Food systems are characterized by interactions among system components and among factors, processes, and outcomes which are often non-linear. They are characterized by time and space lags and thresholds, which make it difficult to implement effective policies for promoting food system sustainability. A sustainable food system therefore embraces the interconnectedness of all the food-related activities and the environment within which these activities occur. The production, distribution and consumption of food are dynamic process involving multiple inputs, outputs and stakeholders. The current global distribution of food is already unequal (about 800 million people worldwide do not have access to sufficient food, while 1.9 billion

Table 1. SDGs and food system challenges

SDGs	Food System Challenges
Goal 2: End hunger, achieve food security and improved nutrition and promote sustainable agriculture	• Access to affordable and nutritious food • Food fortification and vitamin supplementation programmes to needy population
Goal 3: Ensure healthy lives and promote well-being for all at all ages.	• Earlier years' nutrition education for mothers • Encouraging exclusive breast feeding • Regulating marketing and sales obesogenic food • Public education campaigns on optimal diets
Goal 5: Achieve gender equality and empower all women and girls.	• Gender empowerment in food production, processing and distribution • Gender equality in access to agricultural inputs
Goal 6: Ensure availability and sustainable management of water and sanitation for all.	• Sustainable water usage • Ensuring good sanitation for all
Goal 9: Build resilient infrastructures, promote inclusive and sustainable industrialization, and foster innovation.	• Ensuring infrastructure for storage, transport, and logistics • Investment in innovations through R&D
Goal 10: Reduced inequalities	Nutritional deficits in the early years of life can cause lifelong deficits such as stunting and impaired cognitive development, deepening inequality cycles as malnourished children are unable to participate in labour force
Goal 12: Ensure sustainable consumption and production patterns.	• Sustainable use of arable land • Sustainable water management practices • Limiting agriculture- related pollution and emissions
Goal 13: Take urgent action to combat climate change and its impacts	Agriculture is both a cause of climate change through emissions it produces, and a victim as changes in temperature and rainfall impact crop growth and agricultural productivity
Goal 14: Conserve and sustainably use the oceans, seas and marine resources for sustainable development.	Sustainable use of the oceans, seas and marine resources
Goal 15: Protect, restore and promote sustainable use of terrestrial ecosystems, sustainably manage forests, combat deforestation, and halt and reserve land degradation and halt diversity loss.	• Managing deforestation related to food and non-food activities including livestock and biofuels • Lower use of harmful chemicals and related substances in agriculture

adults are overweight or obese), and this is projected to become even worse if we continue with "business as usual". There are obvious health and ethical implications to all of these. One of the major challenges facing many developing countries is how to ensure a food system and a food future that is more secure, sustainable, just and healthy. Lack of nutrition education and poor processing contribute significantly to nutrient losses and food systems in many developing countries. These types of systems are imperative to feed the growing population and minimize the effects of climate change. How food is grown and produced, the types of foods consumed, and quantity of food wasted have major impacts on the sustainability of the food system in developing countries. A given food system therefore offers multiple potentially competing and complementary points for intervention

Integration of food systems, the environment and sustainability require the support from a vast range of stakeholders from the farmers to the final consumer, from governments to private sector to civil society. Investments and innovations are needed. This includes fighting food waste and food loss, promoting knowledge and technology-sharing practices for agricultural producers, investing in farm-to-market infrastructures in developing countries, and advancing research and development (R&D) into new techniques and technologies to improve yield and lower environmental impact. Reducing food loss and waste also requires efforts by retailers and food processors.

Policy makers could help smallholders gain a better position in global food supply chains dominated by agro-food corporations; reduce food losses in line with FAO, (2012); invest in women's agriculture, reform agricultural input markets; improve resource efficiency, participation, and accountability in water and energy systems and invest in nutrition and health. Investment in management practices and research development will enable a more effective food production and process in developing countries. There is need for a regulation to prevent nutrient losses at all stages especially in the livestock sector, proper regulation on access to and use of water, biodiversity, and ecosystems services; removal of subsidies that encourage unsustainable production or practices. Under-nutrition can be reduced significantly when the traditional food systems in developing countries is improved using a combination of strategies including nutrition education.

REFERENCES

Alila, P. O., & Atieno, R. (2006). *Agricultural Policy in Kenya: Issues and Processes*. Paper presented for the Future Agricultures Consortium workshop, Nairobi, Kenya.

De Haan, J., & Rotmans, R. (2011). Patterns in transitions: Understanding complex chains of change. *Technological Forecasting and Social Change*, *78*(1), 90–102. doi:10.1016/j.techfore.2010.10.008

Economist Intelligence Unit. (2015). *Fixing: Towards A More Sustainable Food System*. Available online at http://foodsustainability.eiu.com/whitepaper/

Ericksen, P. (2008). Conceptualizing food systems for global environmental change research. *Global Environmental Change*, *18*(1), 234–245. doi:10.1016/j.gloenvcha.2007.09.002

European Commission. (2014). Impact assessment on measures addressing food waste. Review of EU Waste Management Target. EC.

FAO. (1997). The food system and factors affecting household food security and nutrition. In Agriculture food and nutrition for Africa - A resource book for teachers of agriculture. Rome: FAO.

FAO. (2011). *Global food losses and food waste – Extent, causes and prevention.* Rome: FAO. Available at: http://www.fao.org/docrep/014/mb060e/mb060e00.htm

FAO. (2012). *The state of food insecurity in the World.* Rome: Food and Agriculture Organisation.

FAO. (2013). *The State of Food and Agriculture. Food Systems for Better Nutrition.* Rome: FAO.

FAO, IFAD, & WFP. (2015). The State of Food Insecurity in the World 2015. In *Meeting the 2015 international hunger targets: Taking stock of uneven progress.* Rome: FAO.

FAO and Biodiversity International. (2010). *Biodiversity and Sustainable Diets United against Hunger.* International Scientific Symposium, FAO/Biodiversity International.

Farmar-Bowers, Q. (2013). Framing the research needs for food security in Australia. In Q. Farmar-Bowers, V. Higgins, & J. Millar (Eds.), *Food security in Australia: Fixing Business Media.* doi:10.1007/978-1-4614-4484-8_15

Goodman, D. (1997). World-scale processes and agro-food systems: Critique and research needs. *Review of International Political Economy, 4*(4), 663–687. doi:10.1080/096722997085657 87

Hassen, I. W., Dereje, M., Minten, B., & Hirvonen, K. (2016). *Diet transformation in Africa: the case of Ethiopia.* Ethiopia Strategy Support Program (ESSP). Working Paper no. 87, March. International Food Policy Research Institute.

HLPE. (2014). *Food losses and waste in the context of sustainable food systems. A report by the High Level Panel of Experts on Food Security and Nutrition of the Committee on World Food Security.* Rome: HLPE.

Hollinger, F., & Staatz, J. M. (2015). *Agricultural Growth in West Africa Market and Policy Drivers.* Rome: African Development Bank and Food and Agriculture Organisation.

Ingram, J. A. (2011). food systems approach to researching food security and its interactions with global environmental change. *Food Security*, *3*(4), 417–431. doi:10.1007/s12571-011-0149-9

Kickbusch, I. (2010). *The food system: A prism of present and future challenges for health promotion and sustainable development*. Triggering the debate – white paper. Bern: Promotion Santé Suisse, 2010.

Liverpool-Tasie, L., Adjognon, A., & Reardon, T. (2016). *Transformation of the food system in Nigeria and female participation in the Non-Farm Economy (NFE)*. 2016 Annual Meeting, Boston, MA.

Minten, B., Tamru, S., Engida, E., & Tadesse, K. (2013). *Ethiopia's value chains on the move: the case of teff.* Ethiopia Strategy Support Program (ESSP). Working Paper no. 52, April. International Food Policy Research Institute.

Ng, M., Fleming, T., Robinson, M., Thomson, B., Graetz, N., Margono, C., … Abera, S.F. (2014). Global, regional, and national prevalence of overweight and obesity in children and adults during 1980–2013: a systematic analysis for the Global Burden of Disease Study 2013. *The Lancet*.

Ng, R. (2015). *Making Sustainable Food Choices: Influencing Consumer Behaviour for Food Sustainability in Oxford*. OxPolicy.

Nugent, R. (2011). *Bringing agriculture to the table: How agriculture can play a role in preventing chronic disease*. The Chicago Council on Global Affairs.

Pearson, D., Friel, S., & Lawrence, M. (2014). Building environmentally sustainable food systems on informed citizen choices: Evidence from Australia. *Biological Agriculture and Horticulture*, *30*(3), 183–197. doi:10.1080/01448765.2014.890542

Reardon, T. (2015). The hidden middle: The quiet revolution in the midstream of agrifood value chains in developing countries. *Oxford Review of Economic Policy*, *31*(1), 45–63. doi:10.1093/oxrep/grv011

Regmi, A., & Meade, B. (2013). Demand side drivers of global food security. *Global Food Security*, *2*(3), 166–171. doi:10.1016/j.gfs.2013.08.001

Reisch, L., Eberle U., & Lorek S. (2013). Sustainable food consumption: an overview of contemporary issues and policies. *Sustainability: Science, Practice, & Policy*, *9*(2), 7–25.

Rischke, R., Kimenju, S. C., Klasen, S., & Qaim, M. (2015). Supermarkets and food consumption patterns: The case of small towns in Kenya. *Food Policy*, *52*, 9–21. doi:10.1016/j.foodpol.2015.02.001

Tendall, D. M., Joerin, J., Kopainsky, B., Edwards, P., Shreck, A., Le, Q. B., & Six, J. et al. (2015). Food system resilience: Defining the concept. *Global Food Security*, *6*, 17–23. doi:10.1016/j.gfs.2015.08.001

Tschirley, D., Reardon, T., Dolislager, M., & Snyder, J. (2015a). The rise of a middle class in east and southern Africa: Implications for food system transformation. *Journal of International Development*, *27*(5), 628–646. doi:10.1002/jid.3107

Tschirley, D., Snyder, J., Dolislager, M., & Snyder, J. (2015b). Africa's unfolding diet transformation: Implications for agrifood system employment. *Journal of Agribusiness in Developing and Emerging Economies*, *5*(2), 1–48. doi:10.1108/JADEE-01-2015-0003

UN-DESA. (2015). *World Population Prospects: The 2015 Revision, Key Findings and Advance Tables.* Available online at http://esa.un.org/unpd/wpp/publications/files/key_findings_wpp_2015.pdf

UNEP. (2016). *Food Systems and Natural Resources.* A Report of the Working Group on Food Systems of the International Resource Panel.

Westhoek, H., Ingram, J. S. I., Van Berkum, S., Özay, L., & Hajer, M. (2016). *Food Systems and Natural Resources. A Report of the Working Group on Food Systems of the International Resource Panel.* UNEP Nairobi.

WHO. (2013). *UNICEF-WHO-The World Bank Joint Child Malnutrition Estimates 2013.* Available at: http://www.who.int/nutgrowthdb/summary_jme_2013.pdf?ua=1

WHO. (2015). *Global Health Observatory (GHO) data.* Available at: http://www.who.int/gho/ncd/risk_factors/obesity_text/en/

Chapter 2
Food and Environment:
A Review on the Sustainability of Six Different Dietary Patterns

Pedro Pinheiro Gomes
National Statistics Institute, Portugal

ABSTRACT

Recent studies related the link between food consumption and impacts on environment and health. These may present variations according to the dietary patterns of different populations. This chapter assesses the impacts of six dietary patterns while emphasizing protein overconsumption and sustainability of food systems in a world where one billion people are hungry and several more suffer from conditions related to obesity. The chapter shows the nutritional disparity existent in different dietary patterns and potential to make changes. Changes in dietary patterns are an opportunity to contribute for environmental and health benefits. The analysis was based on a set of environmental indicators such as greenhouse gas (GHG) emissions and land use demand, while providing a nutritional balance. The methodology comprehended a life cycle assessment in order to quantify the GHG emissions and the land use demand for food production. Finally, a review is made to focus on the benefits of shifting from current diet patterns to more sustainable ones, such as the Mediterranean.

DOI: 10.4018/978-1-5225-3631-4.ch002

FOOD AND ENVIRONMENT

With the fast growth of cities and the increase in world population food needs keep increasing in order to face the demands. This demand is responsible for the increase of pressures over the environment. The priorities for the communities' well-being become interlinked with those regarding environmental conservation (Johns and Eyzaguirre, 2001; Chan et al, 2011). Nutrition is the most fundamental aspect for the human needs; deficient nutrition can lead to health problems such as malnutrition, infectious diseases, contamination and also obesity (FAO, 2011). Environmental contamination from industrial sources such as heavy metals and organochlorines can contribute to people's nutrition and health. The main challenge of food for the 21st century is to try to understand how diets can respect the body's needs of nutrients and energy while maintaining the balance of ecosystems and respect cultural differences between communities.

The quality of the environment is vital to the quality of food, since in every region, each species is adapted to the local conditions, if these change, the ecosystems will be impacted, thus, plants and animals will also have to adapt in a direction that may be less productive and even lead to extinction (OECD, 2011). Not every impact is easily predictable, for instance, it is common sense that a forest fire will have severe impacts on the ecosystem, however, what are the effects of climate changes on the productivity? Although several sources point out that many cultivated plants will react positively to the predicted increase of carbon dioxide in the atmosphere, weeds will also react favourably, which can lead to a decrease in soil, thus, leading to the increase in the use of pesticides with potential negative impacts over agriculture. These risks will contribute to a decrease in food safety, especially to those who depend on agriculture for food and income (Oenema et al., 2007; Cassman et al, 2003). The quality of environment in which food is produced also translates, according to FAO (2011) into three classes of potential croplands: prime, good and marginal. About 81% of the cultivated land is classified either as prime or good, which means that 19% have its productivity determined by adding other inputs, this percentage of cultivated lands add negative impacts to the environment and the balance of ecosystems.

The conversion of lands with low productivity in terms of food production will decrease its capacity to provide other goods and services.It is precisely in such situations that land use and its proper management has impact on both food production and environmental sustainability. Regarding water, its quality is essential for populations in terms of consumption and use for agriculture and livestock production. However, it is also used for activities like production and processing of food, its preparation and also for disposal of waste. Since agriculture is highly dependent on water, its scarcity will affect either the quality of produced food, or the water availability needed for livestock. The current patterns of use of water for agriculture are unsustainable with several negative impacts requiring an efficient approach for its use (FAO, 2011). The production or value of crops per volume of water utilized will be a decisive factor into the choices in land management for agriculture. This indicator is currently low (OECD, 2008) since the losses of water are around 50% due to ineffective practices or lack of investment in new technologies such as low pressure sprinklers or drip irrigation.

In a world where it's expected to have less availability of water for agriculture due to the effects of climate changes and with the fertility of soil decreasing with the abundant use of inputs to force high productivities, it is essential to establish efficient pratices for water consumption, while maintaining the soils structure without the use of inputs and pesticides that damage not only surface water but also groundwater. Regarding climate change and its effects on water, it is expected to impact the availability in many regions, affecting the precipitation, hydrologic flows and recharges of groundwater, water quality will also be compromised due to the flooding of fertile coastal regions due to sea level rise (IPCC, 2014). An European Commission study (EC/JRC, 2009), showed that food consumption represented 27% of all environmental impacts in the EU-27, and enhanced a prominent role of meat production on environmental impacts generated along the food chain.

The environment has influence on the quality of food, but the food systems also have impacts over the environment since these use natural resources and depend on human activities. There's still a wide use of fossil fuels to produce, process and transport food which contributes to an increase of carbon dioxide emissions into the atmosphere, the use of imputs lead to the emission of nutrients to water bodies. Food production will always have impacts on the environment since activities such as agriculture and livestock production are open systems based on natural processes, meaning that there's the need of a system in order to manage, control, prevent and if possible avoid effects on the environment (Vermeulen et al., 2012). In Table 1 are synthesized main environmental damages related to each cycle of food, from its production to its waste.

WATER AND SOIL QUALITY

A food system affects the water resources in several manners, through the overuse of inputs which leads to pollution by nutrients, pesticides, other chemicals, bacteria and organic waste. The pollution can be local, mainly from organic sources, such as waste or effluents or assuming regional and global scale with nutrient and chemical pollution, which, not only affects the quality of water used for agriculture but also drinking water, leading to health risks.

Table 1. Negative Impacts of food activities on the environment

Environmental Impact	Food Production	Food Processing and Packaging	Food Distribution	Food Consumption	Waste Management
GHG Emissions	Fertilizers; Irrigation; Machinery; Livestock; Land Conversion	Machinery	Machinery	Cooking	Landfills
Air	Pastures; Livestock	Exhaust	Exhaust	Cooking	Waste
Biodiversity	Land Conversion; Habitat Fragmentation	Paper and Card		Fuel Use	Pollution
Soil	Erosion; Compaction; Salinization	Pollution	Pollution		Pollution
Water	Eutrophication; Pollution	Pollution	Coastal Degradation		Pollution

Source: Adapted from Ingram (2011)

Excessive fertilization, urbanization and livestock production are the main drivers for water pollution with increases in Nitrogen and Phosphorus in the soil which eventually will runoff to the water bodies (Seitzinger et al, 2010; Hall and Richards, 2013). The soils have drivers related to water pollution, the excessive use of inputs such as nutrients and chemicals and emissions from industries will contaminate the soils with heavy metals, copper and zinc, used often in the livestock production. These substances can reduce the productivity of soils, and the effect is persistent, since the removal of contaminants, a process known as "remediation" is very expensive and is often costlier than prevention.

GREENHOUSE GAS EMISSIONS

Emissions occur throughout all activities related to food, such as changes in land use, use of fossil fuels and even energy related diets. Total emissions account for more than 10 gigatonnes of CO2eq in 2010, resulting in around 25% of total greenhouse gas emissions in the reported year (FAO, 2014). Agriculture contributes around 80% of total emissions, due to deforestation and animal sources, such as emission of methane (FAO, 2014). The distribution of food represents an important source of emissions due to the transport, refrigerators leakages and also the preparation of food itself. Mitigation plans for emissions will have to contemplate the food systems in order to have a significant contribution to the environment.

FOOD SYSTEMS DISPARITY

Food security was redefined in the World Summit on Food Security in 2009 (FAO, 2009) depending on four main standards: 1) Food availability, where quantities should be enough to meet the demands; 2) Food Access: Whether physically and economically to support a nutritious diet; 3) Food Use: where the use is related with means and knowledge related to basic nutrition and well being and finally 4) Stability in food availability regarding its access and use. These four pillars should be the back bone of a proper food system while adding a fifth one: Food Sustainability.

Food systems are the main concern of the 2030 Agenda for Sustainable Development, the global commitment is focused to eradicate poverty and hunger while maintaining investments to develop quality of life and economical development. These systems are intended to be "resource-oriented" since they depend largely on natural resources such as land, soil, terrestrial and marine biodiversity (Lang et al, 2009). These resources need to be used efficiently in order to guarantee a sustainable management, since these are also sources of a number of environmental impacts, such as the loss of biodiversity, water degradation and depletion and the emission of greenhouse gas, therefore, the stakeholders related to food systems need to be aware that they influence directly and indirectly the health and quality of life of communities, which also makes them possible agents of change of the current systems (Garnett et al, 2016; Fraser, 2005).

Globalization increased the demand for food, meaning that a system, more than just a way to organize the processes, is a needed requirement since the majority of food that reaches families' plate is no longer produced by families themselves, but it goes (sometimes from long distances) from producers to consumers. Current systems are failing in terms of security, quality of the food itself and threatening sustainability by using more natural resources and inputs to increase productivity to meet the world demands. Food production has more than doubled; diets are energy intense in order to make them affordable, which is often not the equivalent to healthy, or sustainable (Lawrence and Burch, 2007). There is more variety of food, in terms of tastes, presentations, brands and even segments of the brands themselves in the shape of prices, such as the premium brands, well-known brands, low brands and even in terms of presentation with the growth of the "gourmet" sections regarding its marketing. Several local, national and even multinational companies related to food have emerged; some of them used the financial crisis of 2008 to promote these cheaper segments of food to expand their businesses and areas of influence. However, nearly 1 billion people daily are hungry and over 2 billion have nutrient deficiencies and over 2 billion people overweight or even obese, meaning that around 5 billion people have unhealthy patterns which will challenge the sustainability of food systems (Berdegué et al, 2005; Sutton et al, 2013). With the increase in world population, the scenario is not positive; in order to reach the international targets, changes need to be made, to provide access to healthy food to everyone. The main focus should be the markets, at local, regional and even global level to protect the most vulnerable population clusters and rely on information for societies suffering from overconsumption, in order to assist them make the right choices.

It is important to rely on statistics to understand the food systems are and how to assess the equality or inequality between them. The data from Table 2 demonstrate the high dependence of food from nutrients such as Nitrogen and Phosphorus and its impact on the environment and health, especially, natural resources and associated toxicity issues as well as health problems related to deficiencies in nutrition (USGS, 2013; Allen et al, 2011). Land use and the loss of biodiversity still are the main challenges, while the loss of nutrients due to the intensive use of soil leads to the decrease inequality of food (FAO, 2007).. Nonetheless it's also important to highlight the overuse of nutrients as inputs also cause toxicity issues. Higher crop yields have been proposed as an alternative to a more sustainable way of agriculture. However, progress has been slow, since these methods have other problems associated, namely, the pollution of waters due to the use of nutrients and chemicals. Food systems will meet several challenges at short, medium and long term with the population growth, increasing the demand for food, the economy growth in developing countries will change the dietary patterns to more unsustainable ones, with red meat consumption, fish, vegetables and processed foods and drinks (Gustavsson et al, 2011). Climate change is the long term barrier, which will impact weather conditions and the natural resources vital for food production.

The efficient management of food systems will be an imperative, as it is stated in the Sustainable Development Goals, and it goes through the prevention of resource degradation by adapting good practices such as diminishing overexploitation, increasing the efficient use of resources while decreasing the environmental impact of food production, higher agricultural yields without increasing environmental impacts, reducing the use of pesticides, higher productivity of feed conversion, efficient use of energy and water, reduction of food losses throughout the chains and alerting to the impact of overconsumption in diets (Galal, 2002;Verburg et al, 2012).

DIETARY PATTERNS

The six dietary patterns selected for this analysis were based on a study from Auestad and Fulgoni (2015), based on several other studies comparing the method of analysis, the region, the environmental impacts, the conclusions and also the study limitations in order to allow improvement for future analysis. The selection of the studies was subjective, and is intended to cover different global realities. Diet 1 is based on the

Table 2. Minerals needed in food systems and their relation to the environment

Nutrients	Share of Agriculture or Food in use	Deficiency Issues Reported Related to Food	Toxicity Issues
Nitrogen (N)	80%	Protein	Reported
Phosphorus (P)	90%	Reported	Reported
Potassium (K)	85%	Reported	No
Sulphur (S)	60%	Protein	No
Magnesium (Mg)	10%	Reported	No
Calcium (Ca)	10%	Reported	No
Iron (Fe)	1%	Reported	No
Zinc (Zn)	2%	Reported	Reported
Copper (Cu)	1%	Unknown	Reported
Molybdenum (Mo)	1%	Unknown	Reported
Manganese (Mn)	1%	Unknown	No
Boron (B)	12%	Reported	Unknown
Selenium (Se)	10%	Reported	Reported
Iodine (I)	1%	Reported	Unknown
Cobalt (Co)	1%	Reported in deficiencies of B12 vitamin	Unknown

Source: Adapted from USGS (2013)

pattern consumptions of New Zealand, Diet 2 in Indian patterns, Diet 3 of rural China patterns, Diet 4 is be representative of a Northern European pattern, Diet 5 an example of a standard European pattern and Diet 6 should be representative of the American pattern.

Greenhouse gas emissions as an environmental indicator is a measure of the energy-driven procedures applied to the production of food composing the dietary pattern. The more resources are spent and the more processing is applied to the food, more energy will be spent leading to a trend to deforestation patterns for conversion to intensive agriculture. These are the main drivers to increase carbon dioxide emissions, thus, making the first choice indicator in terms of environmental analysis.

Diet 1 has higher costs to the communities; however the tradeoff between food quality, security and price results in higher environmental protection; the reduction of the use of milk increases the cost of food. Diet 2 presents several sources for greenhouse gases; however it also presents threats, since the dietary pattern is varied, and since it refers to a country in its developing stage, if it enters in an industrialized pattern of eating with consumption of meat, probably environmental impacts will increase severely. Diet 3 is in the exact opposite of the industrialized

Table 3. The environmental impacts associated with six dietary patterns

Diet	Study, Year, Reference	Diets	Environmental Impacts	Key Results	Conclusions	Limitations of Study
1	Wilson et al., 2013, New Zealand	Diet modeling study (16 diets) to meet New Zealand nutrient recommendations for men with inputs for energy, macronutrients, and 10 select micronutrients in foods, food prices, food wastage, and food-specific GHGe. Models designed to 1) minimize cost and meet nutritional needs; 2) minimize GHGe and meet nutritional needs; 3) be relatively healthy diets (Mediterranean and Asian style); 4) be familiar New Zealand meals.	GHGe determined for 76 food items. GHGe data were scant for New Zealand foods, so used UK GHGe data; estimates made for some foods. Sensitivity analysis conducted.	Increasing dietary variety and likely acceptability of modeled diets had increased daily cost when optimized for both low-cost and low GHGe. Several diet scenarios had small number of foods (e.g., 9, 10, or 14 foods). All modeled low-cost and low-GHGe diets had likely health advantages over the current New Zealand model. Diets that included "more familiar meals" for New Zealanders had higher costs.	Low-cost, low-GHGe modeled diets are complementary but with trade-offs of higher daily food costs. This is partly because of reduction in higher GHGe foods, such as eggs and milk; pushes food choices to more costly alternative foods containing nutrients such as calcium. Milk is a relatively efficient beverage for nutrient provision (i.e., nutrients per GHGe generated).	Diet-related GHGe limited to 76 food items. LCA for GHGe of foods in New Zealand limited; substituted data from the UK, with approximations for foods not covered.
2	Pathak et al., 2010, India	Five common nutritionally balanced diets in India: 1) vegetarian, 2) lacto-vegetarian (vegetarian with milk), 3) ovo-vegetarian (nonvegetarian with egg), 4) nonvegetarian with poultry meat, 5) nonvegetarian with mutton	GHGe (g CO2eq/d) were determined using LCA (production, processing, transportation, and preparation) based on published data	GHGe were 40% higher for nonvegetarian vs. vegetarian meals. Nonvegetarian meal with mutton had 1.8 times more GHGe than vegetarian, 1.5 times more than nonvegetarian with chicken and ovo-vegetarian, and 1.4 times more than lacto-vegetarian. In nonvegetarian meal with mutton, GHGe from mutton (35%) were similar to those from rice (34%). In lacto-vegetarian meal, 49% GHGe were from rice, 22% were from milk.	A change in food habits could offer a possibility for GHGe mitigation. Some potential options to reduce GHGe from food may be consumption of locally produced foods; less mutton; substitute meat and milk with vegetable protein	Diet-related GHGe limited to 16 food products. LCA for GHGe covered on farm production, transport, processing, and preparation (consumer) but not storage and handling losses during production or storage at households. LCA for GHGe from different sources.
3	Zhen et al., 2010, rural Guyuan, China	Habitual diets	Land requirements for 8 food categories (25 foods/sub categories) determined per capita (m2) and per household (m2 per household)	Food consumption pattern in Guyuan depends on wheat; more mixed for China. Less meat consumed than plant foods; land requirement for meat was only 5.7% of arable land (national average, 8.4%). Animal protein intake was 7.5% of total protein, below recommended 30% protein intake.	In Guyuan, food consumption met only basic energy needs for survival, with protein, especially animal protein, and fat below recommendations. Meat consumption expected to increase with projected increasing incomes of local people, thus toward a more balanced diet.	Several uncertainties and assumptions in estimates of food consumption and land requirements.

continued on following page

Table 3. Continued

Diet	Study, Year, Reference	Diets	Environmental Impacts	Key Results	Conclusions	Limitations of Study
4	Saxe et al., 2012, Denmark	Habitual Danish diet, based on >300 foods/beverages	GHGe based on:	GHGe were lower in diets with less meat and dairy. Type of meat in diet can impact GHGe. Within methodologic constraints, local produce may help reduce GHGe. May be negative effects from organic vs. conventional farming. Alcoholic beverages, sweets, and hot drinks (coffee, tea, cocoa) in habitual diet accounted for 22% of diet-based GHGe (meat at 37%). Theoretical vegetarian diet did not reduce GHGe more than optimized omnivore diet.	A well-designed diet could lead to lower climate impact and improved health. A change to Nordic diets (less animal foods, more fruits and vegetables) could support climate change mitigation, but must be cautious with diet recommendations. Reducing alcoholic drinks, hot drinks, and sweets by 50% would reduce GHGe the same as reducing meat intake by 30%.	Diet-related GHGe limited to 31 food categories. LCA for GHGe only farm to retail and from different sources (Danish LCA food database and from literature).
		Recommended diet:	LCA farm to retail;			
		New Nordic diet (more local foods; 75% organic foods)	LCA ISO 14040 (GHGe/kg food; based on consequential LCA method);			
		Theoretical scenarios, compared to 3 base diets:	Converted to CO2eq (IPCC 2007);			
		1) Conventional farming	Accounted for wasted and spoiled food;			
		2) Conventional farming with factor for transport of imported foods;	31 food categories, derived from >300 foods/beverages			
		3) Organic farming with factor for transport of imported foods;	GHGe reported as:			
		Other scenarios:	kg CO2eq/person per year			
		4, 5) Type of meats varied				
		6, 7) Amounts and types of organic foods varied				
		8) Ovo-lacto version of scenario 7				

continued on following page

Table 3. Continued

Diet	Study, Year, Reference	Diets	Environmental Impacts	Key Results	Conclusions	Limitations of Study
5	Vieux et al., 2012, France	Habitual diet based on a French national diet survey (2006–2007)	GHGe based on:	GHGe were lower with 240 fewer kcal and less meat intake (when kcal not replaced with other foods). GHGe were moderately reduced when meat and deli kcal replaced with kcal from dairy or mixed dishes. GHGe were negated or slightly higher when meat and deli kcal were replaced with kcal from fruits and vegetables.	GHGe linked to amount of food and kcal eaten. Substituting fruits and vegetables for meat (especially deli meat) may be desirable for health but is not necessarily the best approach to decreasing diet-associated GHGe.	Diet-related GHGe limited to 73 foods. LCA for GHGe only farm to retail, via conventional production and distribution.
		Theoretical diets:	LCA farm to retail;			
		Habitual diet with	LCA ISO 14040;			
		1) 240 fewer kcal	g CO_2eq/100 g of edible portion food;			
		2) 20% less meat/deli with a) no replacement of kcal or b) replacement of kcal with fruits/vegetables, dairy, or mixed dishes	73 representative foods			
		3) Red meat limited to 50 g/d; no deli with a) no replacement of kcal or b) replacement of kcal with fruits/vegetables, dairy, or mixed dishes	GHGe reported as:			
			g CO_2eq/d			
6	Eshel and Martin, 2006, USA	Food consumption based on per capita food disappearance	GHGe based on:	Energy efficiency of animal-based portion of diets: lacto-ovo-vegetarian, omnivore with poultry > average US diet > omnivore with red meat, omnivore with fish. GHGe estimates from theoretical diets (tons CO_2eq/person per year): omnivore with red meat > mean US diet > omnivore with fish > lacto-ovo-vegetarian > omnivore with poultry.	A mixed diet at average US calorie intake has 1485 kg CO_2eq higher emissions than the same number of calories from plant foods.	Diet-related GHGe limited to CO_2eq from agricultural production (on farm). GHGe from direct energy (primarily fossil fuel) and non-CO_2emissions from numerous sources. Theoretical diets described as "semirealistic."
		Theoretical diets: semirealistic mixed diets with animal-based foods at 0–50% daily kcal based on:	Farm only ($CO_2$from direct energy use + non-$CO_2$from agricultural production; NH_4, CH_4, N_2O);			
		1) Average, habitual US diet	LCA ISO 14040 (consequential LCA method)			
		2) Lacto-ovo-vegetarian	GHGe reported as:			
		3) Omnivore with fish source	g CO_2eq/kcal;			
		4) Omnivore with red meat	Tons of CO_2/person per year;			
		5) Omnivore with poultry	Energy efficiency (% fossil fuel energy input retrieved as edible energy from protein output)			

Source: Adapted from Auestad and Fulgoni (2015)

pattern, where the dietary pattern is the minimum required for survival in which meat consumption will translate into a more balanced diet. However, the tradeoff will be higher values of greenhouse gas emissions. Diet 4 is an example of what could be a sustainable diet with less consumption of meat and more consumption of fruits and vegetables. Diet 5 and especially Diet 6 are meat oriented, which will reflect in higher emissions and consequently more energy and resources are spent; these are the diets further from sustainability.

SUSTAINABLE DIETS

The main challenge regarding diets is to simultaneously provide enough food, in quantity and in a safe way, while conserving nutritional and cultural values to meet the present needs and for the future generations (Kastner et al, 2012). However, there is another dimension to satisfy, which is the environment in the figure of the protection of natural resources, making the diet fully sustainable. FAO provided estimates referring that for 2050, in order to satisfy the needs of the population, there will be the need to increase food production by 60% focusing on animal products (FAO, 2010). While increasing the efficiency of the food systems and reducing waste can cut this number, the biggest shift must be the promotion, not only of healthier diets, but essentially on sustainable diets. FAO recently begun collecting information in order to design methods to assess in different regions, what could be an harmonized concept for a sustainable diet (FAO, 2010).

Mediterranean Diet

The Mediterranean Diet is very well known by its characteristics as healthier diet, having been subjected to several analyses to assess its impacts not only on health, but also environment. For the latter, it is suggested that it presents a lower environmental impact with other available options. This diet involves several countries (Portugal, Spain, Italy, Greece, France, Morocco, Cyprus and Croatia, the main motive for which is considered cultural heritage. However, recent studies show that there's a decline of this diet as an option in the Mediterranean area, since this region is passing through a transition in terms of diet to a more energy-related one (Gussow, 1995; Belahsen and Rguibi, 2006).

As it is possible to analyze it from Figure 1, the Mediterranean diet is more than just quantity and quality of food, it also incorporated physical exercise into the daily quantities of food. In fact, the basis of the pyramid is daily exercise from moderate to vigorous in order to maintain activity throughout the entire life. Both exercise and correct servings of food allow a proper weight and protect health from

Figure 1. The Mediterranean Diet
Source: Estruch et al, 2013

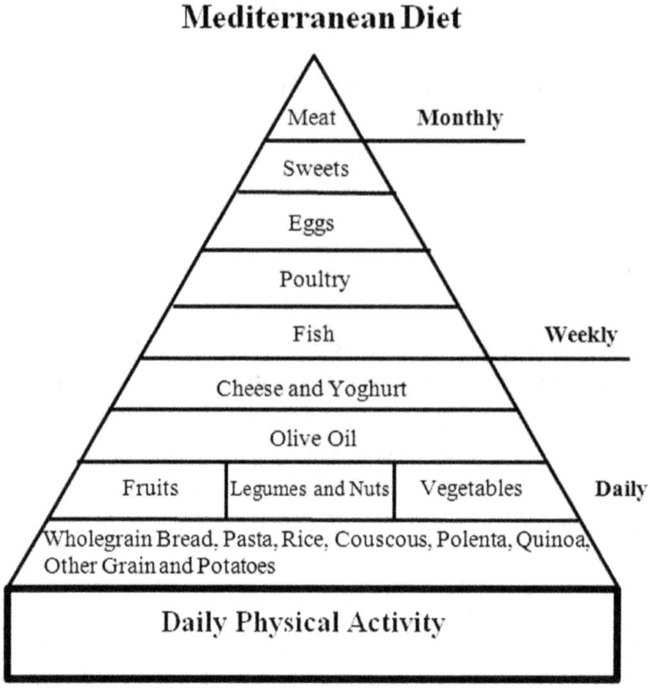

nutrient deficiency and diseases related to overconsumption and obesity, namely heart related diseases (Estruch et al, 2013). Analyzing the environmental component related to food of the Mediterranean Diet Pyramid, the monthly consumption of meat will reduce the amount of energy and water used to produce it, thus, reducing GHG emissions, while maintaining the necessary nutritional values to daily food consumption.

A diet rich in protein and calories has several negative impacts on both health and environment. Eating meat was always regarded as a symbol of economic well-being and after an economic crisis in 2008 which affected the very core of families' consumption, the steady recover verified in the financial and economical systems are re-shifting the diets into ones with patterns of industrialization. While improving the food systems in the Mediterranean area may contribute to enhance its influence in the communities, promoting information among communities and alerting stakeholders in food systems the need to promote sustainable diets should be the steps to take to minimize or even mitigate the negative externalities caused by unsustainable diets. The threat to the Mediterranean Diet was evident in the 2005 Mediterranean Strategy on Sustainable Development (UNEP/MAP, 2005): "Mediterranean agricultural and rural models, which are at the origins of Mediterranean identity, are under increasing threat from the predominance of imported consumption patterns. This trend is illustrated in particular by the decline of the Mediterranean dietary model despite its benefits on health. The prospective scenario for the expected impacts of trade liberalization, climate change and the lack of efficient rural policies offers a gloomy picture in some southern and eastern Mediterranean countries, with the prospect of aggravated regional imbalances, deeper ecological degradation and persistent or accrued social instability."

The price of food is one of the key drivers for its consumption nowadays since it is also related to security and access, two of the main pillars of food systems (Kharas, 2010). In lower income countries, expenditure on food accounts for 40-50% of the household budget (FAO, 2014), meaning that people will condition their food choices on how much energy they can get at the minimum possible. This is one of the reasons that also made fast food popular, since people can get calories at a cheap price; however, it also shifts diets to less sustainable ones.

CONCLUSION AND RECOMMENDATIONS

Feeding the world population will require strategies that comprehend multicultural and transversal communication, while considering health and environment related aspects towards more sustainable patterns to use natural resources while ensuring food security.

The activities surrounding food systems have a considerable impact on the environment, mainly due to the overuse of natural resources. The main impacts are on biodiversity, especially due to changes in land use and ecosystems for intensive agriculture and monocultures. IIts impacts will be on air, soil and water quality with the loss of nutrients and water contamination., Energy use and greenhouse gas emissions not only are a measure of the inefficiency of the food systems, but also the main driver for climate changes, which will impact the systems in all the above referred situations.

The lack of proper food systems can have serious problems, such as the increase of vulnerability of agro-systems to resist climate changes and pests and diseases resulting in loss of productivity. The loss of productivity will only result in higher needs of inputs to face the growing demand due to the population increase and the growth of the food industry as a business characterized by the existence of quantity and variety. The regeneration rates of the natural resources will also be lower since the natural systems are overloaded with the increasing use of inputs and the overconsumption of stocks.

It is important to rethink food systems in order to contemplate sustainability by cutting food losses and waste, alerting regarding the consumption of energy-intensive foods such as red meat, processed food and the way urban environments affect the way people consume further from a sustainable way. Reconnecting to the rural roots could develop in the consumers the consciousness that there's the need to promote healthier lifestyles while empowering small producers. There is a lack of information on how local, regional and national food systems interact between themselves, understanding this dynamic would improve the efficiency between producers and consumers. Reinforce investments in education, technology and environmental services are always good measures to sustainability regarding environmental protection and create the conscience that current practices are negative and leading to food inequalities.

From the Diets analysed it was possible to show that Nordic and Mediterranean diets are possible solutions towards the sustainable pattern since they focus on variety, but especially focus on less consumption of meat. A linear comparison of the different food patterns will have several limitations, the choice of the method by each author is different, thus, the quantification of environmental impacts will have different sources of error. However the added value are the different sources of greenhouse gas registered in each pattern, further studies should be based on a single method, but focusing on the application of the method to different realities in order to establish a common ground to define sustainability.

REFERENCES

Allen C.R., Fontaine J.J., Pope K.L., Garmestani A.S. (2011) Adaptive management for a turbulent future. *Journal of Environmental Management, 92*, 1339-1345.

Auestad, N., & Fulgoni, V. L. (2015). What current literature tells us about sustainable diets: Emerging research linking dietary patterns, environmental sustainability, and economics. *Advances in Nutrition*, *6*(1), 19–36. doi:10.3945/an.114.005694 PMID:25593141

Belahsen, R., & Rguibi, M. (2006). Population health and Mediterranean diet in southern Mediterranean countries. *Public Health Nutrition*, *9*(8A), 1130–1305. doi:10.1017/S1368980007668517 PMID:17378952

Berdegué, J.A., Balsevich, F., Flores, L., & Reardon, T. (2005). Central American supermarkets' private standards of quality and safety in procurement of fresh fruits and vegetables. *Food Policy, 30*, 254-269.

Cassman, K. G., Dobermann, A., Walters, D. T., & Yang, H. S. (2003). Meeting cereal demand while protecting natural resources and improving environmental quality. *Annual Review of Environment and Resources*, *28*(1), 315–358. doi:10.1146/annurev.energy.28.040202.122858

Chan, L., Receveur, O., Sharp, D., Schwartz, H., Ing, A., & Tikhonov, C. (2011). *First Nations Food, Nutrition and Environment Study (FNFNES): Results from British Columbia (2008/2009). Prince George*. University of Northern British Columbia.

EC/JRC. (2009). *Environmental impacts of diet changes in the EU. Technical Report, European Commission (EC), Joint Research Centre (DG JRC)*. Brussels: IPTS.

Estruch, R., Martínez-González, M. A., & Corella, D. (2013). Effects of a Mediterranean-style diet on cardiovascular risk factors: A randomized trial. *Annals of Internal Medicine*, *2006*(145), 1–11. PMID:16818923

FAO. (2007). *Paying farmers for environmental services*. Rome: Food and Agriculture Organization of the United Nations.

FAO. (2009). *Linking people, places and products. A guide for promoting quality linked to geographical origin and sustainable geographical indications*. Rome: FAO/SINER-GI.

FAO. (2010). *Report of the Technical Workshop on Biodiversity in Sustainable Diets*. Rome: FAO.

FAO. (2011). *The state of the world's land and water resources for food and agriculture. Managing systems at risk*. Rome: Food and Agriculture Organization of the United Nations.

FAO. (2014). Agriculture, Forestry and Other Land Use emissions by sources and removals by sinks. Rome: FAO.

Fraser, E.D.G., Mabee, W., & Figge, F. (2005). A framework for assessing the vulnerability of food systems to future shocks. *Futures, 37*, 465-479.

Galal, O. M. (2002). The nutrition transition in Egypt: Obesity, Under-nutrition and the Food Consumption Context. *Public Health Nutrition*, *5*(1a), 141–148. doi:10.1079/PHN2001286 PMID:12027277

Garnett, T., Appleby, M.C., Balmford, A., Bateman, I.J., Benton, T.G., Bloomer, P., … Godfray, H.C.J. (2013). Sustainable intensification in agriculture: Premises and policies. *Science, 341*, 33-34.

Gussow, J. D. (1995). Mediterranean diets: Are they environmentally responsible? *The American Journal of Clinical Nutrition, 61*(suppl), 1383S–1389S. PMID:7754992

Gustavsson, J., Cederberg, C., Van Otterdijk, R., & Meybeck, A. (2011). *Global Food Losses and Food Waste*. Rome: Food and Agriculture Organization of the United Nations.

Hall, A. J., & Richards, R. A. (2013). Prognosis for genetic improvement of yield potential and water-limited yield of major grain crops. *Field Crops Research, 143*, 18–33. doi:10.1016/j.fcr.2012.05.014

Ingram, J. (2011). A food systems approach to researching food security and its interactions with global environmental change. *Food Security, 3*, 417-431.

IPCC. (2014). *Mitigation of Climate Change*. Contribution of Working Group III to the Fifth Assessment Report of the Intergovernmental Panel on Climate Change.

Johns, T., & Eyzaguirre, P. B. (2001). Nutrition for sustainable environments. *SCN News, 21*, 24–29.

Kastner, T., Rivas, M. J. I., Koch, W., & Nonhebel, S. (2012). Global changes in diets and the conseguences for land requirements for food. *Proceedings of the National Academy of Sciences of the United States of America, 109*(18), 6868–6872. doi:10.1073/pnas.1117054109 PMID:22509032

Kharas, H. (2010). *The Emerging Middle Class in Developing Countries*. Development Centre Working Paper No. 285. Organization for Economic Co-operation and Development.

Lang, T., Barling, D., & Caraher, M. (2009). *Food policy: integrating health, environment and society*. Oxford, UK: Oxford University Press. doi:10.1093/acprof:oso/9780198567882.001.0001

Lawrence, G., & Burch, D. (2007). Understanding supermarkets and agri-food supply chains. In D. Burch & G. Lawrence (Eds.), *Supermarkets and agri-food supply chains: transformations in the production and consumption of foods*. Cheltenham, UK: Edward Elgar.

OECD. (2008). *Environmental Performance of Agriculture in OECD Countries Since 1990*. Paris: OECD Publishing.

OECD. (2011). *Evaluation of Agricultural Policy Reforms in the European Union*. Paris: OECD.

Oenema, O., Oudendag, D., & Velthof, G. L. (2007). Nutrient losses from manure management in the European Union. *Livestock Science*, *112*(3), 261–272. doi:10.1016/j.livsci.2007.09.007

Seitzinger, S. P., Mayorga, E., Bouwman, A. F., Kroeze, C., Beusen, A. H. W., Billen, G., & Harrison, J. A. et al. (2010). Global river nutrient export: A scenario analysis of past and future trends. *Global Biogeochemical Cycles*, *24*(4), n/a. doi:10.1029/2009GB003587

Sutton, M.A., Bleeker, A., & Howard, C.M. (2013). *Our nutrient world: the challenge to produce more food and energy with less pollution*. Edinburgh, UK: NERC/Centre for Ecology & Hydrology.

UNEP/MAP. (2005). *Mediterranean strategy for sustainable development: a framework for environmental sustainability and shared prosperity. Tenth Meeting of the Mediterranean Commission on Sustainable Development (MCSD)*, Athens, Greece.

USGS. (2013). *Mineral Commodity Summaries: (various elements)*. Washington, DC: U.S. Geological Survey.

Verburg, P.H., Mertz, O., Erb, K.H., Haberl, H., & Wu, W. (2013). Land system change and food security: towards multi-scale land system solutions. *Current Opinion in Environmental Sustainability, 5*, 494-502. DOI: .cosust.2013.07.00310.1016/j

Vermeulen, S.J., Campbell, B.M., & Ingram, J.S.I. (2012). Climate Change and Food Systems. *Annual Review of Environment and Resources, 37*, 195-222.

Section 2

Sustainable Agricultural Systems, Climate Change, and Other Environmental Issues

This section connects the sustainability of agricultural systems with climate change and other environmental issues.

Chapter 3
Performance of Small-Scale Irrigation Schemes Under Climate Change in Low- and Middle-Income Countries:
A Systematic Review of the Evidence

Edgar Muhoyi
University of Zimbabwe, Zimbabwe

Josue Mbonigaba
University of Kwa-Zulu Natal, South Africa

ABSTRACT

Small-scale irrigation schemes (SSIS) in developing countries have been crucial, but the evidence about their performance has not been sufficiently analyzed. This chapter documents such evidence by reviewing and classifying the performance indicators. It also assesses literature on whether there are discernible trends in the efficiency of SSIS, identifies and classifies SSIS constraints, and characterizes various channels through which SSIS might affect poverty. Objectives are achieved via a systematic review of literature from 1990 to 2017. Results indicate a lack of standardization of irrigation performance indicators, and there is evidence that irrigation has boosted agricultural performance. Even though SSIS were associated with higher productivity than rain-fed agriculture, they performed below their full potential due to undervaluation of irrigation water by irrigation authorities, farmer characteristics, costs, institutional setups, the policy environment, and design, cultural, community, and environmental issues. SSIS are important tools for poverty reduction, and relevant policy implications are outlined.

DOI: 10.4018/978-1-5225-3631-4.ch003

INTRODUCTION AND BACKGROUND

In spite of the fact that small-scale farming contributes significantly to most economies of the developing world, it relies mainly on rain-fed agriculture. This reliance causes this type of farming to be vulnerable to extreme weather events such as droughts, among others (World Bank, 2013). Droughts are in turn likely to perpetuate the incidence of hunger and poverty, compromising the Sustainable Development Goal (SDG) of poverty elimination by 2030.

Unfavourable climate change (droughts), as presented by Maruyama *et al.* (2014), impact negatively on crop productivity. In Africa, and according to Alhassan *et al.* (2013), rain-fed agricultural crop output will have decreased by 50% by the year 2020 due to negative effects of climate change. As a solution to the adverse consequences of drought and the reliance on erratic patterns of rainfall that affect an increasing number of developing countries, irrigation is said to reduce the variance in output, yields and employment (Lipton, Blackman, De Zoysa, Qureshy, & Waddington, 2003; Svendsen, Ewing, Msangi, & others, 2009). As supported also by Akudugu *et al.* (2016), agriculture in Africa is dominated by smallholder farmers whose farms are rain-fed, and with the current trends of unpredictable rainfall patterns, their livelihoods are under threat.

Accordingly, governments across Africa, Asia and Latin America relentlessly invest in irrigation schemes intended to transform agriculture (Yedra, Mesa-Jurado, López-Morales, & Castillo, 2016). The main objective of investing in small-scale irrigation schemes (SSIS) is to reduce poverty amongst farming communities, with the knowledge that such investments have spillover effects on the greater economy. According to Lipton *et al.* (2003), of the 1.2 billion people defined as poor in developing countries, three-quarters of them reside in rural areas, and investment in irrigation is suggested as a solution to poverty. Irrigation has the potential to lead to poverty reduction as a result of increased yields, planting of higher value crops, increased food supplies, higher calorie intakes and better nutrition levels. This potential is also evident in Asia, as argued by Hussain and Hanjira (2004), because improving agriculture and enhancing agricultural productivity through investment in irrigation is an important strategy for rural poverty reduction, as the majority of rural poor people depend on agriculture. It is therefore, imperative that resource-use efficiency, especially in the face of climate change, is supported and achieved through SSIS.

Performance assessment of SSIS is necessary as it makes explicit the level of actual achievement against set production and productivity targets. Understanding the factors contributing to current productivity levels can lead to improving productivity and the more efficient use of resources (Pereira & Marques, 2017). High productivity and efficiency in SSIS production is critical in improving food security, reducing the level of poverty and achieving or maintaining agricultural growth in developing areas. A pertinent question is: how knowledgeable are policy makers and other key stakeholders in irrigation implementation in terms of the link between SSIS and social outcomes (including poverty, food security and nutritional statuses) in developing countries? In trying to understand how SSIS perform and their consequential impact on social outcomes, existing literature and evidence is reviewed and presented in this chapter.

In addition to performance, the sustainability of SSIS is crucial in irrigation sector development. The sustainability of SSIS can be understood by studying the constraints SSIS face and also by how SSIS recipients value irrigation water. These two issues as studied by Mutambara *et al.* (2016), have a bearing on SSIS sustainability in terms of agricultural productivity and production, incomes of farmers, livelihoods of people, and rural and economic development, especially in African and Asian countries. Concurrently, demand for water by other sectors has also been rising, as well as an increased demand for food arising from world population growth, making irrigation water scarcer (Birhane & Geta, 2016; Omondi, Mbogoh, & Munei, 2014; Syaukat, Arifah, & Minha, 2014). Since water is an economic good (Omondi *et al.*, 2014), it follows that how irrigation water is valued and used might contribute to the future sustainability of SSIS. It is worth noting that, in many instances, governments meet the greater proportion of water needs of farmers via subsidies; for instance, in Bangladesh, 50% of the expenses in the irrigation sector are heavily subsided (Akter, 2007). Therefore, the issue of sustainability arises in many developing countries with a paradigm shift away from government-funded to farmer-funded irrigation activities with the aim of reducing pressure on the government fiscus. Since water is increasingly becoming scarce and yet agricultural irrigation is a major consumer of water internationally (Tang, Nan, & Liu, 2013), it is thus worth learning from the literature how farmers value and use irrigation water to sustain the noble development cause.

Small and Svendsen (1990) conceptualized irrigation purposes within a nested *means* and *ends* framework where the provision of water for agricultural crops represents an output from an irrigation system, but an input into an irrigated agriculture system. Likewise, a sustained increase in agricultural productivity is simultaneously an output of the irrigated agriculture system and an input into a broader agricultural economic system. The output of this system – increased incomes in the rural sector – also represents an input into a still broader rural economic system, whose output in turn is an input into the nation's socio-economic system, as presented in Table 1.

Table 1. Irrigation purposes as nested means and ends

Level of End	Means	End
Proximate	Operation of irrigation facilities	Supplying water to crops
Intermediate-1	Supplying to crops	Sustained increase in agricultural productivity
Intermediate-2	Sustained increase in agricultural productivity	Increased incomes in rural sector
Intermediate-3	Increased incomes in rural sector	Rural economic development
Ultimate	Rural economic development	(1) Improved livelihoods of rural people (2) Sustained socio-economicdevelopment for the whole economy

Adopted From Small and Svendsen (1990)

For these results to be achieved, SSIS in low and middle-income countries need to perform at the highest levels. The broad question of the performance of SSIS is, however, less understood and a number of sub-questions can arise from the performance of SSIS. Firstly, can indicators of SSIS performance be standardized? Secondly, are there any discernible patterns of performance among SSIS? Thirdly, how are constraints to SSIS sustainability classified? Finally, what do we know about the linkage between SSIS performance and the provision of food security and poverty reduction? This chapter attempts to provide evidence in answer to these questions.

MATERIALS AND METHODS

This chapter utilized a comprehensive review of the empirical literature on various facets of SSIS performance, based on the standard presentation of a systematic literature review by Pereira and Marques (2017) and Kamwanga-Mthetiwa *et al.* (2016). The attempt to address various issues on SSIS performance in developing countries resulted in extensive literature being gathered, and the exercise was academically exciting and equally challenging. Specific keywords were selected, relevant scientific databases identified, search terms developed and then applied to each bibliographic database. The search period was limited to studies published between 1990 and 2017, a period with increased interest in SSIS development in the low- and middle-income countries. Only studies from low- and middle-income countries were included in the final review. The researchers consulted a wide array of open access websites that provide information that addresses a number of issues under study, and search engines such as Google Scholar and Google.com were utilized. Furthermore, reference lists of identified sources were also examined for literature published in English. Relevant literature was screened based on the title and its abstract and full texts reviewed were only those that met the inclusion criterion of relating to at least one of the research questions. Most studies that compared irrigated with rain-fed agriculture and a few that used 'before and after irrigation' approaches were included for robustness of the results. The studies included applied various methodologies, all robust, within and across topics covered. This method resulted in high quality empirical evidence.

Following all inclusion criteria, finally, 6 papers on performance indicators; 34 papers on SSIS performance assessment; 26 papers on constraints that SSIS face; 27 papers on SSIS irrigation valuation; and 46 papers on the impact of SSIS on social outcomes were considered for full review. Most of the literature reviewed was published in renowned international academic journals, some in books and some sources can be categorized as grey literature (reports, Master's and PhD dissertations).

The authors of the reviewed work came from various research institutions, such as academic institutions, international organizations and agriculture-related institutions. This chapter benefited from the rich expertise arising from various research networks, where most of the work reviewed had a minimum of two authors, depending on the topic studied.

Most of the research works were published in the last decade, reflecting significant interest in the areas studied. The studies were based on data from the developing world, and were largely motivated by the need to reduce and possibly eradicate poverty, which is pronounced in these countries. The highest percentage of studies reviewed arose from the African continent (70.5%), with the rest coming from outside of Africa.

This study extracted data such as year of publication; country of study; performance indicators; performance indicator type used; social outcome used; constraints faced; methodologies used; determinants where applicable and data available; and a studies' final results. The results were analyzed using narrative synthesis categorized by performance indicators, performance assessment, SSIS constraints, irrigation water economic valuation, and SSIS social outcomes.

RESULTS AND DISCUSSION

Irrigation Performance Indicators

Indicators are numbers used to describe the level of actual achievement in respect of the objectives of irrigation schemes (performance assessment), which supports the planning and implementation process so as to achieve the efficient use of resources

(Bos, Burton, & Molden, 2005; D. Molden, 1998). The current status of an irrigation scheme is established via assessing its performance so that underlying factors can be identified. This assessment will in turn signal the means of scheme improvement, since most schemes rarely perform at full capacity.

According to the reviewed literature, the answer to the first research question on standardization of irrigation performance indicators is quite challenging. In fact, studies measuring irrigation performance used various performance indicators. For instance, Rao (1993) and Molden (1990) summarized external performance indicators (EPI) (which reflect the degree of adequacy in resources utilization in obtaining final outputs in irrigation schemes) as output per unit irrigation supply, output per unit water consumed, relative water supply, relative irrigation supply,water delivery capacity, gross return on investment, and financial self-sufficiency. These indicators were then categorized and whether used by the two most cited works, Bos *et al.* (2005) and Greaves (2007), is indicated in Table 2.

Even though there are four studies that used water balance, service and maintenance indicators as irrigation performance indicators, the indicator types applied under this class differ across authors. On one hand, Boss *et al.* (2005) considered the following indicators: overall consumed ratio, field application ratio, depleted fraction, drainage ratio, outflow over inflow ratios, delivery performance ratio, dependability of interval between water applications, canal water level, head-discharge relationship, effectiveness of infrastructure; water level ratio, and modified interquartile ratio. On the other hand, Greaves (2007) considered the following indicators: total annual volume of irrigation water delivery, relative water supply, relative irrigation supply, irrigation efficiency, water delivery capacity ratio, annual irrigation water supply per unit command area, annual irrigation water supply per unit irrigated area, and security of entitlement supply. Furthermore, studies differed by how they treated economic and/or agricultural output indicators. Bos *et al.* (2005) used, for example, water productivity, land productivity, management, operations and maintenance funding ratio, operations and maintenance fraction, fee collection ratio, relative water cost, price ratio, crop yield ratio, cropping intensity, and crop production relative to family food needs as indicators. Greaves (2007) differed from Bos *et al.* (2005) in that the study added the total annual value of total output and the output per unit command area, output per unit irrigation supply, achieved production factor and potential production factor.

However, these two studies concur on the environmental indicators used which were: ground water depth, pollution of water, and sustainability of irrigable area (change in water table depth over time, salt balance). A unique set of indicators were applied by both studies reviewed, where Bos *et al.* (2005) considered a class of emerging indicators as ideal in irrigation performance assessment, these being crop water deficit, relative evapotranspiration, relative soil wetness, and biomass production per cubic meter water support. Whereas Greaves (2007) had financial indicators for performance of irrigation as total management, operations and maintenance cost per unit command area, total cost per person employed, revenue collection performance, staffing numbers per unit command area, and average revenue per unit cubic meter of water supplied.

Gorantiwar and Smout (2005) summarized external performance indicators into *allocation types* (productivity and equity indicators, which need to be attained during allocation of resources at the planning and operation stages); and *scheduling types* (measures of adequacy, reliability, flexibility, efficiency and sustainability regarding the distribution of irrigation water to the users). They further grouped these two categories of performance indicators into: *economic* (productivity), *social* (equity), *environmental* (sustainability), and *management* (reliability, adequacy, efficiency and flexibility) indicators. This system provides a holistic approach to performance assessment of irrigation schemes.

Burt and Styles (1998) argued that external indicators only give an idea of the relative magnitudes of some major inputs and outputs of irrigation projects directly associated with the irrigation project itself, and none provide insight regarding its internal mechanisms such as, management, social, and hardware instruments. It is, therefore, considered necessary to consider both external performance factors and internal performance indicators (IPI) if irrigation project performance is to be improved. Burt and Styles (1998) therefore proposed including internal indicators where no single indicator is capable of gauging a scheme on its own, but when indicators are reviewed holistically and combined with certain external indicators then some insight into the design, operation, and management of an irrigation scheme can

Table 2. Summary of External Irrigation Performance Indicators

Indicator Category	Definition	Studies that Utilized the Indicators
Water Balance, Service and Maintenance	Primary function of irrigation and drainage, provision of water service to users	Bos et al. (2005); Greaves (2007)
Environmental	Extent to which irrigation systems influences the environment and whether they are sustainable or not	Bos et al. (2005); Greaves (2007)
Economic/ Agricultural Output	Quantify crop yield and the related generated funds to manage the irrigation system	Bos et al. (2005); Greaves (2007)
Financial	Concentrate on the costs and returns on monetary value of the irrigation investment	Greaves (2007)
Emerging	Depend on parameters measured via satellite remote sensing	Bos et al. (2005)

be gained. Such internal indicators point to the equity, reliability, and the adequacy of the irrigation supply, and enhance efficient and economic agricultural production. They need to be investigated in order to gauge the potential of an irrigation project.

Irrigation projects' performance is influenced by many factors. Irrigation outputs as identified by Burt and Styles (1998) are cropping intensity, average crop yields, yield per unit of water consumed and downstream environmental impacts. Computation of internal indicators requires information on factors identified by these authors as *physical and institutional constraints,* factors influencing *service quality* (hardware design and management), *service delivered* (actual level and quality), and *symptoms* (percentage collection of water fees, viability of water user associations, condition of structures and canals, water theft). External indicators require some of the information contained under *physical constraints* and *output* categories

EMPIRICAL EVIDENCE ON SSIS AND AGRICULTURAL PRODUCTION

Performance assessment of SSIS is necessary as it makes explicit the level of actual achievement against set production and productivity targets. Understanding the contributing factors to current productivity levels can lead to improving productivity and the more efficient use of resources (Pereira & Marques, 2017). The majority of the studies reviewed (about 92%) as presented in Table 3, provided evidence that SSIS increased crop productivity, using different performance indicators and research methodologies (with stochastic production frontier being the dominant method used). On the other hand, Pereira and Marques (2017) provided a review of the empirical literature on methodologies applied to measure irrigation efficiency where data envelopment analysis was the most popular method. However, mixed results were obtained from a study by Kamwamba-Muthethiwa *et al.* (2016), where yield and profitability indicators showed a positive impact for an irrigated scheme, but pumping and irrigation system efficiency measures produced inconclusive results for SSIS performance. Although no standard performance measures exist (as revealed under Table 3, column 7), which poses a challenge for performance evaluations across schemes, various performance proxies used gave useful insights into SSIS performance that can be used to guide policy formulation. Management style was observed to influence SSIS performance as studied by Makombe and Sampath (1998), Makombe and Sampath (1999), Sharma *et al.* (2001) and Faulkner (2006). Quite a few of the studies utilised both internal and external irrigation performance indicators in their analysis, as presented in Table 3, yet these indicators give a holistic irrigation performance assessment. Furthermore, for a study that considered different irrigation technologies (sprinkler, flood and drip) used in SSIS, it was sprinkler

Table 3. Summary of evidence on the impact of SSIS on agricultural production

No.	Country of Study	Comparator	Performance Indicator Assessed	Methodology	Outcome	Indicator Type Used to Quantify Impact	Source
1	Philippine	Rain-fed vs irrigated agriculture	IPI EPI X	Stochastic Production Frontier (SPF)	Positive	Resource use, Crop productivity	Rola *et al.* (1993)
2	Pakistan	Poor vs non-poor irrigated farmers	X	SPF	Positive	Resource use, crop productivity, economic returns	Ahmed (2003)
3	Ethiopia	Rain-fed vs irrigated agriculture	X	SPF	Positive	Crop productivity	Makombe *et al.* (2007))
4	Iran	Rain-fed vs irrigated agriculture	X	SPF	Positive	Crop productivity	Ghaderzadeh and Rahimi (2008)
5	Ethiopia	Two irrigation schemes with similar settings but contrasting socio-economic performance	X	SPF	Positive	Crop yields; economic returns; irrigation water use efficiencies	Van Halsema *et al.* (2011)
6	Tanzania	Traditional vs modern irrigators; irrigated vs rain-fed farmers	X	SPF	Positive	Crop productivity	Makombe *et al.* (2011)
7	Sri Lanka	Irrigated agriculture	X	SPF	Positive	Crop profitability	Shantha *et al.* (2012)
8	Nigeria	Irrigated agriculture	X	SPF	Positive	Crop productivity	Ahmadu and Alufohai (2012)
9	Ethiopia	Rain-fed vs irrigated agriculture	X	SPF	Positive	Crop productivity	Gebregziabher *et al.* (2012)
10	India	Alternative tenurial contracts	X	SPF	Positive	Crop productivity	Laha and Kuri (2013)
11	Ethiopia	Irrigated agriculture	X	SPF	Positive	Crop productivity	Regassa (2013)
12	Nigeria	Rain-fed vs irrigated agriculture	X	SPF	Positive	Crop productivity and profitability	Usman and Zakari (2014)
13	Ethiopia	Rain-fed vs irrigated agriculture	X	SPF	Positive	Crop productivity	Yuya (2014)
14	India	Rain-fed vs irrigated agriculture	X	SPF	Positive	Crop productivity	Rani and Singh (2015)
15	Uganda	Unfavorable climate Simulations on rain-fed agriculture	X	SPF	Negative	Crop productivity	Maruyama *et al.* (2014)
16	Ethiopia	Rain-fed agriculture	X	SPF	Positive	Crop productivity, economic returns	Debebe *et al.* (2015)

continued on following page

Table 3. Continued

No.	Country of Study	Comparator	Performance Indicator Assessed	Methodology	Outcome	Indicator Type Used to Quantify Impact	Source
17	Nigeria	Rain-fed vs irrigated agriculture	X	SPF	Positive	Crop productivity and profitability	Wakili and Isa (2015)
18	Burkina Faso	Irrigated agriculture	X	SPF	Positive	Crop productivity	Kabore (2016)
19	Ghana	Irrigated agriculture	X	SPF	Positive	Crop productivity	Anang *et al.* (2016)
20	Nigeria	Rain-fed vs irrigated agriculture	X	SPF	Positive	Yield gap, crop productivity and profitability	Babatunde *et al.* (2016)
21	Iran	Rain-fed vs irrigated agriculture	X	Data Envelope Analysis (DEA)	Positive	Crop productivity	Banaeian and Zangeneh (2011)
22	Vietnam	Rain-fed commercial crops vs staple food farming	X	DEA	Positive	Crop productivity	Dao (2013)
23	Pakistan	Tube-well owners vs water buyers under irrigated agriculture	X	DEA	Positive	Crop productivity and irrigation efficiency	Watto and Mugera (2014)
24	Kenya	Rain-fed farming	X	DEA	Positive	Crop productivity	Ogada *et al.* (2014)
25	India	Irrigated agriculture	X	Panel data	Positive	Crop productivity, land prices, crop intensities	Jin *et al.* (2012)
26	Tunisia	Irrigated agriculture	X	Input-specific technical efficiency	Positive	Crop productivity and irrigation efficiency	Naceur and Mongi (2011)
27	Zimbabwe	Different irrigation technologies (Sprinkler, flood and drip) for two schemes	X	Descriptive and inferential statistics	Sprinkler best for farmers	Crop productivity	Mupaso *et al.* (2014)
28	Tanzania	Irrigation scheme cooperative vs private input arrangements	X	SPF and Enterprise budgeting	Cooperative arrangement better	Crop productivity and profitability	Kangile (2015)
29	Sub-Saharan Africa	SSIS	X	Systematic review	Mixed	Technical, crop productivity and profitability	Kamwamba-Muthethiwa *et al.* (2016)
30	Africa, Asia, Europe, North America, Oceania	Methodologies on irrigation performance	X	Systematic review	DEA most used; positive	Irrigation efficiency	Pereira and Marques (2017)
31	Zimbabwe	Three SSIS	X X	Production function, Theil	Farmer-managed better	Crop production, inequity, management performance	Makombe and Sampath (1998)

continued on following page

Table 3. Continued

No.	Country of Study	Comparator	Performance Indicator Assessed	Methodology	Outcome	Indicator Type Used to Quantify Impact	Source
32	Zimbabwe	Three SSIS	X X	Benefit-cost analysis	SSIS community-managed better than government management	Benefit-cost ratios and management styles	Makombe and Sampath (1999)
33	Nepal	Two SSIS	X X	SPF	Farmer-managed better than government-managed	Crop production and management performance	Sharma *et al.* (2001)
33	Ghana	Two SSIS	X X	Benefit-cost analysis	Management style affected performance	Relative water supply, profitability and management influence	Faulkner (2006)
34	Sri Lanka	Rain-fed and irrigated agriculture	X X	DEA	Positive	Water availability and crop productivity	Thibbotuwawa *et al.* (2013)

irrigation that proved to be the best for farmers (Mupaso *et al.*, 2014). Irrigation cooperative input sourcing arrangements for SSIS proved better than own-farmer arrangements in terms of lowering input costs that affected their cost structures and efficiency levels too (Kangile, 2015). It is important to note that all the SSIS studied performed below their full potential, showing that there is much room to improve them so as to increase crop production and productivity, which affect food security and poverty statuses of the people depending on them (directly or otherwise).

MAJOR CONSTRAINTS FACING SSIS: WHAT THE EMPIRICAL LITERATURE SAYS

Based on the evidence on performance of SSIS reviewed in the previous section, it is clear that all systems are operating below their full potential, raising questions on SSIS sustainability. In this section, evidence on constraints SSIS face are presented, followed by an economic valuation of irrigation water as a measure of sustainability.

Quite a number of research studies as presented in Table 4, were undertaken to understand the prevailing challenges facing irrigation farmers. The evidence was gathered and analyzed qualitatively. The general conclusion from the evidence from different countries and continents provided is that the major constraints encountered by SSIS are related to farmer characteristics, costs, institutional setups, the policy

Table 4. Summary of evidence on constraints facing SSIS

No.	Country of Study	Major Constraints	Source
1	Zimbabwe	Poor physical set up, ignorance of socio-economic framework during setup	Manzungu (1995).
2	Ethiopia	Problems related to physical nature of irrigation systems, irrigation water application, marketing, policy related, engineering-related, irrigation economy and levels of farmer participation	Aberra (2004)
3	South Africa	Access to information, training, market access, planning, transport, extension services	Magingxa *et al.* (2006)
4	Ethiopia	Water management, market access, irrigation farming experience, household size, educational levels, training, access to formal credit, extension services, farmer participation during establishment, lack of participation of female household in leadership	Deribe (2008)
5	Zimbabwe	Inadequate and expensive inputs, market access, high and unaffordable water and electricity bills, weak capital base	Chibisa *et al.* (2008)
6	Nigeria	Poor and inadequate irrigation agency services, costs more than revenue collected, lack of irrigation investment, lack of access to farm inputs	Oriola (2009)
7	Ethiopia	Water access and distribution, water seepage problem, lack of servicing for irrigation equipment, market access and facilities, crop disease, input cost	Ayele (2011)
8	Ethiopia	Extension services, poor water institutions, technical and financial capacity, post-harvest storage, processing and marketing services, inadequate planning, design and construction failures, poor operation and management of schemes, soil erosion and sedimentation, lack of maintenance of irrigation infrastructure	Awulachew and Ayana (2011)
9	Ghana	Indigenous and water users' institutions complementarity	Derbile (2012)
10	Nicaragua	Little know-how, lack of technical assistance, low access to credit, low agricultural labor demand	Hallensleben (2012)
11	South Africa	Socio-economic, political, climatic, edaphic and design factors; lack of farmer participation; farmer practice, dilapidated irrigation infrastructure, weak institutional and organizational setups	Fanadzo (2012)
12	Southern Africa	Poor infrastructure, poor crop production knowledge, limited farmer participation, ineffective extension and mechanization services, lack of reliable market access and effective credit services, predominance of subsistence-oriented farming	Mwendera and Chilonda (2013)
13	Swaziland	Inadequate irrigation water access, inadequate knowledge and skills for sustainable agriculture production practices, poor irrigation designs, high debts, poor market environment and inadequate skills for business management, conflicts from poor management, lack of cooperation by members, land tenure issues	Nkambule and Dhlamini (2013)
14	Ghana	High cost of hiring farm machinery, inadequate access to credit, poor water supply, ineffective technical assistance, lack of entrepreneurial skills, inability to store perishable products and inadequate ready market	Dinye (2013)
15	Zimbabwe	Poor asset base, limited access to agricultural inputs, limited access to irrigation water, electricity load shedding, lack of credit facilities, low application rates of fertilizers, low value crops grown in the scheme, and poor agricultural output markets	Mutambara and Munodawafa (2014)
16	Ethiopia	Irrigation water supply unreliability, unfair distribution of water, timeliness problem in water distribution, weak and poor organization of irrigation scheme administration, the water users' associations lacked clear laws/by-laws, no documented administration matters which dealt with users, no scheme auditing, poor irrigation water management institution	Ulsido and Alemu (2014)
17	Ethiopia	Poor irrigation water management, soil salinity, waterlogging, soil erosion and degradation, sedimentation, build-up of pests and diseases, non-existent extension services on irrigation water management	Etissa *et al.* (2014)
18	Cambodia	Poor system design and management, production constraints	de Silva *et al.* (2014)
19	Southern Africa	Management style, irrigation method, crop mix, type of financier and geography	Mutiro and Lautze (2015)
20	Kenya	Size of land under irrigation, per acre operations and maintenance costs, donor funding	Ngenoh *et al.* (2015)
21	Tanzania	Poor access to inputs, farm equipment, transportation, value-adding opportunities and functional markets, lack of finance and the storage of outputs, water supply issues	Mdemu *et al.* (2016)

continued on following page

Table 4. Continued

No.	Country of Study	Major Constraints	Source
22	Zimbabwe	Poor infrastructure and soil fertility, weak input and output markets, high transport costs, lack of farm implements, poor agricultural knowledge, idle and low utilization of irrigated land, uncoordinated marketing of produce, siltation problems, water-leaking canals, land tenure, poor policies, cultivation of low-value crops	Moyo *et al.* (2016)
23	Sub-Sahara	No access to markets; lack of knowledge, farm implements and financing; poor governance; timely access to knowledge; high-quality inputs	Bjornlund *et al.* (2016)
24	Ghana	Poor management of irrigation facility; lack of storage, processing facilities and extension services; inadequate access to credit; poor market prices for farm produce; high cost of farm inputs; lack of transport; high prevalence of diseases and pests; poor maintenance of irrigation infrastructure; high water use levy	Adam *et al.* (2016)
25	Ethiopia	Shortage of water, access to improved seeds, marketing and increment of farm input costs	Mengistie and Kidane (2016)
26	Africa and Asia	Water management challenges; low, peripheral or no farmer engagement by irrigation development agencies; no farmers' ownership and responsibility; failed water reforms; ineffective water management bodies; failure by farmers to meet their utility bills; water and electricity highly subsidized; sudden withdrawal of government support in irrigation schemes before deliberate capacity building of the water users; irrigation management committees remained weak and lacked the necessary legal backing; state intervention in water management and the failure to contextualize the management practices; limited water markets disconnected to rest of markets; limited use of underground water; huge costs of establishing and/or rehabilitating irrigation schemes	Mutambara *et al.* (2016)

environment, design issues, cultural factors, community attributes, and environmental issues. However, as observed from Table 4, the nature and extent of these constraints on SSIS performance differ across schemes within and across countries due to unique scheme characteristics. There is therefore no way they can be generally classified in terms of their nature and extent as they are context-specific. For a comprehensive summary of the constraints affecting most developing countries, see Mutambara *et al.* (2016). Relevant authorities tackling these constraints facing SSIS will help sustain their operations and maximize the benefits they were established for.

EMPIRICAL LITERATURE ON ECONOMIC VALUE OF IRRIGATION WATER

Irrigation systems are truly sustainable if that sustainability stems from both farmers' and government perspectives. Sustained operation of SSIS can be explained by so many factors, some of which were reviewed in the previous section. With climate change challenges, there have been instances of increasing droughts, making irrigation water scarcer. Since water is an economic good, it follows that how irrigation water is valued and used might contribute to the future sustainability of SSIS. It is thus worth learning from the following literature review on how farmers value and use irrigation water to sustain the noble development cause.

Qureshi *et al.* (2010) summarized methods for assessing the value of irrigation water as using either inductive (observation based methods) or deductive techniques. The inductive techniques differ according to type and source of data. The common techniques used to estimate the productivity relationship include, stated preference methods, contingent valuation method (CVM), and land value method. The following methods impute the value of water via land and implementing valuation from land market data: Hedonic Pricing Method (HPM), choice experiment methods (CEM) and econometric valuation of irrigation water based on primary and secondary data. Deductive techniques are used to derive shadow prices of irrigation water and include residual imputation method (RIM) and its variations, value marginal productivity of water, mathematical programing, hydro-ecological models, and computable general equilibrium models. Therefore, the review presented is based on both deductive and inductive methods.

Evidence as presented in Table 5 points to the general conclusion that farmers in SSIS are willing to pay for irrigation water, regardless of the method utilized in the economic valuation of water. In most cases, except in two cases as shown by Zakaria *et al.* (2014) and Omondi *et al.* (2014) where irrigation water was overpriced (overvalued), the overall conclusion was that irrigation water was underpriced (hence undervalued) by respective authorities as against the SSIS farmers who were prepared to pay more for an improvement in irrigation services. This undervaluing undermines SSIS sustainability. If farmers continue paying lower than the economic value of water, they will continue to use it inefficiently. Irrigation water that is priced-right discourages inefficiency in its use and promotes sustainability in the current context of increasing drought. It is important to note that the majority of studies utilized inductive methods in irrigation water evaluation as compared to the deductive methods, or made use of both. This reliance on the use of inductive methods can generally be explained by the complexities in applying the deductive methods, which require strict assumptions and large volumes of agricultural and market data, making it less user friendly. Furthermore, deductive methods tend to yield higher water values than inductive ones (Jaghdani *et al.*, (2012). As revealed by evidence presented in Table 5, the economic values of irrigated water vary by crops, location of farm on the scheme, farm ownership, water management efficiency, farmer characteristics, bid level, and state of irrigation infrastructure, among other issues identified in column 5 of Table 5. These economic values as revealed by SSIS farmers need to be incorporated in water pricing policies if SSIS sustainability is to be achieved.

Table 5. Summary of evidence on economic value of irrigation water

No.	Country	Methodology	Outcome	Determinants	Source
1	Bangladesh	CVM	Water undervalued	bid level, respondents' age, education, family size, number of income sources, ownership of farm land, management system, decision to change cropping patterns	Akter (2007)
2	Iran	HPM	Water undervalued	availability of irrigation water, water quantity, land-structure characteristics neighborhood characteristics	Esmaeili and Shahsavari (2011)
3	Ethiopia	CVM and CE	Water undervalued	practical irrigation experience of households, average annual income, participation in off-farm activities, and market access	Assefa (2012)
4	Ghana	CVM	Water undervalued	location of farm and land lease prices land ownership	Alhassan *et al.* (2013)
5	China	CVM	Water undervalued	Bid level, family size, household's income, area of irrigation land, source of irrigation water, respondents' satisfaction with management and farmers' attitude towards whether current waters price could recover the water supply cost	Tang *et al.* (2013)
6	Tanzania	CVM	Water undervalued	education, farm size, household income, farm distance	Ndetewio *et al.* (2013)
7	Ethiopia	CVM	Water undervalued	sex of the household, household head educational level, total annual income, credit utilization, and perceived trend in rain-fed agricultural productivity, family size and initial bid	Birhane (2014)
8	Kenya	CVM	Water overvalued	Off-farm income, access to credit and satisfaction with the management of water supply system	Omondi *et al.* (2014)
9	Ethiopia	CVM	Water undervalued	Education level, household size, gender, first bid, total family income and cultivated land size	Alemayehu (2014)
10	Ghana	CVM	Water overvalued	Age, maintenance culture of irrigation facilities, on-scheme income as ratio of household income, off-scheme income, farm locations	Zakaria *et al.* (2014)
11	Uganda	CVM	Water undervalued	formal education of the household head; farm size; practical experience in irrigated farming; participation in training related to farming and irrigation water management; access to credit and markets	Namyenya *et al.* (2014)
12	India	CVM	Water undervalued	-	Karthikeyan *et al.* (2009)
13	India	CVM	Water undervalued	Bid price, total area under irrigation, location of a plot, crop income, and age	Biswas and Venkatachalam (2015)
14	Turkey	CVM	Water undervalued	price of irrigation water, education, location, irrigation type, and attitudes toward associations.	Aydogdu (2016)
15	Turkey	Irrigation water productivity	Water undervalued	Primary school graduates, users of modern irrigation technologies, crop pattern, married farmers, property owners, gravity irrigation users and farmers' perceptions about natural resources	Aydogdu and Bilgic (2016)
16	Ethiopia	CVM	Water undervalued	-	Birhane and Geta (2016)
17	Pakistan	RIM	Water values established	-	Ashfaq *et al.* (2005)
18	Pakistan	RIM and Change in net income	Water undervalued	-	Hussain *et al.* (2009)
19	Australia	RIM	Water undervalued	-	Qureshi *et al.* (2010)

continued on following page

Table 5. Continued

No.	Country	Methodology	Outcome	Determinants	Source
20	Spain	RIM	Water undervalued	-	Berbel *et al.* (2011)
21	South Africa	RIM	Water values established	-	Speelman *et al.* (2011)
22	Jordan	RIM	Water undervalued	-	Al-Karablieh *et al.* (2012)
23	Mexico	Net income change	Water values established	-	Yedra *et al.* (2016)
24	Thailand	CVM, marginal value product	Water undervalued	-	Tiwari (1998)
25	Iran	CVM, Marginal value product and Change in net rent	Water undervalued	-	Jaghdani *et al.* (2012)
26	South Africa	RIM and CVM	Water undervalued	extension services, training, use of motorized pumps, farmer perceptions of scheme management, duration of the farmer in the scheme, livestock ownership and road conditions	Njoko (2014)
27	Indonesia	RIM and CVM	Water undervalued	-	Syaukat *et al.* (2014)

IMPACT OF SSIS ON FOOD SECURITY AND POVERTY

The main and final objective of all key stakeholders in the irrigation sector investing in SSIS is to reduce poverty amongst the farming communities, knowing too that such investments have spillover effects on the greater economy. How knowledgeable are policy makers and other key stakeholders in irrigation implementation in terms of the link between SSIS and social outcomes (including poverty, food security and nutritional statuses)? The review evidence with respect to this question is presented here. Many of the studies on the topic that were reviewed concurred that SSIS are important in poverty reduction via so many channels both direct and indirect (see evidence in Table 6). It is important to note that about 63% of the studies presented, compared livelihood outcomes between irrigators and non-irrigators (purely-rain-fed farmers) relative to the potential impact of SSIS, and all utilized a variety of methodologies from simple descriptive analysis to the most complicated econometric methods such as propensity score matching (PSM), and all concurred that irrigation improves the livelihoods of people. The improvements in livelihood outcomes were also captured differently by these studies but were mainly in terms of improvements in food security, increased farm incomes, increased consumption expenditures, increased access to social services, increased nutrients uptake, reduced out-migration, and improved social security networks, among others as presented

Table 6. Summary of evidence on the impact of SSIS on food security and poverty

No.	Country of Study	Comparator	Social Outcome Assessed	Methodology	Outcome	Source
1	Asia	Irrigated agriculture	Poverty	SLR	Positive	Lipton *et al.* (2003)
2	Bangladesh and Nepal	Irrigated vs rain-fed agriculture	Livelihood assets	Sustainable Livelihood Approach	Positive	Brabben *et al.* (2004)
3	International	Irrigated agriculture	Poverty	Poverty-irrigation framework and SLR	Positive	Smith (2004)
4	Asia	Irrigated agriculture	Poverty	SLR	Positive	Hussain and Hanjira (2004)
5	Tanzania	Irrigated vs rain-fed agriculture	Household income	Descriptive statistics	Positive	Kaswamila and Masuruli (2004)
6	China	Irrigated agriculture	Household income	Fixed effects model	Positive	Huang *et al.* (2006)
7	Ethiopia	Irrigated vs rain-fed agriculture	Household expenditures, labor demand and food prices	Ordinary Least Squares (OLS)	Positive	Berg and Ruben (2006)
8	Ethiopia	Irrigated vs rain-fed agriculture	Food security, household income and expenditure	Propensity score matching	Positive	Awlachew *et al.* (2007)
9	Mali	Irrigated vs rain-fed agriculture	Household income, social insurance, nutrient uptake	Propensity score matching	Positive	Dillon (2008)
10	Ethiopia	Irrigated vs rain-fed agriculture	Family employment, household income and consumption	Propensity score matching	Positive	Haile (2008)
11	Zimbabwe	Irrigated agriculture	Household food security and income	Descriptive statistics	Positive	Chibisa *et al.* (2008)
12	Burkina Faso, Mali, Niger and Senegal	Irrigated agriculture	Poverty	Descriptive statistics	Positive	Dittoh *et al.* (2010)
13	Ethiopia	Irrigated agriculture	Household income, nutrition uptake, asset base	Sustainable livelihood approach	Positive	Tucker and Yirgu (2010)
14	Zimbabwe	Irrigated vs rain-fed agriculture	Food security, household income	OLS	Positive	Nhundu *et al.* (2010)
15	Swaziland	Before and after irrigated agriculture	Food security, household income	Descriptive statistics	Positive	Peter (2011)
16	Ethiopia	Irrigated agriculture	Asset base, household income	OLS	Positive	Alemu *et al.* (2011)
17	South Africa	Irrigated agriculture	Poverty and food security	OLS	Positive	Tekana and Oladele (2011)
18	Ethiopia	Irrigated vs rain-fed agriculture	Household income and poverty	Tobit Model	Positive	Ayele (2011)
19	South Africa	Irrigated vs rain-fed agriculture	Poverty profiles	Forster-Greer-Thorbecke index (FGT)	Positive	Meliko and Oni (2011)
20	Ethiopia	Irrigated vs rain-fed agriculture	Poverty profiles	FGT model	Positive	Bacha *et al.* (2011)

continued on following page

Table 6. Continued

No.	Country of Study	Comparator	Social Outcome Assessed	Methodology	Outcome	Source
21	South Africa	Irrigated vs rain-fed agriculture	Food security	Descriptive statistics	Positive	Maliwichi *et al.* (2012)
22	Ghana	Irrigated vs rain-fed agriculture	Household consumption expenditure	PSM, OLS and switching regression	Positive	Kuwornu and Owusu (2012)
23	Ethiopia	Irrigated vs rain-fed agriculture	Household income and consumption expenditure	PSM	Positive	Hagos *et al.* (2012)
24	Ghana	Irrigated vs rain-fed agriculture	Food security and employment	FGT and descriptive statistics	Positive	Apam (2012)
25	Zimbabwe	Irrigated agriculture	Household income, asset base	SLA	Positive	Chazovachii (2012)
26	Ethiopia	Irrigated agriculture	Household income	Descriptive statistics, Heckman's estimation	Positive	Aseyehegu *et al.* (2012)
27	Ethiopia	Irrigated agriculture	Household income	OLS	Positive	Belay and Beyene (2013)
28	Ghana	Irrigated agriculture	Household income	Descriptive statistics	Negative	Dinye (2013)
29	Ghana	Before and after Irrigated agriculture	Household food security and leisure	Descriptive statistics	Mixed	Bagson and Kuuder (2013)
30	Ethiopia	Irrigated vs rain-fed agriculture	Poverty profiles and household consumption expenditure	FGT, descriptive statistics, PSM	Positive	Haji *et al.* (2013)
31	Sri Lanka	Irrigated agriculture	Household income	SLA and regression analysis	Positive	Sellamuttu *et al.* (2014)
32	Malawi	Irrigated vs rain-fed agriculture	Household income, food security	PSM	Positive	Nkhata (2014)
33	Ghana	Irrigated vs rain-fed agriculture	Household consumption, income	FGT	Positive	Banson *et al.* (2014)
34	South Africa	Irrigated vs rain-fed agriculture	Household consumption, food security	Treatment effect model, FGT	Positive	Sinyolo *et al.* (2014)
35	Ethiopia	Irrigated vs rain-fed agriculture	Household income	Descriptive statistics	Positive	Beyan *et al.* (2014)
36	India	Irrigated vs rain-fed agriculture	Household income and consumption	Descriptive statistics	Positive	Beero and Narayanamoorthy (2001)
37	Ethiopia	Irrigated vs rain-fed agriculture	Household income	FGT, descriptive statistics	Positive	Eneyew *et al.* (2014)
38	Ethiopia	Irrigated vs rain-fed agriculture	Household, expenditure income, asset base	PSM, descriptive statistics	Positive	Zeweld *et al.* (2015)
39	Ethiopia	Irrigated ss rain-fed agriculture	Household consumption, poverty profiles	FGT, PSM	Positive	Solomon and Ketema (2015)
40	Ethiopia	Irrigated agriculture	Household income	OLS	Positive	Yihdego *et al.* (2015)

continued on following page

Table 6. Continued

No.	Country of Study	Comparator	Social Outcome Assessed	Methodology	Outcome	Source
41	Ghana	Irrigated agriculture	Household food security, income, social services	Descriptive statistics	Positive	Dawda and Amosah (2015)
42	Ethiopia	Irrigated vs rain-fed agriculture	Household income, asset base, food security	Food balance model, descriptive statistics	Positive	Siraw (2016)
43	Zimbabwe	Irrigated agriculture	Food security, asset base	SLA	Positive	Dube (2016)
44	Ghana	Irrigated vs rain-fed agriculture	Household income, food security	PSM	Mixed	Ukudugu *et al.* (2016)
45	Ethiopia	Irrigated vs rain-fed agriculture	Household income	Descriptive statistics	Positive	Mengistie and Kidane (2016)
46	Ghana	Irrigated vs rain-fed agriculture	Household income, nutritional status, out migration	Descriptive statistics	Positive	Adam *et al.* (2016)

in Table 6, column 4. Robust evidence of the impact of SSIS is attributed to studies that used regression and propensity score matching, as these methods provided rigor and depth in understanding issues investigated, quantitatively. However, there was evidence that suggested that SSIS did not help in reducing poverty, as provided by Dinye (2013). According to the study this outcome was due to the overwhelming challenges faced by particular SSIS. Mixed results were also found regarding SSIS impact on poverty by Bagson and Kuuder (2013) and Akudugu *et al.* (2016), and various arguments were presented.

DISCUSSION OF THE RESULTS AND CONCLUSION

Performance assessment of irrigation projects is critical in that it informs key stakeholders in the sector on how such investments are currently performing so that improvements can be implemented if variations from the objectives that were initially set are observed. Overall, irrigation performance assessment is necessary as the results of such an exercise have a bearing on various levels of irrigation operations as conceptualized by Small and Svendsen (1990). This chapter summarized performance indicators drawn from various sources in the literature that were used to assess irrigation schemes and/or systems performance broadly, using both external and internal indicators. Notably, studies by Burt and Styles (1998) and Gorantiwar and Smout (2005) prescribed a holistic approach to performance assessment. From the reviewed literature, a variety of performance indicators were used and no standardization of assessment was observed. Generally, performance indicators chosen for particular studies depended on the study's objectives. All the same, studies that considered both categories of indicators in their assessments were premised to produce robust and sound performance assessment outcomes.

The chapter reviewed literature on the effect of SSIS on agricultural productivity and/or efficiency, based on performance indicators (internal and external). Results from these studies and from the different methodologies utilized, show that SSIS operate on a higher production frontier than rain-fed farms, indicating that SSIS are technically more efficient and more productive than rain-fed agriculture in terms of resource utilization. However, irrigated agriculture efficiency levels still fall too far below their full capacity in most areas studied, except in Uttarakhand state in India, which recorded the highest TE score of 99% (Rani and Singh, 2015). In short, SSIS were observed to be inefficient, indicating room for further improvement. Furthermore, higher TE levels of modern irrigated farms over traditional irrigation farms did not translate into higher profit levels (Gebregziabher *et al*, 2012 and Babatunde, 2016). The same was also observed between higher TE levels of irrigated and rain-fed farms. Market-oriented crops were observed to have higher TE scores than staple food outputs (Dao, 2013). Irrigation management systems of SSIS impacted on production and distribution of production gains among farmers, where farmer-managed SSIS were observed to have outperformed government-owned irrigation systems (Makombe and Sampath, 1998, 1999; Faulker, 2006).

The study found it equally important to consider literature surrounding SSIS sustainability. In this regard, literature on challenges facing SSIS and irrigation water valuation by small-scale farmers were considered.

Overall, agricultural performance, according to the literature reviewed in the section on challenges facing SSIS, was affected by socio-economic, environmental, geographic, infrastructural set-up, institutional, political and agronomic factors, as well as lack of farmer participation, among other issues. The study by Mutambara *et al.* (2016) reviewed the problems of water management sustainability of farmer-managed smallholder irrigation schemes in Africa and compared their problems with the situation in Asian countries. The literature review yielded a comprehensive summary of these issues. The list of issues includes critical water management challenges, peripheral or no farmer engagement by irrigation development agencies, failure to stimulate the necessary farmers' ownership and responsibility, and failed water reforms that could not sustainably improve the water management. Other issues identified were ineffective water management bodies; failure by farmers to meet their utility bills; highly subsidized water and electricity; sudden withdrawal of government support for irrigation schemes before deliberate capacity building of the water users; irrigation management committees remaining weak and lacking the necessary legal backing to independently solve internal and external challenges; state intervention in water management and the failure to contextualize the management practices that undermined traditional management practices; limited water markets that are disconnected to markets like input supply, and the financial and output market; and, in the face of climate change, use of underground water, which has been critical in Asia but has remained limited in Africa due to high drilling costs. These issues, tied to other factors on the topic as identified in the literature, explain the underperformance of SSIS.

Studies reviewed that addressed irrigation water valuation using various methods, indicate that estimated water charges were much higher than actual charges, indicating that irrigation water was undervalued, which can lead to the inefficient and unstainable use of a vital resource, an exception being a study in Ahero, Kenya, which found that farmers were being overcharged (Omondi *et al.*, 2014). There is also enough evidence on higher willingness to pay (WTP) available for upstream farmers than tail-end farmers, indicating that location of farmland on an irrigation scheme matters in relation to the economic valuation of irrigation water (for instance, Zacharia *et al.*, 2014; Alhassan *et al.*, 2013; Hussain *et al.*; 2009, Ndetewio *et al.*, 2013). However, geographically, higher WTP were revealed by farmers in high water-scarcity zones as compared to farmers in low water-scarcity zones (Biswas and Venkatachalam, 2015). Land ownership also affected WTP where landowners valued irrigation water less than lessees (Alhassan *et al.*, 2013). Irrigation water values as computed from agricultural data differed by crop types (for instance, Al-karablier *et al.*, 2012; Aydogdu and Bilgic, 2016; Ashfaq *et al.*, 2005; Hussain *et al.*, 2009; Qureshi *et al.*, 2010). However, comparison of a variety of methods used to value irrigation water indicates the tendency of deductive approaches yielding very high-water values, and because of these and other strict data requirements, they are less preferred to inductive methods. Water values were mainly explained by almost the same factors that affect SSIS viability with the additions of source of irrigation water, perceived trends in rain-fed agriculture, maintenance culture of irrigation facilities, price of irrigation water, location of farm, attitudes towards water users' associations, bid level and cropping patterns, amongst many issues as reviewed on the topic.

The ultimate objective of all key stakeholders in the irrigation sector investing in SSIS is to reduce poverty amongst the farming communities, knowing too that such investments have spillover effects on the greater economy. It is against this background that impact assessment of SSIS on livelihood outcomes is addressed last in this chapter. Many studies reviewed on the topic concur that SSIS are important in poverty reduction via so many channels, both direct and indirect. Briefly, SSIS are poverty-reducing through increased yields, higher and more reliable outputs (food security), higher incomes, lower producer prices, greater demand for labor and technology, reduction in out-migration, improved and sustained rural livelihoods, reduced dependence on food aid, increased frequency of production, increased nutritional intakes, saving and sharing within communities (informal insurance network), increased base (livestock and non-agricultural assets), increased cultivation of higher value crops (from subsistence to commercial crops), increased resilience to climate change, increased access to social amenities (health and education), increased overall household expenditures, reduced hunger gaps, and lowered poverty incidence, among other gains. These gains can be maximized via the use of small-scale and low-cost irrigation technology (Dittoh *et al.*, 2010, Lipton *et al.*, 2003). However, negative irrigation development externalities were also observed, such as badly designed and managed SSIS that negatively impacted on poverty (Hussain and Hanjira, 2004), and SSIS construction that degraded and altered the environment (Ukudugu et al., 2016). Though SSIS development improved the livelihood outcomes considered, such investment projects have not yet eliminated the food insecurity problem in its totality (Awlachew, 2007; Chazovachii, 2012). Also, such investments have been argued to have significantly reduced leisure and communal intimacy, and degraded cultural heritages (Bagson and Kunder, 2013).

Based on the evidence collated from the systematic review of literature, the following recommendations are made:

- It is critical for policymakers to assess the performance of SSIS from time to time.
- Appropriate policy design and implementation that enables farmers to increase their efficiency in production is necessary as this has multiplier effects ranging from farm productivity growth to economic growth and poverty reduction.
- Among so many factors said to influence irrigation performance, farmers should be encouraged to participate and take ownership of irrigation projects from the initial stages of implementation. This participation promotes and sustains SSIS development.
- Productivity and efficiency targets should be set in conjunction with market access. Market access promotion by relevant authorities may motivate farmers to produce more, whereas aiming to produce more where markets are non-existent will be self-defeating for SSIS farmers. Commercializing SSIS should be the main agenda for policy makers. In short, policy makers should sincerely address all factors said to influence irrigation outcomes;

from irrigation design challenges, implementation challenges, institutional challenges (including markets, inputs, finance, management, tenure, etc.), environmental challenges, and political and socio-economic challenges, amongst many other challenges that affect SSIS operational viability.

- Irrigation water reforms should be guided by the economic value of irrigation water as perceived by SSIS farmers, so as to promote efficient and sustainable use of such a precious environmental good, especially now when the adverse effects of climate change are putting more pressure on its use. Factors behind the economic value as perceived by SSIS farmers should be considered for effective irrigation water pricing.
- It is necessary to invest in new irrigation projects, and rehabilitate and modernize existing ones, as these systems play a significant role in communities where they exist, in terms of improving and sustaining livelihood outcomes. This rehabilitation will enable farmers to move from highly unreliable rain-fed agriculture to irrigated agriculture.

Research gaps exist in understanding SSIS performance, especially now when climate change is threatening the agricultural sector more than before. This concluding section identifies such gaps and looks forward to potential areas for future research into aspects of SSIS. Since irrigated farming differs widely in its form and impacts, and has various local attributes, it will be interesting to understand how unexplored SSIS perform in various parts of the world. Moreover, disaggregating the performance and impact of SSIS into head and tail-end sections can provide more accurate results, as different sites had different values for irrigation water, indicating the varying importance of irrigation water depending on the location of the farm in the irrigation scheme, which could have a bearing on agricultural productivity and efficiency, and hence livelihood outcomes. Studies on SSIS farmers categorized according to their wealth endowments may reveal more robust research outcomes as compared to lumping them into one category, which bring bias to resource use constraints. Challenges facing SSIS vary with the geography of a particular country, and from country to country, and it will be interesting, through performance assessment to ascertain challenges a particular small-scale farmer faces so as to promote an irrigation project's sustainability, as these challenges are not generalizable from elsewhere. In particular, determinants of livelihood outcomes with respect to SSIS are under researched. It will also be interesting to establish which type of irrigation and/or irrigation technology has a more poverty-reducing impact. Future studies can also explore how SSIS farmers are adapting to climate change as this has a bearing on current and future sustainability of their agricultural activities.

REFERENCES

Abebe, M. B., & Geta, E. (2014). *Farmers Willingness To Pay For Irrigation Water Use: The Case Of Agarfa District, Bale Zone, Oromiya National Regional State*. Haramaya University.

Aberra, Y. (2004). Problems Of The Solution: Intervention Into Small-Scale Irrigation For Drought Proofing In The Mekele Plateau Of Northern Ethiopia. *The Geographical Journal*, *170*(3), 226–237. doi:10.1111/j.0016-7398.2004.00122.x

Adam, J. N., Al-Hassan, S., & Akolgo, D. A. (2016). Small-Scale Irrigation And Rural Poverty Reduction In The Upper-East Region Of Ghana. *African Journal of Science and Research*, *2*(5), 38–42.

Adugna, E., Ermias, A., Mekonnen, A., & Mihret, D. (2014). The Role Of Small Scale Irrigation In Poverty Reduction. *Journal of Development and Agricultural Economics*, *6*(1), 12–21. doi:10.5897/JDAE2013.0499

Ahmad, M. (2003). Agricultural Productivity, Efficiency, And Rural Poverty In Irrigated Pakistan: A Stochastic Production Frontier Analysis. *Pakistan Development Review*, *42*(2), 219–248.

Ahmadu, J., & Alufohai, G. O. (2012). Estimation Of Technical Efficiency Of Irrigated Rice Farmers In Niger State, Nigeria. *American-Eurasian J. Agric. & Environ. Sci.*, *12*(12), 1610–1616.

Akter, S. (2007). Farmers' Willingness To Pay For Irrigation Water Under Government Managed Small Scale Irrigation Projects In Bangladesh. *J. Bangladesh Stud*, *9*, 21–31.

Akudugu, M. A., Nyamadi, B. V., & Dittoh, S., & Others. (2016). Transforming Smallholder Agriculture In Africa Through Irrigation: An Assessment Of Irrigation Impact Pathways In Ghana. In *2016 Aaae Fifth International Conference*. African Association Of Agricultural Economists (Aaae).

Al-Karablieh, E. K., Salman, A. Z., Al-Omari, A. S., Wolff, H.-P., Al-Assa'd, T. A., Hunaiti, D. A., & Subah, A. M. (2012). Estimation Of The Economic Value Of Irrigation Water In Jordan. *Journal of Agricultural Science and Technology B*, *2*(5b), 487.

Alemayehu, T. (2014). Smallholder Farmer's Willingness To Pay For Improved Irrigation Water: A Contingent Valuation Study In Koga Irrigation Project, Ethiopia. *Journal Of Economics And Sustainable Development*, *5*(19), 5–15.

Alemu, A. E., Kassa, B., & Yihdego, G. (2011). Determinants Of Land Allocation To Irrigation And Its Wealth Effect: Evidence From Northern Ethiopia. *Journal Of The Drylands*, *4*(2), 310–319.

Alhassan, M., Loomis, J., Frasier, M., Davies, S., & Andales, A. (2013). Estimating Farmers' Willingness To Pay For Improved Irrigation: An Economic Study Of The Bontanga Irrigation Scheme In Northern Ghana. *The Journal of Agricultural Science*, *5*(4).

Anang, B. T., Bäckman, S., & Sipiläinen, T. (2016). Technical Efficiency And Its Determinants In Smallholder Rice Production In Northern Ghana. *Journal of Developing Areas*, *50*(2), 311–328. doi:10.1353/jda.2016.0072

Apam, L. S. (2012). An Assessment Of The Impact Of Small-Scale Irrigation Schemes On Livelihoods And Poverty Reduction. In *The Upper East Region*. Ghana: Kwame Nkrumah University.

Aseyehegu, K., Yirga, C., & Rajan, S. (2012). Effect Of Small-Scale Irrigation On The Income Of Rural Farm Households: The Case Of Laelay Maichew District, Central Tigray, Ethiopia. *Journal of Agricultural Sciences*, *7*(1). doi:10.4038/jas.v7i1.4066

Ashfaq, M., Jabeen, S., & Baig, I. A. (2005). Estimation Of The Economic Value Of Irrigation Water. *Journal Of Agriculture And Soil Science*, *1*(3), 270–272.

Assefa, N. (2012). Valuing The Economic Benefit Of Irrigation Water: Application Of Choice Experiment And Contingent Valuation Methods To Ribb Irrigation And Drainage Project. In *South Gonder, Ethiopia*. Ethiopia: Addis Ababa University.

Awlachew, S. B., Loulseged, M., & Yilma, A. (2007). Impact Of Irrigation On Poverty And Environment In Ethiopia. In *Draft Proceedings Of The Symposium And Exhibition*. Ghion Hotel, Addis Ababa, Ethiopia.

Awulachew, S. B., & Ayana, M. (2011). Performance Of Irrigation: An Assessment At Different Scales In Ethiopia. *Experimental Agriculture*, *47*(S1), 57–69. doi:10.1017/S0014479710000955

Aydogdu, M. H. (2016). Evaluation Of Willingness To Pay For Irrigation Water: Harran Plain Sampling In Gap Region - Turkey. *Applied Ecology and Environmental Research*, *14*(1), 349–365. doi:10.15666/aeer/1401_349365

Aydogdu, M. H., & Bilgic, A. (2016). An Evaluation Of Farmers' Willingness To Pay For Efficient Irrigation For Sustainable Usage Of Resources: The Gap-Harran Plain Case, Turkey. *Journal of Integrative Environmental Sciences*, 1–12. doi:10.1080/1943815X.2016.1241808

Ayele, G. K. (2011). The Impact Of Selected Small-Scale Irrigation Schemes On Household Income And The Likelihood Of Poverty. In *The Lake Tana Basin Of Ethiopia (Master Of Professional Studies)*. Cornell University.

Babatunde, R. O., Salami, M. F., & Mohammed, B. A. (2016). Determinants Of Yield Gap In Rain Fed And Irrigated Rice Production: Evidence From Household Survey In Kwara State. In *2016 Aaae Fifth International Conference*. African Association Of Agricultural Economists (Aaae).

Bacha, D., Namara, R., Bogale, A., & Tesfaye, A. (2011). Impact Of Small-Scale Irrigation On Household Poverty: Empirical Evidence From The Ambo District In Ethiopia. *Irrigation and Drainage*, *60*(1), 1–10. doi:10.1002/ird.550

Bagson, E., & Kuuder, C.-J. W. (2013). Assessment Of A Small-Scale Irrigation Scheme On Household Food Security And Leisure In Kokoligu. Academic Press.

Banaeian, N., & Zangeneh, M. (2011). Assessment Of Productive Efficiency For Irrigated And Dryland Wheat Farming In Iran Using Data Envelopment Analysis. *Or Insight, 24*(4), 215–230.

Banson, K. E., Asare, D., Heng, L., Cobbinah, J. F., & Adu-Sarkodieh, A. (2014). Impact Of Small Scale Irrigation Technologies On Poverty Alleviation Among Peri-Urban And Urban Farmers. *Journal of Life Science*, *8*(2), 142–151.

Berbel, J., Mesa-Jurado, M. A., & Pistón, J. M. (2011). Value Of Irrigation Water In Guadalquivir Basin (Spain) By Residual Value Method. *Water Resources Management*, *25*(6), 1565–1579. doi:10.1007/s11269-010-9761-2

Beyan, A., Jema, H., & Adem, K. (2014). Effect Of Small-Scale Irrigation On The Farm Householdsincome Of Rural Farmers: The Case Of Girawa District, East Hararghe, Oromia, Ethiopia. *Asian Journal of Agriculture and Rural Development*, *4*(3), 257.

Birhane, M., & Geta, E. (2016). Farmers' Willingness To Pay For Irrigation Water Use: The Case Of Agarfa District, Bale Zone, Oromia National Regional State. *International Journal Of Agricultural Economics*, *1*(2), 35–39.

Biswas, D., & Venkatachalam, L. (2015). Farmers' Willingness To Pay For Improved Irrigation Water — A Case Study Of Malaprabha Irrigation Project In Karnataka, India. *Water Economics And Policy, 01*(01), 1450004. doi:10.1142/S2382624X14500040

Bjornlund, H., Van Rooyen, A., & Stirzaker, R. (2016). Profitability And Productivity Barriers And Opportunities In Small-Scale Irrigation Schemes. *International Journal of Water Resources Development*, 1–15.

Bos, M. G., Burton, M. A. S., & Molden, D. J. (2005). *Irrigation And Drainage Performance Assessment: Practical Guidelines*. Wallingford, UK: Cabi Pub.

Brabben, T. E., Angood, C., Skutsch, J. C., & Smith, L. (2004). *Irrigation Can Sustain Rural Livehoods: Evidence From Bangladesh And Nepal*. Academic Press.

Burt, C. M., & Styles, S. W. (1998). Modern Water Control And Management Practices. In *Irrigation: Impact On Performance (Water Reports No. 16)*. Rome: The World Bank.

Chandrasekaran, K., Devarajulu, S., & Kuppannan, P. (2009). Farmers' Willingness To Pay For Irrigation Water: A Case Of Tank Irrigation Systems In South India. *Water (Basel), 1*(1), 5–18. doi:10.3390/w1010005

Chazovachii, B. (2012). The Impact Of Small Scale Irrigation Schemes On Rural Livelihoods: The Case Of Panganai Irrigation Scheme Bikita District Zimbabwe. *Journal of Sustainable Development in Africa, 14*(4), 217–231.

Dao, G. T. N. (2013). *An Analysis Of Technical Efficiency Of Crop Farms In The Northern Region Of Vietnam* (Unpublished PhD thesis). University of Canberra, Canberra, Australia.

Dawda, T. D., & Amosah, J. (2015). An Analysis Of Small Scale Irrigation Farming And Poverty Reduction In North-Eastern Ghana. *International Journal Of Engineering Research & Technology (Ijert), 4*(10), 473–481.

De Silva, S., Johnston, R., & Sellamuttu, S. (2014). *Agriculture, Irrigation And Poverty Reduction In Cambodia: Policy Narratives And Ground Realities Compared*. Cgiar Research Program On Aquatic Agricultural Systems, Working Paper: Aas-2014-13.

Derbile, E. K. (2012). Water Users Associations And Indigenous Institutions In The Management Of Community-Based Irrigation Schemes In Northeastern Ghana. *European Scientific Journal. Esj, 8*(26), 118–135.

Deribe, R. (2008). *Institutional Analysis Of Water Management On Communal Irrigation Systems In Ethiopia: The Case Of Atsbi Wemberta, Tigray Region And Ada'a Woreda, Oromiya Region* (Msc Thesis). Addis Ababa University, Ethiopia.

Dillon, A. (2008). *Access To Irrigation And The Escape From Poverty: Evidence From Northern Mali*. Ifpri Discussion Paper 00782.

Dinye, R. (2013). Irrigated Agriculture And Poverty Reduction In Kassena Nankana District In The Upper-East Region, Ghana. *Journal Of Science And Technology (Ghana)*, *33*(2), 59–72. doi:10.4314/just.v33i2.6

Dittoh, S., Akuriba, M. A., Issaka, B. Y., & Bhattarai, M. (2010). *Sustainable Micro-Irrigation Systems For Poverty Alleviation In The Sahel: A Case For "Micro" Public-Private Partnerships*. Poster Presentation Presented At The Joint 3rd African Association Of Agricultural Economists (Aaae) And 48th Agricultural Economists Association Of South Africa (Aeasa) Conference, Cape Town, South Africa.

Dube, K. (2016). Implications Of Rural Irrigation Schemes On Household Economy. A Case Of Lower Gweru Irrigation Scheme, Zimbabwe. *South African Journal Of Agricultural Extension (Sajae)*, *44*(1), 75–90. doi:10.17159/2413-3221/2016/v44n1a371

Esmaeili, A., & Shahsavari, Z. (2011). Valuation Of Irrigation Water In South-Western Iran Using A Hedonic Pricing Model. *Applied Water Science*, *1*(3–4), 119–124. doi:10.1007/s13201-011-0015-0

Etissa, E., Dechassa, N., Alamirew, T., Alemayehu, Y., & Desalegne, L. (2014). Irrigation Water Management Practices In Smallholder Vegetable Crops Production: The Case Of The Central Rift Valley Of Ethiopia. *Science. Technology And Arts Research Journal*, *3*(1), 74–83. doi:10.4314/star.v3i1.13

Fanadzo, M. (2012). Revitalisation Of Smallholder Irrigation Schemes For Poverty Alleviation And Household Food Security In South Africa: A Review. *African Journal of Agricultural Research*, *7*(13), 1956–1969. doi:10.5897/AJARX11.051

Faulkner, J. W. (2006). Water Use, Productivity, And Profitability Of Small Scale Irrigation Schemes. In *Ghana's Upper East Region*. Cornell University.

Gebregziabher, G., Namara, R. E., & Holden, S., & Others. (2012). Technical Efficiency Of Irrigated And Rain-Fed Smallholder Agriculture In Tigray, Ethiopia: A Comparative Stochastic Frontier Production Function Analysis. *Zeitschrift für Ausländische Landwirtschaft*, *51*(3), 203–226.

Gebrehiwot Yihdego, A. (2015). The Impact Of Small-Scale Irrigation On Income Of Rural Farm Households: Evidence From Ahferom Woreda In Tigray, Ethiopia. *International Journal Of Business And Economics Research*, *4*(4), 217–228. doi:10.11648/j.ijber.20150404.14

Ghaderzadeh, H., & Rahimi, M. H. (2008). Estimation Of Technical Efficiency Of Wheat Farms: A Case Study In Kurdistan Province, Iran. *American-Eurasian J. Agric. & Environ.*, *4*(1), 104–109.

Gorantiwar, S. D., & Smout, I. K. (2005). Performance Assessment Of Irrigation Water Management Of Heterogeneous Irrigation Schemes: 1. A Framework For Evaluation. *Irrigation and Drainage Systems*, *19*(1), 1–36. doi:10.1007/s10795-005-2970-9

Greaves, K. R. (2007). *Quantifying And Benchmarking Irrigation Scheme Performance With Water Balances And Performance Indicators* (MSc thesis). University of Kwazulu-Natal, South Africa.

Hagos, F., Jayasinghe, G., Awulachew, S. B., Loulseged, M., & Yilma, A. D. (2012). Agricultural Water Management And Poverty In Ethiopia. *Agricultural Economics*, *43*, 99–111. doi:10.1111/j.1574-0862.2012.00623.x

Haile, H. T. (2008). *Impact Of Irrigation Development On Poverty Reduction In Northern Ethiopia* (PhD thesis). National University of Ireland, Cork, Ireland.

Haji, J., Aman, M., & Hailu, T. (2013). Impact Analysis Of Mede Telila Small Scale Irrigation Scheme On House Poverty Alleviation: Case Of Gorogutu District In Eastern Haratghe Oromia National Regional State Ethiopia. *International Journal Of Development And Economic Sustainability*, *1*(1), 15–30.

Hallensleben, N. (2012). *Characterizing Poor Smallholder Farmers In Nicaragua: Factors Influencing The Success Or Failure Of Micro-Irrigation Users* (Masters' Thesis). University Of St. Gallen (Hsg).

Huang, Q., Rozelle, S., Lohmar, B., Huang, J., & Wang, J. (2006). Irrigation, Agricultural Performance And Poverty Reduction In China. *Food Policy*, *31*(1), 30–52. doi:10.1016/j.foodpol.2005.06.004

Hussain, I., Biltonen, E., & International Water Management Institute. (Eds.). (2001). Irrigation Against Rural Poverty: An Overview Of Issues And Pro-Poor Intervention Strategies In Irrigated Agriculture In Asia: Proceedings Of National Workshops On Pro-Poor Intervention Strategies In Irrigated Agriculture In Asia, Bangladesh, China, India, Indonesia, Pakistan, And Vietnam. Colombo, Sri Lanka: International Water Management Institute.

Hussain, I., Sial, M. H., Hussain, Z., & Akram, W. (2009). Economic Value Of Irrigation Water: Evidence From A Punjab Canal. *The Lahore Journal Of Economics*, *14*(1), 69.

Hussain, I., & Hanjra, M. A. (2004). Irrigation And Poverty Alleviation: Review Of The Empirical Evidence. *Irrigation and Drainage*, *53*(1), 1–15. doi:10.1002/ird.114

Jaghdani, T. J., Brümmer, B., & Barkmann, J. (2012). Comparison Of Methods For The Valuation Of Irrigation Water: Case Study From Qazvin, Iran: Comparison Of Methods For The Valuation Of Irrigation Water. *Irrigation and Drainage*, *61*(3), 375–385. doi:10.1002/ird.683

Jin, S., Jansen, H. G., Muraoka, R., & Yu, W. (2012). The Impact Of Irrigation On Agricultural Productivity: Evidence From India. International Association Of Agricultural Economists (Iaee).

Kabore, D. P. (2016). Technical Efficiency Of Irrigated Rice Production In Burkina Faso Water Control Infrastructures: A Stochastic Frontier Approach. *Asian Journal of Agriculture and Rural Development*, *6*(10), 188–200.

Kamwamba-Mtethiwa, J., Weatherhead, K., & Knox, J. (2016). Assessing Performance Of Small-Scale Pumped Irrigation Systems In Sub-Saharan Africa: Evidence From A Systematic Review: Performance Of Small Pumped Irrigation Systems In Sub Saharan Africa. *Irrigation and Drainage*, *65*(3), 308–318. doi:10.1002/ird.1950

Kangile, R. J. (2015). *Efficiency In Production By Smallholder Rice Farmers Under Cooperative Irrigation Schemes In Pwani And Morogoro Regions, Tanzania* (Msc Thesis). Sokoine University Of Agriculture, Tanzania.

Kaswamila, A. L., & Masuruli, B. M. (2004). The Role Of Traditional Irrigation Systems. In *Poverty Alleviation In Semi-Arid Areas: The Case Of Chamanzi In Lushoto District, Tanzania*. Dar Es Salaam, Tanzania: Mkuki Na Nyota Publishers.

Kuwornu, J. K. M. (2012). Irrigation Access And Per Capita Consumption Expenditure In Farm Households: Evidence From Ghana. *Journal of Development and Agricultural Economics*, *4*(3), 78–92.

Laha, A., & Kuri, P. K. (2013). Measurement Of Technical Efficiency Under Alternative Institutions Of Tenurial Contracts And Its Determinants: Evidence From Rural West Bengal, India. *Discourse Journal Of Agriculture And Food Sciences*, *1*(11), 160–170.

Lipton, M., Blackman, R., De Zoysa, D., Qureshy, L., & Waddington, H. (2003). Preliminary Review Of The Impact Of Irrigation On Poverty With Special Emphasis On Asia. Rome: Academic Press.

Machethe, C. L. (2011). Smallholder Irrigation And Agricultural Development In The Olifants River Basin Of Limpopo Province: Management Transfer, Productivity, Profitability And Food Security Issues. Gezina, South Africa: Water Research Commission.

Magingxa, L. L., Alemu, Z. G., & Van Schalkwyk, H. D. (2006). Factors Influencing The Success Potential In Smallholder Irrigation Projects Of South Africa: A Principal Component Regression. *Presented At The International Association Of Agricultural Economists Conference*, Gold Coast, Australia.

Mahdhi, N., Sghaier, M., & Bachta, M. S. (2011). Water Use And Technical Efficiencies In Private Irrigated Perimeters In South-Eastern Of Tunisia. *Emirates Journal of Food and Agriculture*, *23*(5), 440.

Makombe, G. (2011). *A Comparative Analysis Of The Technical Efficiency Of Rain-Fed And Smallholder Irrigation In Ethiopia*. Working Paper 143. International Water Management Institute.

Makombe, G., Kelemework, D., & Aredo, D. (2007). A Comparative Analysis Of Rainfed And Irrigated Agricultural Production In Ethiopia. *Irrigation and Drainage Systems*, *21*(1), 35–44. doi:10.1007/s10795-007-9018-2

Makombe, G., & Sampath, R. K. (1998). An Economic Evaluation Of Smallholder Irrigation Systems In Zimbabwe. *International Journal of Water Resources Development*, *14*(1), 77–90. doi:10.1080/07900629849510

Makombe, G., & Sampath, R. K. (1999). A Benefit-Cost Analysis Of Smallholder Irrigated Farms In Zimbabwe. *Irrigation and Drainage Systems*, *13*(1), 1–12. doi:10.1023/A:1006133220120

Maliwichi, L. L., Oni, S. A., & Obadire, O. S. (2012). An Investigation Into The Factors Affecting Food Availability, Choices And Nutritional Adequacy Of Smallholder Farming Households Under Irrigation And Dryland Farming In Vhembe District Of Limpopo Province, South Africa. *African Journal of Agricultural Research*, *7*(25), 3653–3664. doi:10.5897/AJAR11.1062

Manzungu, E. (1995). Engineering Or Domineering? The Politics Of Water Control In Mutambara Irrigation Scheme, Zimbabwe. *Zambezia*, *22*(2), 115–136.

Maruyama, A., Haneishi, Y., Okello, S. E., Asea, G., Tsuboi, T., Takagaki, M., & Kikuchi, M. (2014). Rice Green Revolution And Climatic Change In East Africa: An Approach From The Technical Efficiency Of Rainfed Rice Farmers In Uganda. *Agricultural Sciences*, *05*(04), 330–341. doi:10.4236/as.2014.54035

Mdemu, M. V., Mziray, N., Bjornlund, H., & Kashaigili, J. J. (2016). Barriers To And Opportunities For Improving Productivity And Profitability Of The Kiwere And Magozi Irrigation Schemes In Tanzania. *International Journal of Water Resources Development*, 1–15.

Meliko, M. O., & Oni, A., S. (2011). Effect Of Irrigation On Poverty Among Small-Scale Farmers In Limpopo Province Of South Africa. *The Journal of Agricultural Science*, *3*(3), 190–195.

Mengistie, D., & Kidane, D. (2016). Assessment Of The Impact Of Small-Scale Irrigation On Household Livelihood Improvement At Gubalafto District, North Wollo, Ethiopia. *Agriculture*, *6*(3), 1–22. doi:10.3390/agriculture6030027

Mission, E. (2008). Smallholder Irrigation Schemes In Nyanga North As Strategies For Poverty Reduction And Sustainable Rural Livelihoods. *Journal of Sustainable Development in Africa*, *10*(2), 653–676.

Molden, D. (1998). *Indicators For Comparing Performance Of Irrigated Agricultural Systems*. Colombo: Iimi.

Molden, D. J., & Gates, T. K. (1990). Performance Measures For Evaluation Of Irrigation-Water-Delivery Systems. *Journal of Irrigation and Drainage Engineering*, *116*(6), 804–823. doi:10.1061/(ASCE)0733-9437(1990)116:6(804)

Moyo, M., Van Rooyen, A., Moyo, M., Chivenge, P., & Bjornlund, H. (2016). Irrigation Development In Zimbabwe: Understanding Productivity Barriers And Opportunities At Mkoba And Silalatshani Irrigation Schemes. *International Journal of Water Resources Development*, 1–15. Retrieved from Https://Doi.Org/10.1080/07900627.2016.1175339

Mujwahuzi, M., Ndetewio, P. I., Mwakaje, A. G., & Ngana, J. (2013). Factors Influencing Willingness To Pay For Watershed Services. In *Lower Moshi*. Pangani Basin.

Mupaso, N., Manzungu, E., Mutambara, J., & Hanyani-Mlambo, B. (2014). The Impact Of Irrigation Technology On The Financial And Economic Performance Of Smallholder Irrigation In Zimbabwe: Impact Of Irrigation Technology On Smallholder Irrigation. *Irrigation and Drainage*, *63*(4), 430–439. doi:10.1002/ird.1820

Mutambara, S., Darkoh, M. B. K., & Atlhopheng, J. R. (2016). A Comparative Review Of Water Management Sustainability Challenges In Smallholder Irrigation Schemes In Africa And Asia. *Agricultural Water Management*, *171*, 63–72. doi:10.1016/j.agwat.2016.03.010

Mutambara, S., & Munodawafa, A. (2014). Production Challenges And Sustainability Of Smallholder Irrigation Schemes In Zimbabwe. *Journal Of Biology Agriculture And Healthcare*, *4*(15), 87–96.

Mutiro, J., & Lautze, J. (2015). Irrigation In Southern Africa: Success Or Failure? *Irrigation and Drainage*, *64*(2), 180–192. doi:10.1002/ird.1892

Mwendera, E., & Chilonda, P. (2013). Conceptual Framework For Revitalisation Of Small-Scale Irrigation Schemes In Southern Africa. *Irrigation and Drainage*, *62*(2), 208–220. doi:10.1002/ird.1723

Namyenya, A., Sserunkuuma, D., & Bagamba, F. (2014). Willingness To Pay For Irrigation Water And Its Determinants Among Farmers At Doho Rice Irrigation Scheme (Dris) In Uganda. *Journal of Development and Agricultural Economics*, *6*(8), 345–355. doi:10.5897/JDAE2014.0580

Ngenoh, E. K., Kirui, L. K., Mutai, B. C., Maina, M., & Koech, W. (2015). Economic Determinants Of The Performance Of Public Irrigation Schemes In Kenya. *Journal of Development and Agricultural Economics*, *7*(10), 344–352. doi:10.5897/JDAE2015.0676

Nhundu, K., Gwata, C., & Mushunje, A. (2010). Impacts Of Zimbabwe European Union Micro-Project Programme (Zim/Eu Mpp) In Funding Smallholder Irrigation Projects On Food Security And Income Levels: A Case Study Of Mopane Irrigation Scheme In Zvishavane, Midlands Province, Zimbabwe. *African Journal of Agricultural Research*, *5*(14), 1759–1771.

Njoko, S. L. (2014). *Smallholder Farmers' Willingness And Ability To Pay For Improved Irrigation: A Case Of Msinga Local Municipality, Kwazulu-Natal Province* (Master thesis). University Of Kwazulu-Natal, South Africa.

Nkambule, B. L., & Dlamini, C. S. (2013). Towards Sustainable Smallholder Irrigation Development Projects: A Case Study Of The Maplotini Irrigation Scheme, Swaziland. *Journal of Agricultural Extension and Rural Development*, *5*(10), 216–224.

Nkhata, R. (2014). *Does Irrigation Have An Impact On Food Security And Poverty: Evidence From Bwanje Valley Irrigation Scheme In Malawi*. Ifpri Working Paper 04.

Ogada, M. J., Muchai, D., Mwabu, G., & Mathenge, M. (2014). Technical Efficiency Of Kenya's Smallholder Food Crop Farmers: Do Environmental Factors Matter? *Environment, Development and Sustainability*, *16*(5), 1065–1076. doi:10.1007/s10668-014-9513-1

Oladimeji, Y. U., & Abdulsalam, Z. (2014). An Economic Analysis Of Dry Season Irrigated Farming In Asa River, Kwara State, Nigeria: Implications For Poverty Reduction. *Journalof Sustainable Development In Africa*, *16*(7), 1–15.

Omondi, S. O., Mbogoh, S. G., & Munei, K. (2014). An Evaluation Of The Factors Influencing Farmers'willingness To Pay (Wtp) For Irrigation Water: The Case Of Ahero Irrigation Scheme In Kenya. *International Journal Of Science. Environmental Technology*, *3*(5), 1778–1789.

Oriola, E. O. (2009). Breaking The Vicious Cycle In Irrigation Farming System For Sustainable Food Security In Nigeria. *African Research Review*, *1*(3), 234–245.

Pereira, H., & Marques, R. C. (2017). An Analytical Review Of Irrigation Efficiency Measured Using Deterministic And Stochastic Models. *Agricultural Water Management*, *184*, 28–35. doi:10.1016/j.agwat.2016.12.019

Peter, G. (2011). The Impact Of Small-Scale Irrigation Schemes On Household Food Security In Swaziland. *Journal of Sustainable Development in Africa*, *13*(6), 102–117.

Qureshi, M. E., Ranjan, R., & Qureshi, S. E. (2010). An Empirical Assessment Of The Value Of Irrigation Water: The Case Study Of Murrumbidgee Catchment. *The Australian Journal of Agricultural and Resource Economics*, *54*(1), 99–118. doi:10.1111/j.1467-8489.2009.00476.x

Rani, R., & Singh, H. N. (2015). A Comparative Study Of Technical Efficiency Of Rice Production In Irrigated And Rainfed Environment Of Uttrakhand. *Indian Journal Of Hill Farming*, *28*(2), 102–106.

Rao, P. S. (1993). *Review Of Selected Literature On Indicators Of Irrigation Performance*. Colombo, Sri Lanka: International Irrigation Management Institute.

Regassa, B. A. (2013). *Technical Efficiency Variation For Smallholder Irrigated Maize Producers: Stochastic Frontier Approach (A Case Study Of Tibila Surface Water Irrigation Scheme)* (Master thesis). Mekelle, Ethiopia.

Rola, A. C., & Quintana-Alejandrino Jocelyn, T. (1993). Technical Efficiency Of Philippine Rice Farmers In Irrigated, Rainfed Low-Land And Upland Environments: A Frontier Production Function Analysis. *Philippine Journal of Crop Science*, *18*(2), 59–69.

Senaratna Sellamuttu, S., Aida, T., Kasahara, R., Sawada, Y., & Wijerathna, D. (2014). How Access To Irrigation Influences Poverty And Livelihoods: A Case Study From Sri Lanka. *The Journal of Development Studies*, *50*(5), 748–768. doi: 10.1080/00220388.2013.841887

Shantha, A., Ali, B. G. H., & Bandara, R. A. G. (2012). Efficiency And Managerial Ability Of Paddy Farming Under Minor Irrigation Conditions: A Frontier Production Function Approach. *Journal of Agricultural Sciences*, *7*(3), 145–158. doi:10.4038/jas.v7i3.4878

Sharm, K. R., Pradha, N. C., & Leung, P. (2001). Stochastic Frontier Approach To Measuring Irrigation Performance: An Application To Rice Production Under The Two Systems In The Tarai Of Nepal. *Water Resources Research*, *37*(7), 2009–2018. doi:10.1029/2000WR900407

Sinyolo, S., Mudhara, M., & Wale, E. (2014). The Impact Of Smallholder Irrigation On Household Welfare: The Case Of Tugela Ferry Irrigation Scheme In Kwazulu-Natal, South Africa. *Water S.A.*, *40*(1), 145–156. doi:10.4314/wsa.v40i1.18

Sisay, B., & Fekadu, B. (2013). Small-Scale Irrigation And Household Income Linkage: Evidence From Deder District, Ethiopia. *African Journal of Agricultural Research*, *8*(34), 4441–4451. doi:10.5897/AJAR12.1793

Sisay, D., Jema, H., Degye, G., & Abdi Khalil, E. (2015). Technical, Allocative, And Economic Efficiency Among Smallholder Maize Farmers In Southwestern Ethiopia: Parametric Approach. *Journal of Development and Agricultural Economics*, *7*(8), 282–291. doi:10.5897/JDAE2015.0652

Small, L. E., & Svendsen, M. (1990). A Framework For Assessing Irrigation Performance. *Irrigation and Drainage Systems*, *4*(4), 283–312. doi:10.1007/BF01103710

Smith, L. E. D. (2004). Assessment Of The Contribution Of Irrigation To Poverty Reduction And Sustainable Livelihoods. *International Journal of Water Resources Development*, *20*(2), 243–257. doi:10.1080/0790062042000206084

Solomon, S., & Ketema, M. (2015). Impact Of Irrigation Technologies On Rural Households' Poverty Status: The Case Of Fogera District, North-Western Ethiopia. *AGRIS On-Line Papers in Economics and Informatics*, *7*(1), 59–67.

Speelman, S., Frija, A., Perret, S., D'haese, M., Farolfi, S., & D'haese, L. (2011). Variability In Smallholders' Irrigation Water Values: Study In North-West Province, South Africa. *Irrigation and Drainage*, *60*(1), 11–19. doi:10.1002/ird.539

Svendsen, M., Ewing, M., & Msangi, S. (2009). *Measuring Irrigation Performance In Africa*. International Food Policy Research Institute (IFPRI).

Syaukat, Y., Arifah, F. N., & Minha, F. (2014). Economic Value And Service Fee Of Irrigation Water In The Districts Borgor And Kudus: Indonesia. *J. Issaas*, *20*(2), 157–172.

Tang, Z., Nan, Z., & Liu, J. (2013). The Willingness To Pay For Irrigation Water: A Case Study In Northwest China. *Global NEST Journal*, *15*(1), 76–84.

Thibbotuwawa, M., Mugera, A., & White, B. (2013). *Production Efficiency And Technology Gap In Irrigated And Rain-Fed Rice Farming Systems In Sri Lanka: Non Parametric Approach*. In *57th Aares Annual Conference*, Sydney, Australia.

Tiwari, D. N. (1998). *Detemining Economic Value Of Irrigation Water: Comparison Of Willingness To Pay And Indirect Valuation Approaches As A Measure Of Sustainable Resource Use*. Cserge Working Paper Gec 98-05, Centre For Social And Economic Research On The Global Environment University College London And University Of East Anglia.

Ulsido, M. D., & Alemu, E. (2014). Irrigation Water Management In Small Scale Irrigation Schemes: The Case Of The Ethiopian Rift Valley Lake Basin. *Environmental Research, Engineering and Management*, *67*(1), 5–15. doi:10.5755/j01.erem.67.1.6240

Van Den Berg, M., & Ruben, R. (2006). Small-Scale Irrigation And Income Distribution In Ethiopia. *The Journal of Development Studies*, *42*(5), 868–880. doi:10.1080/00220380600742142

Van Halsema, G. E., Keddi Lencha, B., Assefa, M., Hengsdijk, H., & Wesseler, J. (2011). Performance Assessment Of Smallholder Irrigation In The Central Rift Valley Of Ethiopia. *Irrigation and Drainage*, *60*(5), 622–634. doi:10.1002/ird.613

Wakili, A. M., & Isa, A. H. M. (2015). *Technical Efficiency Of Small Scale Rice Production In Adamawa State, Nigeria*. Presented At The Chemical, Food And Environment Engineering (Iccfee'15), Dubai, UAE.

Watto, M. A., & Mugera, A. W. (2014). Measuring Production And Irrigation Efficiencies Of Rice Farms: Evidence From The Punjab Province, Pakistan: Irrigation Efficiencies Of Rice Farms. *Asian Economic Journal*, *28*(3), 301–322. doi:10.1111/asej.12038

World Bank. (2013). Turn Down The Heat: Climate Extremes, Regional Impacts, And The Case For Resilience (A Report For The World Bank By The Potsdam Institute For Climate Impact Research And Climate Analytics). Washington, DC: World Bank.

Yedra, H., Mesa-Jurado, M. A., López-Morales, C. A., & Castillo, M. M. (2016). Economic Valuation Of Irrigation Water In South-Eastern Mexico. *International Journal of Water Resources Development*, *32*(6), 931–943. doi:10.1080/07900627.2015.1133404

Yuya, B. T. (2014). Determinants Of Technical, Allocative And Economic Efficiencies Among Onion Producing Farmers In Kobo District, Amhara Region, Ethiopia. *Journal Of Agricultural Economics, Extension And Rural Development*, *2*(5), 54–62.

Zakaria, H. (2014). Factors Affecting Farmers Willingness To Pay For Improved Irrigation Service. *International Journal Of Agricultural Economics And Extension*, *2*(1), 68–76.

Zewdu, S. (2016). The Role Of Irrigation In Household Food Security In Upper Blue Nile Basin: The Case Of Jedeb Irrigation Scheme, Amhara Region, Ethiopia. *International Journal Of Agricultural Economics*, *1*(4), 108–116.

Zeweld, W., Huylenbroeck, G. V., Hidgot, A., Chandrakanth, M. G., & Speelman, S. (2015). Adoption of Small-scale Irrigation and Its Livelihood Impacts in Northern Ethiopia. *Irrigation and Drainage*.

KEY TERMS AND DEFINITIONS

Constraints: Limitations, often regarded as challenges.

Economic Value: Monetary measure of benefits obtained from consumption of a certain commodity.

Food Security: Internationally regarded condition of having enough food for quality human life.

Irrigation: Artificial application of water to land by farmers to aid the production of crops.

Performance Indicators: A measurable quantitative value that give the extent at which set objectives have been achieved.

Poverty: State of inadequacy of people's basic needs in their various internationally recognized standard forms.

Productivity: Output realized per unit of inputs utilized in the production process.

Small-Scale Irrigation Schemes: Irrigation on small plots where farmers should have greater control over all aspects of their operations.

Chapter 4
Biofuel Production and Its Implications in a Transitive Low Carbon Development Country:
The Case of South Africa

Constant Labintan
Centre de Partenariat et d'Expertise pour le Développement Durable (CePED), Benin

Harald Winkler
University of Cape Town, South Africa

Abiodun Elijah Obayelu
Federal University of Agriculture – Abeokuta (FUNAAB), Nigeria

ABSTRACT

This chapter explains the implication of South Africa's transport fuel 2% blending. Using dry grain sorghum as feedstock with guaranteed food security has lower emission of 24.93kg/ha with emerging farmers who constituted 30% of the suppliers with a 3-year payback period. Using irrigated sorghum with food security as a priority has a relatively lower emission level of 11.47kg/ha from emerging farmers with a 9-year payback period. Using sugar beet has lower emission level of 0.12kg/ha with emerging farmers and a 3-year payback period. Soil organic content has significant influence on emissions from land use practices. Commercial sugar beet ethanol production caused high emission (4.84kg/ha) but has a short payback period of only 2 years which enhanced household food consumption by 12.5% and 31.50%

DOI: 10.4018/978-1-5225-3631-4.ch004

under food security not a priority and food security as a priority, respectively. In all, grain sorghum food and beverage gross domestic product (GDP) increased, respectively, by 8%, 0.19%, and 0.23% under food security as not a priority, and increased by 20.83%, 0.44%, and 0.61% in opposite scenario, respectively.

INTRODUCTION

In a transitive economy country like South Africa, energy is a key factor for achieving economic growth and a major contributor to greenhouse gas (GHG) emission. In South Africa, the main sources of energy are: coal (56.69%), crude oil (11.9%), electricity (8.8%), gas (1.35%), hydro-power (0.16%), nuclear power (1.5%), petroleum products (14.8%), and renewable energy and waste (4.8%) (Statistics South Africa, 2009). The energy sector contributes about 15% of the country's Gross Domestic Products (GDP) followed by the mining industry with 8.8% (GCIS, 2012) and creates over 86000 jobs (Statistics South Africa, 2010). Over the past 20 years, the country's energy sector has been characterized by low investment and low research and development (R&D), and accounts for 0.4% of GDP (Department of Science and Technique (DST), 2010). With economic growth, the energy demand has increased significantly especially in the transport and industrial sectors. In addition, over 30% of the population does not have access to electricity and 13.7% lives on less than 1.25 USD a day (World Bank, 2013).

Between 2000 and 2010 South Africa's Green House Gas (GHG) emission increased significantly by as much as 32% and yielded 4,879,334 Gg Carbone dioxide(CO_2 equivalents. Of this quantity, the energy sector contributed 67% of the total emission, followed by transport sector with 9.24% (Department of Environmental Affairs(DEA), 2011). In 2009, the country contributed 1.49% of global emission that yielded 9.18 tons of CO_2 per capita and was the 14th emitter of the intensive energy user countries (International Energy Atomic Agency (IEA), 2009). The electricity sector contributes a large share of the country's energy related emission of 53%, followed by the transport sector (DEA, 2011).

In order to address the global climate change challenges, the country acceded to GHG emission reduction by responding to several international emission reduction commitments. Following Conference of Parties (COP) 15th, several reforms were undertaken to reduce the country's GHG emission intensity, principally from the energy sector. This included the launch of National Climate Change Response White Paper by the Department of Environmental Affairs (DEA) (DEA, 2011), the 2012 Integrated Energy Plan (IEP)(DOE,2013), the National Transport Master Plan (NATMAP 2050)(DOT,2010), the 2010 Integrated Resources Plan (IRP), which was later revised to adjusted IRP and launched by the Department of Energy (DoE), and the Carbon Tax Policy Paper launched by the South African Department of National Treasury (DNT)(2013). These different policy mechanisms will assist in realizing the country's vision in managing the impacts of climate change effectively and translate to a climate-resilient, low-carbon economy.

The proposed Adjusted IRP plan aims at improving the country's electricity production by 2030 to 9,6 GW nuclear; 6,3 GW coal; 17,8 GW renewable; and 8,9 GW other generation sources using 65% coal, 20% nuclear, 5%, hydro power, 1,1% gas, and 9% renewable energy (IRP, 2012). The new plan is expected to reduce the country's greenhouse-gas emissions by 34% by 2020 and 43% by 2025, equivalent to 275 million tons of carbon dioxide per year after 2024 (DoE, 2012). In related development, the IEP, recognized the potentiality of biofuel production to meet transport sector liquid fuel demand while improving the sector's emission intensity. This programme will support the country's biofuel industrial strategy initiated in 2007 by the DoE intended to achieve 2% blending in transport sector by 2013, produce 400 millions of ethanol as biofuel, and create over 25 000 jobs (DoE, 2013).

Notwithstanding the controversial debate of biofuel production and the country's goal of achieving food security, the use of maize as fuel feedstock was banned in South Africa. In addition, the biodiversity invasion impact of Jatropha Curcas constrains its growth as biofuel feedstock (African Biodiversity Network, 2007;von Maltitz et al, 2010; von Maltitz et al., 2011; Henning, 2012). Therefore, sorghum and sugar beet became the major crops under consideration for ethanol production in South Africa. Indeed, in 2012, the country's total grain sorghum output was 158000 tons, representing considerable decreased by 21% between 2010/2011 (DAFF, 2012). With an average yield of 2t/ha of rain-fed sorghum, 10t/ha of irrigated sorghum and 110t/ha of irrigated sugar beet, the 2% blending goal will directly require 675000 tons of grain sorghum, 1980000 tons of sugar beet and respectively 151500 ha and 18000 ha of land to meet grain sorghum and sugar beet feedstock demand. This is illustrated in Tables 1 and 2. This is to be achieved through 70% of commercial farmers' and 30% of emerging farmers' supply (Agrarian Research and Development Agency (ARDA), 2013). Given the country water availability, grain sorghum market, arable land availability and bio-crop technology adoption, effectiveness of biofuel production is a great concern in South Africa. In addition, food security is also a concern while rotation cropping system is an option to contribute to address food crops supply. This is expected to contribute a total production 2374500 tons and 7755000 tons of grain sorghum and sugar beet respectively and 573450 ha, for grain sorghum 70500 ha sugar beet of land. This is illustrated in Table 1 and Table 2.

Although, the bio-fuel strategy considers the country resources availability and other possible risk of growing biofuel, very limited research do investigate deeply the economics and environmental implication of biofuel production. Most studies which assess the socioeconomics and environmental impact of first and second

Table 1. Direct and Indirect Land Use Change for Dry and Irrigated Grain Sorghum Production

Direct land Use	Commercial Farmer's=70%	Emerging Farmer's=30%	Indirect land Use	Commercial Farmer's=70%	Emerging Farmer's=30%
Dry Land(Hectare)	73500	31500	Dry Land(Ha)	294000	126000
Irrigated land (Ha)	32550	13950	Irrigated (Ha)	97650	55800
Total land (Hectare)	106050	45450	Total(Ha)	391650	181800

Sources: Author Own Computation based on ARDA 2014 database

Table 2. Direct Sugar beet and Indirect Land Use Change Demand for Sugar Beet Production

Indirect Land Use	Commercial Farmer's=70%	Emerging Farmer's=30%	Direct Land Use	Commercial Farmer's=70%	Emerging Farmer's=30%
Irrigated Land(Hectare)	66000	4500	Irrigated Land(Ha)	16500	1500

Sources: Author Own Computation based on ARDA 2014 database

generation biofuels reveal that their production effectiveness depends considerably on a country's resources constraints, the impact of indirect land use emission from biofuel production and feedstock supply chain, land conflict, national food security, country technology adoption and R&D, market and biodiversity (International Assessment of Agricultural Knowledge, Science and Technology for Development (IAAKSTD), 2008; IAE, 2010a; International Energy Forum, 2010; Gasparatos *et al.,* 2013a). This means that biofuel production impact varies across countries and should be assessed case by case.

In related development, this paper aims to assess the environmental and economic impact of South Africa 2% blending goals. This will contribute to the potentiality of the blending policy in achieving the country environmental goals while addressing other development goals such as food security, income distribution and employment. Doing so, the paper applies an integrated assessment framework in quantifying the emission abatement potential of the 2% blending and its economic impacts. Further details of applied integrated assessment framework are presented in methodologies section while the remaining part presents the simulation results and conclusion on the effectiveness of the blending policy.

METHODOLOGY

This paper assesses the mitigation potentiality of the 2% blending mandate in several phases. The first part quantifies GHG abatement of direct and indirect land use for production, transportation and processing of both grain sorghum and sugar beet feedstocks. This stage used the Ex-Act model (FAO, 2013). This method is considered more appropriate for quantifying the emissions during the production of the feedstock under specific farming practices and agro-ecological zones. In addition, ASPEN(AspenTech,2011) and Greet (CEPA, 2010a, 2010b; Argonne National Laboratory, 2013) calibration were used for assessing the emission from the processing and transport activities in the production cycle. This is an accurate estimation of the emission in biofuel production because it includes farming behaviour instead of using Life Cycle Assessment. Land use change data, farming practices, and processing data were obtained in October 2013 from the Agrarian Research and Development Agency (ARDA). Land use change data were from Tables 1 and 2. Land use change is characterized by grassland clearing from the first year and land rotation from second year for both commercial and emerging farmers. However, commercial farming system is characterized by high demand of water; liquid fuel and electricity for land preparation, irrigation and harvesting and high demand of fertilizer and herbicide during crop for crop grow.

The second part applied input-output (I/O) table multiplier decomposition analysis approach to the social and economic impacts of the expected additional demand for agricultural sorghum and sugar beet on the agricultural GDP as well as the country total GDP, and changes in household consumption. It used the SimSIP SAM model (Parra, Juan and Wodon, 2009) and the South Africa Social Accounting Matrix 2005 to compute leontief matrix inverse.

1. **The Social Accounting Matrix:** In SimSIP model, the 2005 SAM serves as the core database. It is made up of 57 industries and 50 commodities including energy and non-energy industries and commodities;7 production factors (capital, 4 labors, crop land and energy land use); 4 institutions (government, exported and imported company and rest of the world) and 9 households income groups. The agriculture sector was disaggregated into 4 sub-sectors namely Sorghum (asorg), Sugarcane (asug), livestock (allve), fishery(afishery), forestry(aforestry), food(afood), beverage (abeverage) and other agricultural commodities (aaother). The South African 2005 Social Accounting Matrix was obtained from International Food Policy Research Institute (IFPRI). Data use for decomposing the 2005 SAM agricultural industry were obtain from Abstract of Agricultural Statistics (DAFF, 2012) and trade data obtain from trade TradeMap (ITC,2013).

2. **Factors of Production:** The mobility of factors of production is a central feature of any general equilibrium. A factor is mobile when it has the capacity to be relocated across industries in response to changes in rate of return. The greater the factor mobility in the model, the greater is the economy's simulated capacity to respond to changes in the economic environment.

Four type of labor were identified, 'flab-p; flab-m; flab-s and flab-t;Flab-p represents primary-educated labor ie between grade1 and grade7. Flab-m represents middle –educated labor (that is, between grade 8 and grade10). Flab-s represents secondary –educated labor (that is between grade 11 and grade12) while flab-t represents tertiary–educated labor. In South Africa labor market flab-t are mobile and fully employed while flab-p, flab-m and flab-s are mobile and not fully employed)

Two types of capital were identified. Fcap represent a capital use in non-energy industry while fegy represent the capital allocated for energy industries Fcap is mobile and fully employed.

However, one type of land was identified Flnd represent land allocated mainly for agricultural activities and is also mobile and fully employed.

3. **Agent:** The model identified 4 agents respectively government, private enterprise, household and the rest of the world. Household was characterized by 9 household income groups. Household income was distributed in percentile. hhd-o represent a group of household which income share is between 0 and 10% of total national income, hhd-1 represents household with income share between 10 and 20% of total national income, hhd-2; hhd-3;hhd-4;hhd-5;hhd-6;hhd-6;hhd-7;hhd-8;hhd-91;hhd-92;hhd-93;hhd-94;hhd-95 represents household which income share are between 20 and 30%; 30-40%; 40-50%; 50-60%; 60-70%; 70-80%; 80-90%; 91-100%; 92-100%; 93-100%; 93-100%; 94-100% and 95-100% of total national income respectively. From this categorization of South African households, hhd-0 up to hhd-4 belong to low income household category while hhd-5 up to hhd-6 belong to middle income household group and from hhd-7 up to hhd-95 belong to high income household group.

Parametric Assumptions

In order to capture household commodities consumption behavior, substitution between production inputs and trade characteristics in the model, different parameter assumptions were used. On demand side, household Sorghum, sugar, livestock, forestry, food consumption or income elasticity values were each equal to 0.76, while their fish and beverage consumption or income elasticity value were 0.66 and 0.67

respectively (Anne Case,1998; TIPS,2002). This was relatively close to the value used by Ramos Mabugu *et al.,* (2013).

However, on the supply side, the production function used in the model was a leontif function at the top nest.

On trade assumption, elasticity of transformation was presumed to be 4.2, for Sorghum, sugar, livestock, forestry, fishery and 0.89, 2 .80 respectively for food and beverage (DBSA,2005).

Core Equation

Indeed, considering the economy n industries producing n homogenous products and F an $n \times 1$ vectors of sectorial final demands, A an $n \times n$ matrix coefficient of intermediate inputs per unit of gross output and X an $n \times 1$ vectors of sectoral gross outputs, the well-known leontief function(Wassily Leontief,1970) is expressed as follow:

$$X = (I - A)^{-1} \times F \tag{1}$$

where I is an $n \times n$ identity matrix and $(I - A)^{-1}$ is the leontief inverse with j^{th} column the gross output of each sector i directly and indirectly required to supply one unit of final demand of product j.

Following equation (1), it was deduced that:

$$\Delta X = (I - A)^{-1} \times \Delta F \tag{2}$$

Simulation Scenario

The model simulation consisted of two scenarios. The first scenario is a scenario when food security is not a concern and this mean only direct land use change (DLUC) is taken into account. The scenario assumed an increase of 4.81% in total agricultural land demand, reduction of 2% of demand of petrol in transport sector and increased of 4900% in biofuel demand.

However, in the second scenario when food security is a concern there is consideration of indirect land use (ILUC) for improving food supply. In this scenario, it assumed an increase of 9% in total agricultural land demand, reduction of 2% of demand of petrol in transport sector and increased of 4900% in biofuel demand. The meso-micro linkage is therefore used to assess the blending target impact on sectorial growth and on household consumption and their income generation.

RESULTS AND DISCUSSION

Analysis of the GHG emission per litre of ethanol produced revealed that in producing bio-ethanol using dry sorghum, total emission is much higher with commercial farmers (77.88kg C02 eq/l) under indirect land use change (ILUC) than total emission with emerging farmers at 24.93 kg C02 eq/l under the same ILUC. In addition, under the ILUC scenario, commercial farmers' land clearing emission is also much higher at 22.66kg C02 eq/l than land clearing emission from emerging farmers at 9.71 C02 eq/l.

However, total emission for producing bio-ethanol using commercial farmers under direct land use change (DLUC) is lower at a value of 5.87 kg $C0_2$ eq/l than using commercial farmers with a value of 9.16 kg $C0_2$ eq/l. Similarly, commercial farmers' land clearing emission of 7.55kg $C0_2$ eq/l is also much lower than for emerging farmers at a value of3.2kg/l $C0_2$eq (Table 3).

Indeed, achieving the 2% blending goal while also maintaining food security creates a significant increase in demand for land, which is much higher with commercial farming. This justifies the relative higher land clearing emission for commercial farmers.

In producing bio-ethanol using irrigated sorghum, total emission under ILUC with commercial farmers was found to be 26.79 kg C02 eq/l). This is relatively higher than total emission using emerging farmers with a value of 11.47 Kg C02 eq/l. Similarly, under DLUC, total emission for commercial farmers, which stands at 7.3 kg C02 eq/l is relatively higher than for emerging farmers at 3.27 kg C02 equation l. In addition, under ILUC and DLUC scenarios, commercial farmers' land

Table 3. Indirect and Direct Emission of Production of a litter of Bio-ethanol, Clearing Land and Dry Sorghum Production

Indirect Emission (Kg $C0_2$ eq/l)	**Commercial Farmers=70%**	**Emerging Farmers=30%**
Production (Rain-fed land)	77.88	24.93
Sorghum growth	-71.86	-11.55
Land Clearing	22.66	9.71
Processing and Transport	3.60	3.60
Direct Emission (Ton $C0_2$ eq/l)	**Commercial Farmer's=70%**	**Emerging Farmer's=30%**
Production (Dry land)	5.87	9.16
Sorghum growth	-1.17	-0.7
Land Clearing	7.55	3.23
Processing and Transport	3.60	3.60

Sources: Author Own Computation based on EX-ACT-ASPEN-GREET Results

clearing emission is also relatively higher than emerging farmers' land clearing emission (Table 4). Indeed, similar to dry sorghum scenario, commercial farmers' land use demand is also higher than emerging farmers'. This justifies their higher land clearing emission.

Using dry grain sorghum yields lower than irrigated sorghum with large scale farmers, having relatively more significant mitigation potential under direct LUC than using emerging farmers. However, considering food security and the payback period due to soil organic carbon (SOC) ethanol production using dry sorghum will have higher emission but with a relatively shorter payback period of 3-9years depending on farming scale. However, with irrigated sorghum under ILUC scenario, the emission will be significant but the payback period will be slightly longer in the range of 4-9years depending on farming scale. This is because land clearing is a once off exercise while the ethanol production will continue and SOC will depend on farmers' land management techniques. Sustainable land use management will contribute to mitigation of the emission from land clearing that occurs during the first year of production.

Analysis of sugarcane ethanol producing scenario under ILUC and DLUC scenarios reveals that the total emissions using commercial farmers is more significant than using emerging farmers (Table 5). In addition, commercial farmers' land clearing emission in both scenarios is also higher than emerging farmers. This is mainly due to the large demand for land by commercial farmer's. Furthermore, when considering the goal of food security, the emission intensity increases under ILUC.

Indeed, sugar beet ethanol production with the goal of achieving food security has relatively lower emission than grain sorghum bio-ethanol. In addition, sugar

Table 4. Indirect and Direct Emission of Production 1 litre of Bio-ethanol clearing land and from Irrigated sorghum production

Indirect Emission (KgCO2eq/l)	**Commercial Farmer's=70%**	**Emerging Farmer's=30%**
Production (Irrigated land)	26.79	11.47
Sorghum grow	-10.99	-4.71
Land Clearing	10.03	4.3
Proccssing and Transport	3.6	3.6
Direct Emission (KgC02 eq/l)	**Commercial Farmer's=70%**	**Emerging Farmer's=30%**
Production (Irrigated land)	7.53	3.27
Sorghum growth	-2.75	-1.118
Land Clearing	3.34	1.43
Processing and Transport	3.6	3.6

Sources: Author Own Computation based on EX-ACT-ASPEN-GREET Results

Table 5. Indirect and Direct Emission of Production 1 litter of Bio-ethanol clearing land and from Irrigated Sugar beet production

Indirect Emission (KgCO_2 eq/l)	Commercial Farmer's=70%	Emerging Farmer's=30%
Production (Irrigated land)	4.84	0.47
Sugar beet grow	-1.85	-1.68
Land Clearing	1.81	0.47
Processing and Transport	0.56	0.56
Direct Emission (KgCO_2 eq/l)	**Commercial Farmer's=70%**	**Emerging Farmer's=30%**
Production (Irrigated land)	1.36	0.12
Sugar beet grow	-0.49	-0.13
Land Clearing	0.60	0.05
Processing and Transport	0.56	0.56

Source: *Author Own Computation based on EX-ACT-ASPEN-GREET Results*

beet production emission payback period ranges between 2 to 3 years depending on farmer's scale (Table 7).

Sugar beet yield is higher than sorghum yield; requires less land clearing; and needs less input than grain sorghum. Moreover, the carbon content of sugar beet ethanol is much lower than for grain sorghum ethanol. Also, sugar beet ethanol production total emission using emerging farmers under ILUC and direct LUC scenarios for both farming systems are relatively lower than emissions from burning a liter of gasoline, which emits 2.7 kg C02/l (Carbon Trust, 2013). In summary, sugar beet ethanol production GHG abatement potentiality is important than in grain sorghum ethanol production scenario. However, analysis of both feedstock and fuel cost reveal that sugar beet feedstock production cost in relatively higher

Table 6. Emission Payback Period

Bio-fuel Fuel Technology	Indirect Emission Payback Period (Years)		Bio-fuel Fuel Technology	Direct Land Use	
	Commercial Farmers (Proportion=70%)	Emerging Farmers=30%		Commercial Farmers (Proportion=70%)	Emerging Farmers (proportion=30%)
Dry land	9	3	Dry Land	3	3
Irrigated land	4	9	Irrigated land	3	4
Direct Sugar beet land use	2	3	Sugar beet Land	3	3

Source: Author Own Computation based on EX-ACT-ASPEN-GREET Results

that grain sorghum while sugar beet ethanol production is much more cheap. This is illustrated by Table 7.

The results of the analysis of the social economic impact of the 2% blending target under food security different scenarios reveal a relatively positive change in agriculture and food industry. This is illustrated by Table 8. It shows that the grain sorghum industry GDP under food security as a priority scenario will increased by 20.83% while the food industry will GDP will have increased by 0.44% and the beverage industry by 0.61%. This is relatively higher compare to the scenarios without food security as priority which respectively grain sorghum, food and betterave industry GDP will slightly increase respectively by 8%, 0.19% and 0.23%. Nonetheless the fuel policy will contribute to improve food and crops industry production.

In related development, the labor market will also experience positive changes relatively higher while food security is priority. The Table 9 below show that the labor market will improve by an average of 0.125% in scenario without food security set up as priority while under food security scenario the employment will increase by an average of 0.30%.

Analysis of changes induce by the blending on household income improvement also shows a relative improvement in household income with the highest changes in higher income household in both scenarios. In a context of higher inequality, this will become a concern and will constitute other challenges to be address.

Notwithstanding, the increased in industry production, household income will in turn contribute to improve commodities availability and household consumption. The Table10 below shows that in both scenarios, household food consumption will improve. This improvement is significantly higher in middle income household with 17% increase in scenario without food security and 41% increase in scenario with food security is a priority.

Although the blending policy will contribute to relatively improve high income household income, it will also contribute to improve significantly low income household food consumption. Lower income household consumption and income will also increase but alike relatively lower compare to middle and higher income household. This may be considering in implementation of the policy in order to reduce the inequality gap and other social challenges.

Table 7. Biofuel Production Cost per liter and per ha of Land Used

Feedstock	Feedstock Production Cost	Ethanol Production Cost
Dry Sorghum	R 3200/ha	R4/l
Irrigated Sorghum	R 8700/ha	R2.175/l
Sugar beet	R20000/ha	R1.42/l

Source: Author Own Computation based on ARDA Data

Table 8. Impact of the 2% blending target on Agricultural and Food Industries GDP Without and With Food Security Scenarios (% change from 2005 baseline)

Industries	Changes in GDP Without Food Security	Changes in GDP With Food Security
aSorghum	8.077570796	20.83279279
asugar	0.084932753	0.213896792
alivestock	0.050209556	0.130141066
aother	0.317220976	0.748000366
aforestry	0.164778636	0.40079126
afishery	0.180696545	0.434383378
afood	0.191804593	0.445847003
abeverage	0.238075966	0.618652025

Sources: Simulation Results from SimSIP

Table 9. Impact of the 2% blending target on Employment Without and With Food Security Scenarios (% changes from 2005 baseline)

Labor	Changes in Labor Without Food Security	Changes in Labor With Food Security
flab-p	0.151934326	0.356556007
flab-m	0.126868438	0.297731954
flab-s	0.156183575	0.366528047
flab-t	0.150256535	0.352618605

Sources: Simulation Results from SimSIP

CONCLUSION AND RECOMMENDATIONS

South Africa in attempts to mainstream climate change mitigation in its development has adopted biofuel production policy using different production strategies in the bid to move from high carbon emission energy intensive to low carbon emission country. The country owes this paradigm shift to the Department of Energy Bio-energy strategy of 2007. Given the country's development challenges that stride between simultaneous achievement of sustainability development, food security, reduced social and income inequality, and increased employment, the country adopts bio-fuel production from sugar beet and grain sorghum to reduce carbon emission as climate change mitigation measure. The biofuel development scheme embarked on the use of commercial farmers as emerging farmers and feedstock suppliers. The scheme invariably increased the demand for land, especially grassland considerably.

Table 10. Impact of the 2% blending target on Household Income Without and With Food Security Scenarios

Household	Changes in Income Without Food Security	Changes in Income With Food Security
hhd-0	0.108931535	0.252132683
hhd-1	0.10954875	0.256112384
hhd-2	0.110906036	0.263045941
hhd-3	0.110872983	0.266330055
hhd-4	0.11408383	0.27602576
hhd-5	0.116732622	0.284779106
hhd-6	0.119892541	0.297619543
hhd-7	0.118288053	0.296194911
hhd-8	0.116070759	0.293045478
hhd-91	0.11572399	0.296467784
hhd-92	0.127092435	0.341341206
hhd-93	0.130139518	0.352675622
hhd-94	0.156714819	0.447134707
hhd-95	0.189860675	0.567706311

Sources: Simulation Results from SimSIP

Ethanol production from sugar beet as alternative energy source is found to have relatively lower carbon emission compared to grain sorghum production using emerging farming system. Given the lack of sufficient farming experience in the adoption of the sugar beet technology, irrigated grain sorghum is therefore preferred despite its relatively lower ethanol production emission and a slightly longer payback period of between 4 to 9 years.

Given the country's resource constraints and the competition for limited agricultural resources, achieving the 2% blending is expected to enhance the country household food consumption pattern as well as to boost the agricultural sector production and perhaps enhance the sectoral contribution to the country's economy. With the biofuel project, household consumption is expected to changes by 12.5% in scenario where food security is not a priority and increase by 31.50% in scenario of food security as a priority. This means in both scenarios, food security has improved, and significantly improved in the second scenario. This positive impact of biofuel production in South Africa is in line with von Maltitz and Brent (2008), Haywood *et al.,* (2008) and Maltsoglou and Khwaja (2010) findings but contrary to FAO (2008) finding which highlight that biofuel production could lead to high food prices and thus food insecurity. The type of adopted cropping ensemble and farming system are

Table 11. Impact of the 2% blending target on Household Consumption Without and With Food Security Scenarios

Without Food Security								
	csorghum	csugar	clivestock	cother	Cforestry	cfishery	cfood	cbeterave
hhd-0	0.215538631	0.053884658	0.366416	0.4418542	0.44529197	0.00580824	8.032001	0.8781012
hhd-1	0.379826967	0.094956742	0.645706	0.7786453	0.68474043	0.01314259	12.49955	1.7276296
hhd-2	0.471103185	0.117775796	0.800875	0.9657615	0.55352504	0.02069732	14.215	2.3400463
hhd-3	0.605811104	0.151452776	1.029879	1.2419128	0.44005302	0.01921322	14.50069	2.9661744
hhd-4	0.675128892	0.168782223	1.147719	1.3840142	0.32265162	0.0228111	15.9252	3.9531918
hhd-5	0.722589453	0.180647363	1.228402	1.4813084	0.27743103	0.03890122	16.35217	5.0143818
hhd-6	0.687739941	0.171934985	1.169158	1.4098669	0.15956697	0.05913231	16.76618	6.5317009
hhd-7	0.762279492	0.190569873	1.295875	1.562673	0.07871576	0.04414992	17.47063	8.2873429
hhd-8	1.167286836	0.291821709	1.984388	2.392938	0.10206747	0.09592802	23.60064	12.668962
hhd-91	0.239482206	0.059870551	0.40712	0.4909385	0.01925237	0.03393728	4.895799	4.2573266
hhd-92	0.352984904	0.088246226	0.600074	0.7236191	0.0241642	0.02880047	6.195795	4.5896083
hhd-93	0.498467271	0.124616818	0.847394	1.0218579	0.0216506	0.03469526	5.78113	4.2455226
hhd-94	0.577498823	0.144374706	0.981748	1.1838726	0.04841904	0.08147149	8.124848	7.6178568
hhd-95	1.475107904	0.368776976	2.507683	3.0239712	0.06768448	0.12148068	10.70587	12.197569
With Food Security								
	csorghum	csugar	clivestock	cother	Cforestry	cfishery	cfood	cbeterave
hhd-0	0.498885226	0.124721307	0.848105	1.0227147	1.03067179	0.01344374	18.59085	2.0324509
hhd-1	0.887991782	0.221997945	1.509586	1.8203832	1.60084441	0.03072586	29.22251	4.0389994
hhd-2	1.117358307	0.279339577	1.899509	2.2905845	1.31284573	0.04908973	33.71501	5.5501008
hhd-3	1.45523012	0.36380753	2.473891	2.9832217	1.05705953	0.04615242	34.83237	7.1251027
hhd-4	1.633473957	0.408368489	2.776906	3.3486216	0.78065541	0.05519145	38.531	9.5647452
hhd-5	1.762818101	0.440704525	2.996791	3.6137771	0.67681647	0.09490282	39.89249	12.233009
hhd-6	1.707235874	0.426808968	2.902301	3.4998335	0.39610677	0.14678921	41.62013	16.214202
hhd-7	1.908758331	0.477189583	3.244889	3.9129546	0.19710535	0.110552	43.74669	20.751621
hhd-8	2.947065501	0.736766375	5.010011	6.0414843	0.25769117	0.24219082	59.58487	31.985506
hhd-91	0.61351807	0.153379518	1.042981	1.257712	0.04932173	0.08694231	12.54231	10.906642
hhd-92	0.9480367	0.237009175	1.611662	1.9434752	0.06489952	0.07735148	16.64049	12.326638
hhd-93	1.350836841	0.33770921	2.296423	2.7692155	0.05867273	0.0940235	15.66675	11.505286
hhd-94	1.647704847	0.411926212	2.801098	3.3777949	0.13814798	0.23245236	23.18161	21.735074
hhd-95	4.410750494	1.102687623	7.498276	9.0420385	0.20238475	0.36324189	32.01185	36.472203

Sources: Simulation Results from SimSIP

the main two factors which would significantly impact on the decision making with regards to the implication of the proposed biofuel blending policy on food security.

In related development the grain sorghum industry, food and beverage industries GDP will increased respectively by 8%, 0.19% and 0.23% under scenarios of food security is not a priority. While under a scenario where food security is a priority, grain sorghum industry, food and beverage industries GDP will increase respectively by 20.83, 0.44% and 0.61%. Moreover, the results show that, biofuel production in South Africa will contribute to the improvement of household income especially the middle-income household. In addition, employment will be significantly being improved in middle and higher income household than in lower income household. The positive impact of biofuel in improving household income is also supported by other scholars (African Biodiversity Network (ABN), 2007; Arndt *et al.,* 2010). However, the contribution of biofuel production in improving household income generation and reducing inequality will depend on the fuel production value chain and mainly the feedstock supply sources. This is in line with the findings of Gasparatos *et al.,* (2013b). Indeed, the effective implementation of the country biofuel policy with respect to social economics and environmental benefit would have to adopt integrated cropping system.

ACKNOWLEDGMENT

The Lead Author of this paper thanks the Energy Research Center (ERC) of University of Cape Town for their support while on Post-doctorate research at the Institute.

REFERENCES

Agrarian Research and Development Agency (ARDA). (2012). Sugar Beet Report. ARDA.

Alexandros, Lee, von Maltitz, Mathai, de Oliveira, & Willis. (2013). Biofuels in Africa Impacts on Ecosystem Services, Biodiversity and Human Well-being. UNU-IAS Policy Report.

Argonne National Laboratory. (2013). *Greenhouse Gases, Regulated Emissions, and Energy Use in Transportation Model (GREET). GREET version 2.7.* Author.

Arndt, C., Benfica, R., Tarp, F., Thurlow, J., & Uaiene, R. (2010). Biofuels, poverty and growth: A computable general equilibrium analysis of Mozambique. *Environment and Development Economics, 15*(1), 81–105. doi:10.1017/S1355770X09990027

AspenTech. (2011). *AspenEngineeringSuiteV7.3*. Author.

California Environmental Protection Agency (CEPA). (2010). First Draft Detailed of The California-Modified GREET Pathway for Sugar beet Ethanol. California Air Resources Board.

CarbonTrust. (2013). Energy and carbon conversions. Carbon Trust.

Department of Agricultural Forestry and Fishery (DAFF). (2012). *Abstract Agricultural Statistics 2012*. DAFF.

Department of Energy (DOE). (2012). *Draft of Integrated Resources Plan For Electricity-2010-2030*. DOE.

Department of Energy (DOE). (2013). *Draft 2012of Integrated Energy Planning*. Pretoria, South Africa: DOE.

Department of Environment (DEA). (2011). *National Climate Change Response*. White Paper. DEA.

Department of Environmental Affaire (DEA). (2011). *South Africa's Second National Communication under the United Nations Framework Convention on Climate Change*. DEA.

Department of Government Communication and Information Systems (GCIS). (2012). *Energy, South Africa Yearbook 2011/12*. GCIS.

Department of Minerals and Energy (DOE). (2007). Biofuels Industrial Strategy of the Republic of South Africa. DoE.

Department of National Treasury (DNT). (2013). *Reducing Greenhouse Gas Emissions and Facilitating the Transition to a Green Economy*. Policy Paper For Public Comment Carbon Tax Policy Paper. DNT.

Department of Science and Technology (DST). (2010). *National Survey ofResearch & Experimental Development Main Results 2009/10*. DST.

Department of Transport (DOT). (2010). *National Transport Master Plan*. Pretoria, South Africa: DOT.

FAO. (2008). *From Subsistence farming to sugar-cane monoculture: impacts on agrobiodiversity, local knowledge, and food security*. Rome: Food and Agriculture Organization.

Food and Agriculture Organization of the United Nations (FAO). (2013). EX-ACT Version 3.4.1. FAO.

Gasparatos, A., Stromberg, P., & Takeuchi, K. (2013, April). Sustainability impacts of first-generation biofuels. *Animal frontiers*, *3*(2), 12–26. doi:10.2527/af.2013-0011

Haywood, L., von Maltitz, G., Setzkorn, K., & Ngepah, N. (2008). *Biofuel production in South Africa, Mozambique, Malawi and Zambia: A status quo analysis of the social, economic and biophysical elements of the biofuel industry in Southern Africa*. Pretoria: Council for Scientific and Industrial Research.

Henning, R. K. (2013). *Jatropha curcas in Africa - an Evaluation*. Weissenberg: Bagani. Available at http://www.underutilized-species.org/Documents/PUBLICATIONS/jatropha_curcas_africa.pdf

International Assessment of Agricultural Knowledge, Science and Technology for Development (IAAKSTD). (2008). *Bio-energy and Biofuel: Opportunity and Constraints. Issue in brief.* Island Press.

International Atomic Energy Agency (IAEA). (2009). International Energy Statistics Report 2009. Washington, DC: IAEA.

International Atomic Energy Agency (IAEA). (2011). Technology Roadmaps: Biofuel for Transport. OECD/IEA Report.

International Food Policy Research Institute (IFPRI). (2013). *South Africa Social Accounting Matrix 2005*. Washington, DC: IFPRI.

International Trade Centre (ITC). (2013). *Trade Map Statistics Database*. ITC.

Leontief, W. (1970). Environmental Repercussions and the Economic Structure: An Input-Output Approach. *The Review of Economics and Statistics*, *52*(3), 262–271. doi:10.2307/1926294

Maltsoglou, I., & Khwaja, Y. (2010). *Bioenergy and food security: The BEFS analysis for Tanzania*. Rome: Food and Agriculture Organization.

Parra, J. C., & Wodon, Q. (2009). *SimSIP-SAM: A Tools for the Analysis of Input-Output Tables and Social Accounting Matrices*. World Bank.

Statistics South Africa. (2012). *Energy Accounts for South Africa 2002/2009*. Statistics South Africa.

Statistics South Africa. (2012). *South Africa Quarterly Labour Force Survey*. Pretoria: Statistics South Africa.

Von Malititz, G. P., & Brent, A. (2008). Assessing the biofuel options for Southern Africa. In *CSIR (Council for Scientific and Industrial Research), Science real and relevant: 2nd CSIR Biennial Conference*. Pretoria, South Africa: CSIR.

von Maltitz, G., & Stafford, W. (2011). *Assessing opportunities and constraints for biofuel development in sub-Saharan Africa*. Bogor: Center for International Forestry Research.

von Maltitz, G. P., Nickless, A., & Blanchard, R. 2010. Maintaining biodiversity during biofuel development. In Projects in developing countries: A framework for policy evaluation. Newcastle upon Tyne, UK: Newcastle University.

World Bank. (2013). World Databank Statistics. World Bank.

Section 3

Sustainable Consumption and Health: Contemporary Issues and Policies

This section discusses the contemporary issues and policies on sustainable consumption and health.

Chapter 5
Sustainable Food Consumption in the Neoliberal Order:
Challenges and Policy Implications

Henry E. Alapiki
University of Port Harcourt, Nigeria

Luke Amadi
University of Port Harcourt, Nigeria

ABSTRACT

In recent decades, we have seen the rise of the sustainable food consumption field and its push for disciplinary space in development studies. This chapter turns to the original impetus of sustainable food consumption and the question of how neoliberal order can be reconciled with the need to save the ecology. Beyond the fundamental objectives, there is a need to assess the links between the global food system, as influenced by neoliberal order, and the signs that it leads to adversity for low-income countries. A review of relevant literature in the sustainable consumption field is explored using content analysis to examine links between neoliberal food consumption dynamics, the logic of global food politics, and the emerging terminological shifts from food consumption to food system. The world systems theory and the Marxian political ecology framework are used to show that sustainability is notable for emphasizing resource efficiency and equitability, which can be useful when sustainability challenges are matched with ecological policies. This chapter makes some policy recommendations.

DOI: 10.4018/978-1-5225-3631-4.ch005

INTRODUCTION

Since the end of the Cold War sustainable food consumption has been a major concern in development studies. In the Earth Summit of 1992 in Rio de Janeiro, world leaders came to a conclusion that "the major cause of the continued deterioration of the global environment is the unsustainable pattern of consumption and production" (UN, 1992). The phases of food production encompass agriculture, food processing, warehouse/retail, consumption (including storage and preparation) and waste management (Åström, et al., 2013).

This process provides some lucid explication of the patterns of sustainable or unsustainable food consumption. In particular, the inverse relationship between increasing food consumption in the affluent North and ecological breakdown in the poor South has resulted in the destruction of the environment. This partly includes the emergence of genetically modified(GM) seeds or genetically engineered(GE) foods which suggests that "genetic engineering is one type of genetic modification that involves the intention to introduce a targeted change in a plant, animal or microbial gene sequence to effect a specific result" (NRC, 2004). The notion is that "genetic engineering has increased the number and type of substances that can be intentionally introduced into the food supply"(NRC,2004).

A critical perspective suggests that the reliance on technology by the affluent societies including globalization, free trade and the methods of production impact directly on climate change variables like green house /carbon emission, deforestation, land grab and degradation arising from capitalist farming, plantation agriculture and the use of organic fertilizers (Wise, 2015).

For neo liberal proponents, consumption represents the liberal ideals of freedom of choice in a market society. The consumer is the King of modern freedoms according to which he/she freely chooses from a broad offering of goods and services (Lock & Ikeda, 2005). The U.S. food system provides a remarkably varied food supply to the U.S. consumer at lower cost than nearly anywhere else in the world' (National Research Council, 2015). However, freedom of choice results in lower costs but leads to choices that might not be sustainable. With billions of people on earth, freedom of choice might not comply necessarily with ecological requirements.

By 2050, the world's population is projected to grow by one-third, reaching between 9 billion and 10 billion people (FAO, 2010). Meting the food needs of this population draws policy attention to sustainable food consumption. This has pressured scholarly engagement with the question of sustainable food consumption as food is inevitably at the center of both human survival and sustainable development.

The objective of this chapter is to stimulate synergies to mitigate unsustainable consumption and related efforts to strengthen proactive policy initiatives. In particular, the chapter suggests ways to provide food for an increasing population on sustainable basis. The central argument is that unsustainable consumption and lifestyles of the affluent societies framed in the context of the neo liberal policies, may lead to increasing and unsustainable food demands which consequently have deleterious effects on both food system and the environment.

Sustainable Food Consumption

Recent trends suggest that debates on sustainable food consumption have moved to the center of development discourse (Lock & Ikeda, 2005). A number of factors account for this emphasis. First is that food is essential for human life and secondly its consumption is interwoven with the natural environment which has implications for sustainability. In their views, Jongen and Meerdink (1998) and Vitterso et al., (1999) recount that "close to half of all human impact on the environment, such as loss of biodiversity, is directly and/or indirectly related to food production and consumption".

Since the end of the Cold War, neo liberal ideology has dominated much of the thinking in most of the powerful industrialized societies and in supranational institutions such as the World Bank and International Monetary Fund, which in turn have imposed this ideology on countries in the global South. Such trajectories claim that the exercise of "freedom of choice" and "consumer sovereignty" creates a dynamic and democratic society (Lock & Ikeda, 2005).

Neo liberalism is first and foremost a belief in the ability of unrestricted market forces to achieve the best possible economic outcomes for all people. Other neo liberal doctrine includes the primacy of economic growth, reduction of regulatory capacity and size of governments, the importance of free trade to economic growth and individual choice (Steger, 2002). In this worldview, the whole notion of democracy is reduced to the freedom to choose between various goods and services in the market place. Conway and Heynen (2006:20) argue that neoliberalism emerged as a shift from the Keynesian economic model. They pointed out that a critical element of the neoliberal order is the fact that it represents a similar bout of suffering and impoverishment for the poor in the global South just like colonialism and post colonialism, modernization which results in dependency and "development of underdevelopment". Agnew (2005) recounts that neoliberalism is propagated by America as its global project including institutionalization of globalization. This has been increasingly linked to the rise in global food system and the politics of sustainable food consumption.

Within consumption and neoliberalism, there are a number of insightful debates which underscore the increasing effects of unsustainable consumption. Debates on finite resources suggest that global resources are in a decline. Hinrichsen, Salem & Blackburn, (2002) argue for instance that global fresh water is in decline and this exerts pressure on water resource use following the rise of global water politics in the West. This is accompanied by global environmental insecurity threats including climate change, ecological breakdown and the need for ecological security (Roggers,1995). With the rise in environmental insecurity, meeting future food security challenges requires a shift in thinking from "food production and consumption" to sustainable food consumption (Schor, 2005). Milner (2012) had identified the consequences of consumption suggesting the basis of sustainable consumption. The recently emergent field of sustainable consumption attempts to remedy deficiencies in the overall food consumption patterns particularly in the context of equitable and resourceful consumption. Lucia Reisch, Ulrike Eberle, and Sylvia Lorek (2013) argue that the unsustainable consumption dynamics of current arrangements arises from the industrialization and globalization of agriculture and food processing. This suggests that the shift of consumption patterns to dietary animal protein, the resurgence of processed products, results in increasing gap on a global scale between rich and poor, and the paradoxical lack of food security amid an abundance of food.

Against this background, the chapter puts the domain of sustainability frontal in the food consumption agenda as the foundational basis for alternative policy options.

THEORETICAL FRAMEWORK

Food consumption has been explored from divergent theoretical perspectives. To live sustainably in a highly capitalist society driven largely by profit motives remains a development challenge. Since the end of the Cold War, the global food system has become a powerful paradigm for understanding the link between human activities and sustainable consumption.

The rise in capitalist food corporations which exploit the poor societies as sources of raw materials, point out the need to investigate inequality in capitalist resource extraction. Luke Amadi (2012) recounts that pioneering work in this field was undertaken by Harry Braverman in his work *Labour and Monopoly Capital*. Baravcrman notes that capitalism is founded on the ability of capitalists to extract surplus labor from production activities and that in the process of development of capitalism, any vestiges of worker control over production that might give them the ability to regulate how much surplus value is produced have been removed (Amadi, 2012). As Braverman argued that the capitalist mode of production systematically destroys all-around skills where they exist, and brings into being skills and occupations that correspond to its needs (Amadi, 2012). This forms part of the wider theoretical basis of "development of underdevelopment" a strand of dependency debate advanced in the World Systems theory popularized by Wallerstein (1976). The systems theory argues that any country's development conditions and prospects are primarily shaped by economic processes, commodity chains, division of labour and geo-political relationships operating at the global scale (Klak, 2013:121).

Another key relevance of this theory is its exploration of the dynamics of entrenched economic interests of the global food corporations and patterns of food chains linked to capitalism.

In a distinct manner, capitalism has been an integral part of the food consumption and underdevelopment logic. Nesheim, et al; (2015) argue that the food system is woven together as a supply chain that operates within broader economic, biophysical, and sociopolitical contexts. This suggests a complex interplay of cross sectorial entities such as health, environmental, social, and economic effects linked to food distribution chain in a globalizing world. The food system is both advantageous and disadvantageous, the system supplies food across all other sectors but in an increasingly unequal manner with divergent implications. The major detrimental effect as Nesheim, et al;(2015:6) suggest is unhealthy dietary patterns considered as a risk factor resulting in mortality and morbidity. They identified additional effects of the food system to include "climate, land, and water resources depletion". They identify the depletion of resources (e.g., water) and flow of outputs (e.g., nitrogen from fertilization, pesticides, and greenhouse gases) to the environment as a result of food system activities" (Nesheim, et al., 2015:6). This remains a persistent challenge to sustainable consumption.

The less developed societies such as the sub- Saharan Africa(SSA), Latin America and South Asia have been vulnerable to undernourishment including challenges posed by food insecurity (Shopouri, et al; 2010). This partly arises from alienating land owners from their lands particularly in agrarian societies, while degradation results in the washing away of top soil nutrients (leaching), use of organic chemicals and depletion of soil nutrients, due to intensive or unsustainable agricultural production methods (Moomaw, et al., 2012). This is a strand of the imperialism debate. This capitalist trend also forms the basis of the emerging ecology and world systems debate (O'Connor, 1998). This is a central basis of mitigating capitalism's war on nature (Harvey, 2005).

In particular, the ecological effects which are expressed in varying degrees between the high income societies and the poor societies have been less lucid. This has equally made the Marxian political ecology theory a useful framework to examine the consequences of unsustainable food consumption.

Since the 1970s following the rise in environmental movements, unsustainable consumption has been under serious attack by the ecological Marxist scholars. These scholars argue that the traditional role of the State, in protecting its citizenry from environmental degradation is seriously put into question (Peet & Watts, 1996; Bryant & Bailey, 1997; O'Connor, 1998; Robbins, 2004; Amadi, Igwe & Wordu, 2014). This is premised on the notion that capitalism is implicitly profit motivated.

Critical of ecological factors is to capture relations between food consumption and natural environment, Akenji & Begsson (2014) argue that beyond being an "overarching objectives of, and essential requirements for sustainable development", consumption is central to the actualization of environmental protection and poverty reduction. Political ecology as a strand of Marxism argues that capitalist resource extraction taints nature (Peet & Watts, 1996; Amadi & Igwe, 2015). They point out the role of capitalists in the global food system and the attendant decimation of the natural resources as 'every-thing' in nature becomes a commodity, resulting in nature marketization and commodification. This is equally linked to the dialectical materialistic contradictions of food systems between the affluent and poor societies in which the material conditions of resource expropriation are largely exploitative (Ake, 1981). This creates a complex environmental system one which is at variance with ecological justice and socially responsible consumption.

Whereas the World systems perspective in which core/periphery asymmetry exists—suggests a regulatory dynamic in which the periphery provides the raw materials which are exploited by the advanced /core societies in an unequal scale, resulting in "development of underdevelopment" (Wallerstein, 1976; Amadi, 2012). The ecological Marxists clearly explore the patterns of deleterious effects including inequality in natural resource exploitation. Thus, the world systems theory and Marxian political ecology perspectives provide useful theoretical tools for this study.

METHODOLOGY

The methodology to achieve integrated analysis of sustainable food consumption requires a deepened exploration of the conceptual and theoretical issues raised in the literature. And how the issues relate to both consumption patterns of the affluent societies of the North and associated deleterious effects.

This at the same time aims at providing on the ground evidence which suggests patterns of such effects of unsustainable consumption. This includes anthropogenic problems linked to consumption notably economic resource exploitation, green-house emission, health implications of consumption and related ecological breakdown.

On these bases, this study is framed within the content analysis methodology. Content analysis according to Holsti (1969:14) is, 'any technique for making inferences by objectively and systematically identifying specified characteristics of messages'. Content analysis transcends "identifying characteristics of messages" rather reviews and analyses key issues in the literature. This becomes suitable for a broader elucidation of the literature as the aim is to deepen the investigation and in a distinct manner advance new knowledge.

Amadi and Imoh-Ita (2017) argue that content analysis examines and analyses contents or set of data for relevant research objectives. This methodology is important as it critically identifies sets of data in relation to the subjects under investigation and in particular, provides evidence of how the neoliberal food system results in unequal resource access and extraction between the North and South.

The ensemble of divergent perspectives into a unified whole provide a coherent account of the challenges associated with unsustainable consumption patterns. This helps to strengthen the arguments advanced in the study in line with various perspectives on the subject of sustainable food consumption.

Specific methods of selecting relevant data for the study was adopted which includes both online and direct library resource materials. The online search of secondary data such as seminal conceptual, empirical and theoretical data that provide on the ground evidence of the research problems was identified and selected. This is premised on identifying and explaining how such patterns of consumption have posed challenges to sustainable food consumption. The extensive search was specifically based on scholarly journals published on the proquest and Ebsco data bases since the post -Cold War era. This timeline of relevant publications from different schools of thought on neo liberal food consumption dynamics, sets of data on conceptual issues related to the study including global corporation, food politics, sustainable consumption patterns etc was relevant to access a robust set of data for possible generalization. This helps to deepen knowledge of the neo liberal consumption patterns, understand the logic of capitalist consumption and need for policy response to redress the problems identified.

The search yielded 300 results on themes related to the study. However, specific themes directly linked to the research was chosen and a total of 120 works were found useful. A number of similar relevant data were sourced from institutional publications which included the United Nations Food and Agriculture Organization (FAO) reports which for instance suggests that in order to ensure food security needs, additional 70% food must be available by 2050, from an already severely depleted natural resource base (Moomaw, et al., 2012). Others institutional literature included the FAOSTATS,World Bank,UN etc.

This aims at exploring causal connections between food and sustainable consumption and to explore prevailing perspectives in the on-going debate. And to explain how these identified problems could be explored within the political ecology framework to advance a new scholarship and policy agenda.

The rest of the chapter is structured as follows; conceptual clarifications, global corporation, food politics and sustainable consumption, conclusion, recommendations and future research directions.

Figure 1. Per capita consumption of major food items in developing countries, 1961-2005

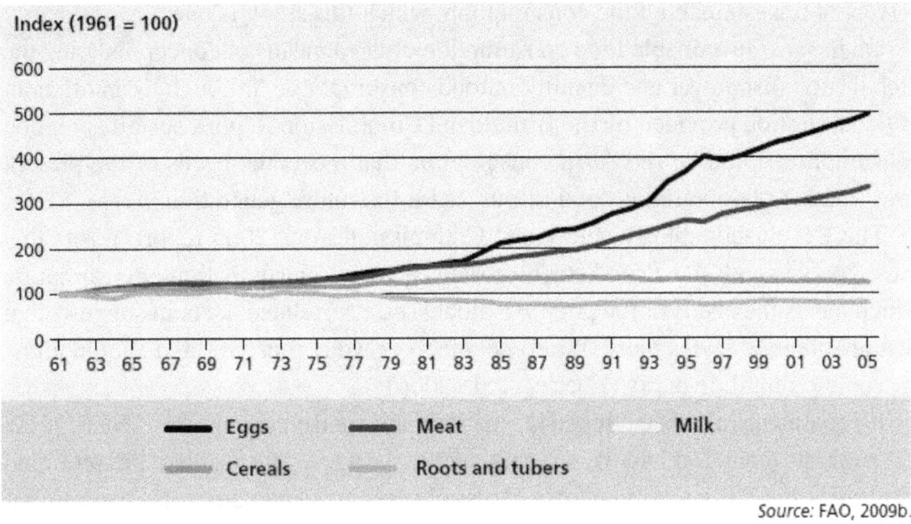

**For a more accurate representation see the electronic version.*

Conceptual Clarifications

The conceptual literature builds on relevant studies in the broad field of consumption. It corroborates dominant debates on sustainable consumption (Koten,1995; Harvey, 2005; Schor, 2005; Davidson & Hatt, 2005; Stiglitz,2010; Milner, 2012; Amadi, Igwe & Wordu, 2014) and advances a novel argument which suggests the need for sustainable food consumption.

There is a widespread assumption that the term food is a common place to everybody. However the conceptual exploration of the term sustainable food consumption is important to understand the interwoven contextual use of the concept and beyond this, to offer a way of understanding the linkages between food system and unsustainable consumption. Thus, concepts like "food system" and "sustainable consumption" form part of broader elucidation of consumption dynamics deployed to explore the processes, practices and social relations which shape capitalist food production and consumption.

The contemporary food system is a dynamic, fast-changing, multidimensional enterprise, particularly influenced by technological advancement (IOS & NRC,2015). In the global food system dynamics and in scholarly debates, there have been absence of consensus on common theoretical approach to the study of sustainable food consumption. Thus, theoretical exploration of what contexts and ways consumption could be classified as sustainable and how it could be classified as unsustainable is less clear. This has increasingly stimulated global concern on collective efforts at institutionalizing effective mechanisms and common grounds to understand the key drivers of unsustainable food consumption, which this study is one.

On its part, sustainable food consumption offers a means of conceptualizing the global capitalist power and equality in food consumption. Although as an offshoot of the definition provided by the Brundtland Commission Report, sustainable food consumption could be seen as a consumption that meets the needs of the present consumers without tainting food resources for the future generation.

The UK Sustainable Development Commission in its 2005 report, posits that "sustainable food and drink" include safe, healthy, and nutritious consumption which meets the needs of the poor at a global scale. A related account suggests that sustainable food styles must fit into people's everyday lifestyles and should allow for socio-cultural diversity (Eberle, et al., 2006).

Inequality has also been linked to unsustainable food consumption. UNEP (2012: 12) suggests that, "food demand is only met in the aggregate, as there are profound disparities in access to food across geographic regions and across the spectrum of incomes at both the household and country levels". There are variations in access to food including cultural and religious barriers, government policies and control mechanisms, demographic factors linked to food consumption dynamics.

The neo Malthusian perspective argues that population growth partly accounts for low food availability and consumption function among the poor societies. Beyond this perspective are the implications of capitalist consumption patterns which is les explored. This gives rise to various analytical, theoretical and conceptual difficulties. For example, while capitalist system is riddled with inequality (Harvey, 2005; Muller, 2013), the same capitalist system and its protagonists propagate sustainable and equitable resource consumption (Lock & Ikeda, 2005). Where these propositions are concerned with addressing the relationship between capitalist production, exploitation and equality, natural resource extraction and ecological breakdown are less critiqued. For instance, since the emergence of the green consumerism paradigm in the late 1980s and early 1990s, in the United States, its impressive response during the decade, has deepened. However, it appears the state of green consumption and business transactions have been less revolutionary.

The critical implications of unsustainable consumption is ever scant in development studies as several people in the poor societies are marginalized in the event of asymmetrical appropriation of food resources including the deleterious food consumption patterns of the affluent societies. This includes disproportionate food consumption of the industrialized societies like the United States (Schor, 2005). There is introduction of genetically modified (GM)foods or genetically Engineered (GE) foods (Paalberg, 2013), groceries, burger etc, which could have carcinogenic effects including the rise in obesity.

This reaffirms the importance of inclusive dynamics in food consumption and a critique of the new paradigm of global capitalist food system. According to Marx (1978), capitalism breeds both love and crisis. This points out the inevitable contradictions surrounding food consumption including recent conceptual and terminological shifts associated with food consumption such as ecological conscious consumer behavior (ECCB)' (Straughan & Roberts, 1999), green purchase behavior and the green consumer (Jansson, Marell, & Nordlund, 2010; Akehurst & Akenso, 2012),, food security (Shopouri,etal;2011), 'pro-environmental consumer behaviour'(Riley, Kohlbacher & Hofmeister, 2012), 'green consumer values(Haws, Winterich & Naylor, 2014),environmental sustainability measurement (Amadi & Imohita, 2017) etc, have attracted novel scholarly interest as issues of sustainable food consumption have increasingly become a globalized phenomenon. For instance, in 2015 a committee was appointed by the Food and Nutrition Board of the Institute of Medicine (IOM) in collaboration with the Board on Agriculture and Natural Resources of the National Research Council (NRC) to develop a report and an analytical framework to assess the health, environmental, social, and economic aspects of the U.S. food system. The committee took into account the complexity of the system and recognized that "the U.S. food system is embedded in a global system that is broadly interconnected" (IOM and NRC,2015).

In particular, the report "provided insights into how aspects of the food system influences modern life" (IOM and NRC,2015). This includes changes and adaptation to Western consumption patterns of packaged foods, lifestyles, beverages, blended drinks including alcohol at the expense of traditional or indigenous patterns of production and consumption and the rise in GM foods often liked to food modernization.

Proponents of modernization of agriculture argue that there is need to encourage GM foods in the South particularly in Africa.Robert Paalberg (2008) and similar advocates argue that African farmers need to grow more crops through GM seeds and subsequently increase the potential productivity of African farmers. Paalberg (2008) contends that Africa is 'starved of science' as they are 'kept out of biotechnology' and that the central basis for advancing food production in Africa should be revamping agriculture in line with modernization models, that this has been a key to Africa's economic growth.

Right or wrong, much of Paalberg's debate perhaps falls within Gidden's (1990) thesis on "societal consequences of modernization".Critiques argue that GM foods and similar modernization thesis may not bring the poor out of poverty nor check unsustainable consumption. Masters, et al;(2004:8) have explored the unintended effects of genetically engineered foods on human health and found that "all evidence evaluated to date indicates that unexpected and unintended compositional changes arise with all forms of genetic modification, including genetic engineering".

Thus, food consumption has become a central trend in economic advancement, identity, lifestyle, interaction between one another and the increasing consumption dichotomy among the societies of the global North and South (Davidson & Hatt, 2005; Milner, 2012).

Among the developed societies, Milner (2012) had identified the relevance of consumption in contemporary development dynamics and argued that Norway has been an example of a society with a modest consumption pattern unlike the United States. This has increasingly resulted in complexities in the study of contemporary food system. Such complexities include the issues of world hunger, sustainable production, inequality and poverty.

FAO (2015) reports that 'world hunger is on the rise: the estimated number of undernourished people increased from 777 million in 2015 to 815 million in 2016". Global food politics which needs to be reconciled with freedom of choice and the imbalance between food consumption in the high and low income countries. FAO estimates for 2016 indicate that the global prevalence of undernourishment in 2016 may have actually risen to 11 percent (FAO, 2017). According to the food first and World Watch institute, global hunger hits an increasing percentage since the 1990s to the 2000s particularly in food insecure areas.

Beyond food consumption, food production has also been a core challenge in understanding the underlying triggers of unsustainable consumption. Sustainable production entails the mode of manufacturing food from 'process to production' including packaging such as canned foods or similar modes of food processing mechanisms etc. A central issue has been the dynamics of equitable and responsible use of resources in the cause of food production. This has been more critical in the neo liberal order where ecological factors which should have been given priority attention appears less prioritized (Korten, 1995).

This disproportionate mode of food consumption results in inequality (Schor, 2005).

The rise of the consumer culture is typified in the capitalist system (Schor, 2005). Equally from the perspectives on environmentally significant consumption, Stern (1997:20) argues that 'consumption consists of human and human-induced transformations of materials and energy'. He recounts that this transformation is critical to the environment and humans to the extent that it makes materials or energy less available for future use.

Growing poverty remains a central concern in global food politics as most people of the South are poorer today than 100 and 50years ago(Weinsten,2008). In both urban and rural contexts poverty remains pervasive. The urban poor live in slums and squatter settlements, without food and adequate access to clean water, sanitation, and health care (Hinrichsen, et al., 2002). Thus,the poor who are predominantly in the less developed societies have been at the margins of popular discourse on global

Figure 2. Number of People and shares of population living in poverty in low-and-middle-income countries

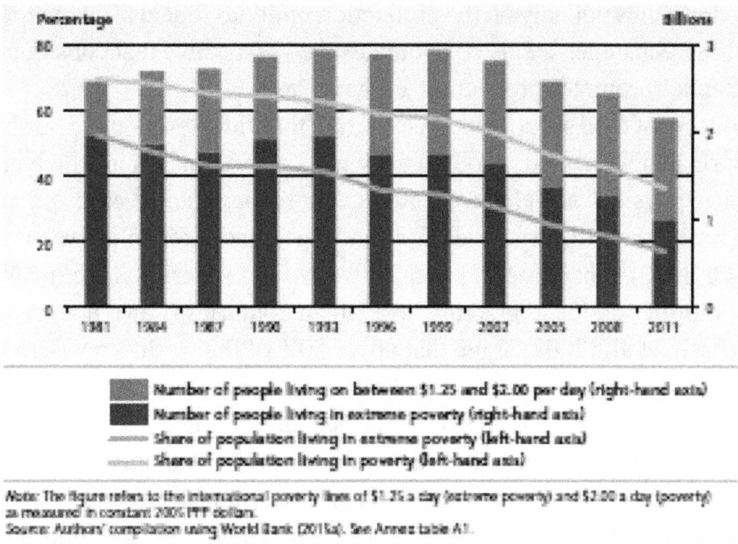

**For a more accurate representation see the electronic version.*

food politics. Martin Ravallion (2001) estimates that, worldwide, 30% of poor people live in urban areas. By 2020 the proportion is projected to reach 40%, and by 2035 half of the world's poor people are projected to live in urban areas.

In 1988, the World Bank estimated that some 330 million urban poor in the developing world were living on less than US$1 a day (World Bank 1991). In 2000 the estimate had increased to 495 million (Hinrichsen et al., 2002). Sub-Saharan Africa has some of the world's highest levels of urban poverty, reaching over 50% of the urban populations in Chad, Niger, and Sierra Leone. Chen and Ravallion (2001) measured absolute poverty in terms of household consumption expenditure percapita in the Middle East, North Africa, South Asia and Sub Saharan Africa. Their findings suggest the prevalence of inequality in income and consumption distribution. The study pointed out that the region with highest poverty incidence relative to $1perday line is sub Saharan Africa(SSA) followed by South Asia, that this group accounted for 70% of those living below 1$ per day in 1998.East Asia came Third in terms of incidence of poverty with Latin America.

Green consumerism which results in green economy has often been advocated as it legitimizes sustainable consumption. However, it has also been criticized in capitalist contexts as it results in "ecological calamity", or "disaster capitalism" (Klein, 2008). It becomes "an opportunity for corporations to turn the very crisis that they generate through their accumulation of capital via the exploitation of nature" (Klein, 2008). This according to Estavo and Prakash (1998) has resulted in the waste of nature and production of rot and decay. Eastwood (2006:118-119), has created nexus between capitalist production and consumption in the contexts that "capital accumulation relies not only on the production of goods, but also on the production of the willing consumer". This corroborates the perspective that consumption in a distinct manner promotes production and vice vasa.

The concept of food security has been part of the wider terminological shifts food consumption had witnessed. Food security includes a situation in which all people always 'have physical, social and economic access to sufficient, safe and nutritious food that meets their dietary needs and food preferences for an active and healthy life" (FAO, 2010). There are linkages between food security and climate change including biofules and land grab by the affluent countries (Lawrence, et al;2011)

Hinrichsen, et al; (2002) argue that about 50% of the world's poor are living in food insecure areas with low subsistent level. There are issues of resource depletion. For instance fresh water shortages soars and accounts for resurgence of several water related diseases where access to clean water is scarce, sanitation is poor. This is reinforced in related accounts on ecological breakdown while related environmental consumption challenges arising from fossil fuel and industrial pollution are critical. The pollution rate recorded in the high income countries between 1900 to 2015, suggest the imminent dangers of unsustainable food consumpton.

Global Food Politics and Sustainable Consumption

Food in global contexts has divergent interpretations. At the aftermath of the second world war food became a major instrument of foreign policy. However following the end of the Cold War a period of competitive world food prices and a return to a more "free market" food policy remerged. The scenario has persisted resulting to a rise in Western food hegemony. The contention remains the need to understand how sustainable a food system is and in particular, the failure of the neo liberal food system to deliver sustainable and equitable consumption.

Christopher Rosin,Paul Stock and Hugh Campbell (2011) analyze the contemporary global food system in the context of its failure to actualize food security. This includes complexities and contradictions surrounding dominant dynamics of production linked with industrial capitalism. Thus, world food systems cannot be discussed in isolation of globalization, free trade and the logic of capitalist mode of production as well as persistent inequality.

Global food politics linked to the neoliberal policy prescriptions of the World Trade Organization (WTO), the World Bank, and the International Monetary Fund (IMF)have resulted in persistent and unmitigated unequal natural resource consumption which has been at variance with sustainable and equitable development. Capitalist consumption is perhaps contradictory as it includes the politics of market and competition informed by the logic of liberalism which preaches freedom and capitalism which is riddled with inequality and exploitation.

In the United States, Natural Resources Defense Council (2010) contends that a typical American meal contains ingredients from five foreign countries, and even domestically grown produce travels an average of 1,500 miles before it is sold. In Germany,260 million tons of CO^2 equivalents are emitted per year, i.e. 3.2 tons per inhabitant to feed the country's 80 million persons (Fuchs & Lorek, 2001).

Capitalist exploitation is linked to the patterns of economic relationship between the North and South. By the latter part of the nineteenth century, places as diverse as Malaysia (rubber producer), India (cotton producer), Egypt (cotton producer),Argentina (beef and wheat producer), Ghana (cocoa producer), and Cuba (sugar producer) had become specialized in the production of one or more export crops for European (and later U.S.) markets(Wolf, 1997).This results in unequal exchange and poverty on a grand scale.This is reinforced in the unequal labour relationships which shape the economic relationship of the 'North' and ' South' and have become issues of scholarly concern in contemporary sustainable consumption debate(Amin, 1972).

Paalberg (2008) argues that because Western consumers derive no benefits from biotechnological advances in agriculture they inflict a frivolous "imperialism of rich tastes" on African farmers.

This points out that the prevailing food system ignored the divergent effects of unsustainable consumption. This compels the need for provision of food for an increasing population on sustainable basis. In the 'North', increasing food consumption must take adequate audit of the environment which is perhaps less examined. Paul Stern (1997) argued that consumption is central to development of every society especially within ecological contexts. This includes chemical residues in crops and livestock, aflatoxin/ mycotoxin contaminants which reduce the quality and acceptability of food, evidence of contaminants which in turn, adversely affects the poor societies as they rely on 'finished' food products from the high income societies.

Equally, the rise in Western food corporations and capitalist resource consumption have increasingly become an opportunity for the corporations to accumulate and exploit nature in a bid for profit and investment revenue. In this particular case, as the global food system makes food available, the affluent societies benefit to the exclusion of the poor, through access and hegemonic control mechanisms. A sustainable food system entails meeting present food consumption and nutrition security needs which should not undermine food and nutrition security for future generations. The sustainability of food system is influenced by anthropogenic and non- anthropogenic factors. These complexities are situated within the interwoven contexts of dynamics of 'food production, distribution, consumption, nutritional

Figure 3. Shares of the population in low- and middle-income countries living in extreme poverty, by region

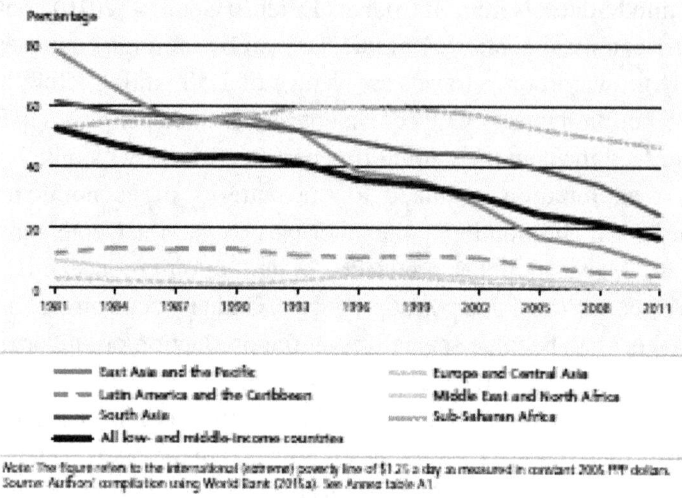

**For a more accurate representation see the electronic version.*

health, socio-economic and environmental factors related to the quantity, quality and affordability of food, as well as health and wellbeing'(UN, 2015).

Globalization of food consumption gained relevance at the instance of the US hegemony. Globalization fosters unequal wealth distribution, favors a small economic elite, leaving out the majority who are economically vulnerable (Rees, 1998).This is perhaps evident in the rise of global food corporations-food supply chains, stores and outlets that consume disproportionate resources notably, Nestlé (Switzerland), PepsiCo (USA), Kraft (USA), ABinBev (Brazil), ADM (USA), Coca-Cola (USA), Mars Inc. (USA), Unilever (Netherlands), Tyson Foods (USA), Cargill (USA) (Berne Declaration,2013), have resulted in dependency linked to globalization as they extract natural resources from the developing economies at a relatively cheap rate and export finished products at a higher economic rate (Bello, 2004;Amadi, 2012; Wise, 2015).

A recent beach clean -up and audit at Freedom Island in Manila Bay, Philippines by the Greenpeace showed that Nestlé, Unilever and Indonesian company PT Torabika Mayora are the top three contributors of plastic waste discovered in the area, contributing to the 1.88 million metric tonnes of mismanaged plastic waste in the Philippines per year(Tiu, 2017).Thus, food corporations seem not to have fully recognized the challenges of environmental degradation and complexities of ecological breakdown and social distortions .Despite the alleged claim on green corporate social responsibility(Amadi & Imohita, 2017).

This provides useful insights on contemporary global food dynamics. On this basis, corporate interest dominates much of the mainstream burgeoning thinking on neo liberal food consumption with less emphasis on the imminent long term dangers of the ecological consequences(Korten, 1995).

Another critical dimension contemporary global food politics had taken is necessitated by the rise in global integration of national economies as globalization takes consumption to a new level. Paehlke (2003) argues that the result is a fully integrated global system dominated by corporate actors. Essentially global food politics dominates much of the advanced industrialized societies. Wapner, (1996) argues that the emergence of international and global economic institutions diminishes state sovereignty and reduces the state's capacity to intervene in economic activities.

In this regard, global food politics remains a complex matrix of interactions advancing from colonial and post -colonial era to the neo colonial era including imperialism and globalization. In most post-colonial societies of the Third World a circle of economic exploitation through trade is linked to a systemic exploitation through the colonial marketing boards (Ekekwe, 1986). Accordingly, in Latin America, Janvry (1981) recounts the unequal patterns of resource extraction through, agro-export production by the Spanish and Portuguese colonizers which extended to the post- independence era. Pointing out that the newly independent states of Latin America re-entered the global economy as suppliers of agricultural commodities to Europe and the United States and as purchasers of manufactured goods.

Figure 4. Meat consumption and share of net imports in consumption, least developed countries, 1961-2005

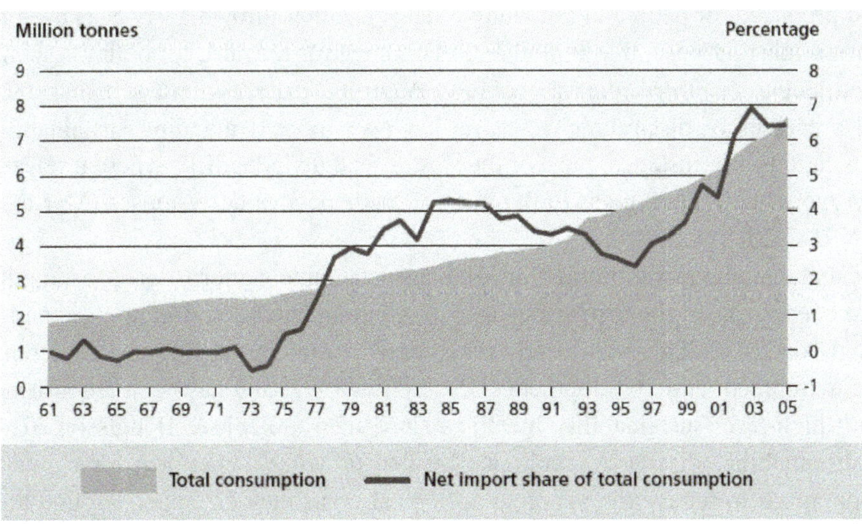

**For a more accurate representation see the electronic version.*

The implications of the free market debate has equally favored the economies of the developed capitalist societies as most developed countries, including the United Kingdom, the United States, France, Germany, and Japan, used tariffs, subsidies, and other interventionist measures in order to promote industrialization (Chang, 2003:426).

Paalberg (2013) argues that farm subsidies shape contemporary international agricultural trade. However, highly indebted developing countries are precluded from using these measures as they are subject to the neoliberal policy prescriptions of the IMF and the World Bank leading to structural adjustment programs designed to ensure loan repayment (Chang, 2003). In Africa, for example, exports diversification and development of manufacturing sector have been negatively impacted by the neoliberal reforms imposed by the IMF and the World Bank which emphasizes comparative advantage in raw material and primary product exports (Stein, 2003).

There are issues of consequences of overconsumption in the industrialized countries(Schor,2005) This is in line with the disproportionate consumption debate by the industrialized societies which have been examined in seminal writings (Shove,2003;Hobson, 2003, Schor, 2005; Milner, 2012). In the United States the NRDC (2010)estimates that 'in California, which imports food distributed throughout the nation, the smog-forming emissions from importing fruits and vegetables are equivalent to the annual emissions from 1.5 million cars". This is linked to the need for ecological footprints of the North (Rees, 1998). The several limitations posed by capitalist international trade particular in relation to the rual peoples of the tropics have been examined (Carlsson-Kanyama, 1997)

Another key component of unsustainable consumption is food wastage. This includes water and similar consumables. The USDA estimates 'that 27% of all food produced for people in the United States is either thrown away or is used for a lower-value purpose, like animal feed. This suggests the need for policy response and strategies to mitigate the waste of food. Again, the patterns of waste management and processes of discarding the waste has ever more demanding sustainability implications. For instance, most of the wastes 'end up in landfills where it releases even more heat-trapping gas in the form of methane as it decomposes'(Martin,2008;NRDC,2010).

Capitalist mode of production and unsustainable patterns of consumption have been declared the primary causes of environmental deterioration (Akenji & Bengtsson,2014). This was clearly recognized at the United Nations Conference on Environment and Development (UNCED) in 1992, and has been reconfirmed in all high-level sustainability meetings since then (Akenji & Bengtsson,2014). In sustainability discourse, ecological justice provides a platform for capitalist corporations to retrace the perverse ecological crisis they engender through their accumulation ethos in the context of exploitation of nature for profit maximization.

Table 1. Per capita consumption of livestock products by region, country group and country, 1980 and 2005

REGION/COUNTRY GROUP/ COUNTRY	MEAT		MILK		EGGS	
	1980	2005	1980	2005	1980	2005
	(kg/capita/year)		(kg/capita/year)		(kg/capita/year)	
DEVELOPED COUNTRIES	**76.3**	**82.1**	**197.6**	**207.7**	**14.3**	**13.0**
Former centrally planned economies	63.1	51.5	181.2	176.0	13.2	11.4
Other developed countries	82.4	95.8	205.3	221.8	14.8	13.8
DEVELOPING COUNTRIES	**14.1**	**30.9**	**33.9**	**50.5**	**2.5**	**8.0**
East and Southeast Asia	12.8	48.2	4.5	21.0	2.7	15.4
China	13.7	59.5	2.3	23.2	2.5	20.2
Rest of East and Southeast Asia	10.7	24.1	9.9	16.4	3.3	5.1
Latin America and the Caribbean	41.1	61.9	101.1	109.7	6.2	8.6
Brazil	41.0	80.8	85.9	120.8	5.6	6.8
Rest of Latin America and the Caribbean	41.1	52.4	109.0	104.1	6.5	9.4
South Asia	4.2	5.8	41.5	69.5	0.8	1.7
India	3.7	5.1	38.5	65.2	0.7	1.8
Rest of South Asia	5.7	8.0	52.0	83.1	0.9	1.5
Near East and North Africa	17.9	27.3	86.1	81.6	3.7	6.3
Sub-Saharan Africa	14.4	13.3	33.6	30.1	1.6	1.6
WORLD	**30.0**	**41.2**	**75.7**	**82.1**	**5.5**	**9.0**

Source: FAO, 2009b.

Panitch and Leys (2007) argue, that there is need for multiple moves in ‘political education: from blind consumerism to a mobilization against specific corporations to an organized understanding of the unsustainable logic of the capitalist system in toto”.

Technological advancement is another issue of relevance in sustainable food consumption. In recognition of the interwoven contexts of the natural environment and food, the challenge to sustainable food consumption results in the novel turn technological advancement had taken in recent decades pointing to the ecological modernization perspectives which are “confident” of the ability of capitalism to transform itself in the face of ecological crisis(Sparaagen,1997). This partly accounts for the rise in GE foods in the industrialized societies since the early 2000s. Genetically modified foods (GMs) or genetically engineered crops(GE) and its products derive from genetically engineered organisms. They are among a number of biotechnological developments intended to improve shelf life, nutritional content, flavor, color, and texture, as well as agronomic and processing characteristics.

Scholarly literature suggests two key perspectives on the GEs namely the proposing perspective which argue for the technological, nutritional and economic relevance of GE foods (Sears, 1981, Gupta & Tsuchiya, 1991; Heyman, et al., 1998, Georges, 2001) and the opposing perspectives which hold a contrary view on the sustainability of the GE foods (Gepts, 2006; Wang, et al. 2006; Altieri, 2008; Frison, 2009;Jacobsen, et al. 2013). Altieri (2008) argued on resilience and sustainability of non GE crops suggesting that agro-ecological farms contain a high level of biodiversity and are self-supporting systems in harmony with their environment.

Similarly, Frison (2009) argued that contemporary agro-systems with low biodiversity offer low-diverse diets which in turn lead to high incidence of lifestyle illnesses such as obesity, Type II diabetes, heart diseases and cancer. There is a debate which contends that the persistence of GE has resulted in the loss of crop and livestock biodiversity including deteriorated genetic loss and vulnerability (Gepts, 2006).

In many instances, states in industrial societies are constrained by the logic of capital or the need to subsidize industry so that the corporate organizations can continue to make a profit, keep people employed, and in turn generate revenues for the state through taxation. The constraints placed on capitalist states are only exacerbated when the priorities of economic growth become codified in international trade agreements. The regulations of NAFTA, for example, make it much more difficult for states to protect natural resources, or to support locals in doing so (Schnaiberg & Gould, 1994; O'Connor,1998; Roberts & Grimes, 2002, Moore, 2000, Chew, 1995; Chase-Dunn, 1989; Davidson & Hatt, 2005, p.238).

Carmen Gonzalez (2004) recounts that inequality remains a core neo liberal threat to sustainable development. For instance, close to a billion people go hungry in the world every day, with 10 children dying of starvation every minute (Pinstrup-Andersen, 2010b). This not only suggests the urgency of sustainable food consumption rather points to food equality which underscores the primacy of equal access to food consumption across regions and in gender contexts. Data below suggests that women are most affected with food insecurity issues.

Globalization has been a central factor to consumption disparity creating a wide gap between the affluent and poor societies. This includes issues of international trade networks, globalization of capital concentration, technological advancement and innovation, Western hegemonic power, information flow and diffusion of values. This relates to the expansion of capitalism replicated in global land grab, plantation agriculture (capitalist farming) and land degradation which have been at issue in unsustainable consumption debates (Wise, 2015). The global convergence of consumption pattern has given rise to emphasis on thc North-South divide with

Figure 5. Women are slightly more likely to be food insecure than men in every region of the world

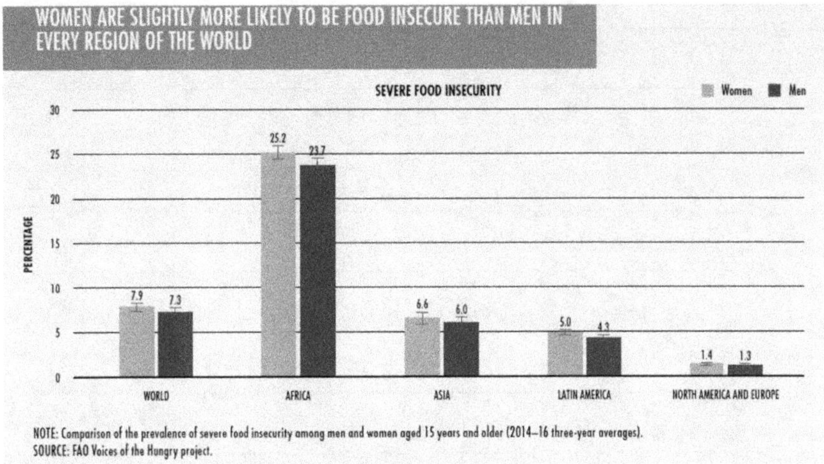

**For a more accurate representation see the electronic version.*

emphasis on influence of the North on consumption response and behavior of the South as the latter adapt to the behavior of the former.

Contemporary society is confronted with the challenge of attaining sustainable food consumption for its growing population. Sustainable food consumption is no longer optional, rather mandatory. Since the Rio summit of 1992 sustainable consumption has become a global issue in development discourse. Similarly Goal 2 of the Sustainable development Goal(SDGs)which is Zero Hunger, points to the saliency of sustainable and equitable consumption. To strengthen this paradigm, there have been growing literature on aspects of sustainable consumption most notably green consumerism, eco- friendly consumption, ecological justice, consumption culture etc.

To redirect the prevailing global food politics requires collective policy engagement. Thus, while sustainable food consumption discourse plays a key role in recreating the dominant notion of consumption and patterns of lifestyle, the affluent societies typically fail to distinctively, attenuate the deleterious consumption patterns. In Sweden, the Swedish Environmental Protection Agency pointed out that the 'use and misuse of nature's resources account for environmental unsustainability. For instance, food consumption alone accounts for 20% of the greenhouse emission in Sweden (Lönnroth, 2010).

Figure 6. Per capita and meat consumption by country, 2005

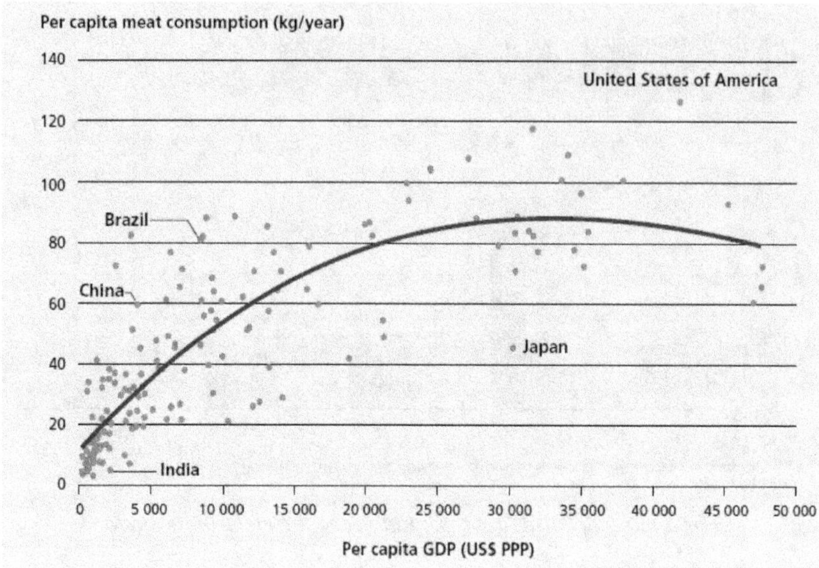

Note: GDP per capita is measured at purchasing power parity (PPP) in constant 2005 international US dollars.
Source: Based on data from FAOSTAT (FAO, 2009b) for per capita meat consumption and the World Bank for per capita GDP.

RECOMMENDATIONS

Although sustainable consumption policies and practices are still in its early stages, there is urgent need for food reform. This chapter has attempted to chronicle some of the scholarly evidence of unsustainable consumption. Since we cannot retain the natural environment amidst unsustainable consumption and marketization of nature and since capitalism is built on profitability, how can the overarching sustainability and consumption be framed to resolve the challenges of deleterious environmental implications? The logic of sustainable consumption helps to resolve the problem and allows for a better policy intervention more generally linked to equality in food consumption. The important question now becomes, how to overcome the challenges of both inequality and deleterious effects of unsustainable food consumption.

The emergence of the field of sustainable food consumption has provided useful insights pushing for disciplinary space in addressing these problems. Beyond this, a return to the original impetus of sustainable food consumption one which seeks for equality and resource efficiency within the capitalist system has not been satisfactorily resolved particularly in ecological contexts as sustainable consumption remains contestable.

In this regard, no single set of policies can meet all the challenges of global food subsystem particularly in the international capitalist system. A broader reconstructive consumption which goes beyond greening project is necessary one which links sustainable consumption with production for an equitable democratization is suggested.

Policy instruments to promote sustainable food systems which range from wide perspectives and strategies, should include the need to return to local food systems as they are eco- friendly, healthy and largely interconnected with green consumerism. The eco- friendly debate on food consumption contends that the natural environment cannot be divorced from mode of consumption and suggests instead the preservation the ecosystem. Thus, "eco-friendly consumption", is a strand of sustainability" that "is seen as an act of equitable development, which aims to meet the material needs and at the same time create a healthy environment (Vlaemincka, et al;2014). However a critical dimension has been the notion that 'food is heavily politicised and considered a global industry worth approximately USD 4.8 trillion yearly' (World Bank, 2006).

The ecological justice thesis had emerged to offer a direct and serious challenge to unsustainable capitalist consumption (Korten, 1995; Hobson, 2003; Harvey, 2005; Eastwood, 2006; Akenji & Begsson, 2014; Amadi, et al., 2014). Korten (1995) had called for an ecological revolution one which includes environmental sustainability, economic justice, biological and cultural diversity. Specifically, to serve the needs of the people and not corporations or governments.

Thus, the entire line of debate in this work emphasizes the need for pro poor, inclusive and eco-friendly consumption pattern.

Eco friendly consumption is a pro-environment consumption pattern which forms part of the alternative consumption debate. It is akin to green consumerism which includes reduction in food toxins, global warming emissions, reduction in biodiversity loss, wildlife preservation, protection of the ecosystem and natural habitats etc.

Green consumerism aims to promote consumption in "natural state" namely eco-efficient access and extraction of natural resources, eating fresh and healthy meals including fruits or vegetables. Thus natural resources if sustainably consumed can add value in many ways to the environment.

The "ecosystem first" consumption dynamics is a pro-environment research agenda which aims to protect the ecosystem needs of the present and future generations in the cause of consumption without tainting the environment. It protects species that inhabit the system. As it aims at ecological conscious consumer behaviour (ECCB) (Straughan & Roberts, 1999)., green purchase behaviour and the green consumer (Jansson, Marell, & Nordlund, 2010; Akehurst & Akenso,2012; Riley, Kohlbacher & Hofmeister, 2012).

Ecological foot print in consumption maintains a clean and healthy consumption dynamic. The aim is to check emission and related green -house effects associated with capitalist consumption.

Strict state regulation based on command-and-control policies to weak state intervention relying on market-based instruments and to private voluntary initiatives could prove helpful in mitigating unsustainable choices and 'consumer freedom' that guides capitalist consumption patterns (Reisch et al., 2013).

Sustainability measurement (Amadi & Imohita, 2017) suggests "ecological accounting" to check unequal and deleterious resource consumption. The point this chapter continues to argue is that sustainability is notable for its critical emphasis on resource efficiency and equitability. And that such hugely significant insights could only be useful where sustainability challenges are matched with ecological policies as some of the foundational questions of development, particularly the relationship of food consumption to sustainability have not been adequately resolved.

The fundamental objectives of this chapter require collaborative policy framings involving relevant stakeholders including the global food giants, policy makers and corporations. Getting the capitalist corporations to account for their mode of natural resource extraction remains at issue. Korten (1995) suggests that neo liberal consumption patterns are linked to global consolidation of corporate power. He provides useful critique of neo liberal consumerism, market deregulation, free trade and privatization. Bakan (1995) had explored this within the context of 'pathological pursuit of profits' pointing out the evidence of capitalist exploitation that drives corporations.

The chapter re-emphasizes the global dichotomy in food system among the affluent North and the poor South. In particular, there is need to rethink the consumption culture and the patterns of lifestyles of the affluent societies. Policy response in this direction should argue instead for resource renewal and ecological justice. This orientation has given rise to the need for new policy initiative with the question of food and consumption policies in the modern societies including global and regional conferences, the involvement of international food and research institutes and NGOs.

The underlying practices and understanding of what policies that are driving these trends in the is critical. Much of these require new policy intervention which this chapter emphasizes at intervals.

Perhaps one of the greatest contributions of sustainable food consumption is the need for resource accountability and equality to check the access and resource control dynamics of the powerful nations. Against the idealization of sustainable consumption, terms such as green consumerism as explicated, consistently emphasize the socially just and equitable mode of consumption. Thus, sustainable consumption is not a set of abstract term insulated from ecological and social reality. Rather, it constitutes the reality of everyday living, "food is life" and should be protected in equitable contexts.

Thus, idealist orientation which conceives food as mere "thing" within abstract contexts, devalues food from the actual development realm and rather suggests a mundane and naïve rhetoric. In particular, while sustainable food consumption is critical to resource efficient development, it points out that humanity must assimilate resource renewal, ecological justice and eco-friendly dynamics. This implicates the capitalist material accumulation which appears to be at variance with the ideal values and norms of sustainable food consumption. This points out the inevitability of new constitutive relationships between 'food' and 'consumption', in the liberal order in a manner which puts sustainability at the center of policy discourse.

So, while the dominant neo liberal consumption patterns remain largely materialistic in orientation, this arguably vitiates sustainable consumption. Harvey (2005)provides one of the most influential treatise in this direction leading to his contention on "creative destruction" .Thus, pattern of food consumption that meets generational needs of humanity and preserves the natural environment is inevitable.

FUTURE RESEARCH DIRECTIONS

Although this chapter tackled an important aspect of sustainable development that could result in further thought provoking debates, several important research themes are worth investigating for a future study. The point the chapter has been aiming at is that the dilemma surrounding international political economy of food consumption may likely worsen the existing inequalities and gaps in well-being across the North and South. Emphasis should be laid on strategies to bridge this gap. Future research should critically examine modernization of agriculture to either provide a nuanced critique or otherwise. A new research engagement with "green revolution" is suggested to help the developing societies to boost food production in a sustainable manner. Novel trends such as GM or GE seeds or crops should be re-examined in further research to understand the future policy direction of food system.

Against, the rise in capitalist farming and plantation agriculture are themes of scholarly relevance and should be given research attention including dynamics and patterns of the use of organic fertilizers. Thus, future research direction should insist on theorizing the possible strategies or alternative policy options for inclusive and green consumption.

CONCLUSION

Based on the key issues raised in this chapter in line with the methodological tools and over- all objective of the study, a number of evidence suggesting how sustainable food consumption has been compromised particular since the post 1992 Rio summit and its agenda 21 the plan of action for implementation of sustainable development was provided.

The neoliberal consumption patterns and trade regimes are not entirely free with regards to the poor societies. While it fosters protectionism in the developed countries in which they open their markets to highly subsidized foreign competition, the poor societies are increasingly restricted with highly stringent conditions. This impedes economic diversification, on the contrary, fosters dependency and inequality. Thus the developing societies are tailored in line with Western food systems in the international capitalist system. This has dominated the ideals of the Breton Woods institution.

There is need for urgent policy response. Thus, food provision alone as propagated by neo liberal proponents is not enough but creating the enabling policy framework to mitigate the adverse environmental challenges is important.

The mainstream notion of consumption reinforces the liberal ideals of freedom of choice in a market society seemingly at the expense of the ecosystem. Capitalism encourages profit maximization regardless of ecological and environmental costs. The chapter demonstrates that this increasingly taints the ecosystem, creates global disparity and inequality. Yet the dominant notion is that there is no alternative. While the chapter does not suggest alternative order, are there no alternative ways of mitigating the rise in unsustainable consumption? This forms the basis for the argument on sustainable food consumption.

REFERENCES

Agnew, J. (2005). *Hegemony The New Shape of Global Power*. Philadelphia: Temple University Press.

Ake, C. (1981). *A Political Economy of Africa*. London: Longman.

Akehurst, G., Afonso, C., & Gonçalves, H. (2012). Re-examining green purchase behavior and the green consumer profile: New evidences. *Management Decision*, *50*(5), 972–988. doi:10.1108/00251741211227726

Akenji, L., & Begsson, M. (2014). *Making Sustainable Consumption and Production the Core of the Sustainable Development Goals*. Discussion Paper Institute for Environmental Global Strategies.

Altieri, M. (2008). *Small farms as a planetary ecological asset: Five key reasons why we should support the revitalization of small farms in the Global South*. Third World Network.

Amadi, L. (2012). Africa: Beyond the "new" dependency: A political economy. *African Journal of Political Science and International Relations*, *6*(8), 191–203. doi:10.5897/AJPSIR12.022

Amadi, L., & Igwe, P. (2016). Maximizing the Eco Tourism Potentials of the Wetland Regions through Sustainable Environmental Consumption: A Case of the Niger Delta, Nigeria. *The Journal of Social Sciences Research*, *2*(1), 13–22.

Amadi, L., Igwe, P., & Wordu, S. (2014). Sustainable Development, Greening and Eco-efficiency. A Political Ecology. *Journal of Sustainable Development Studies*, *7*(2), 161–196.

Amadi, L., & Imoh-Ita, I. (2017). Intellectual capital and environmental sustainability measurement nexus: A review of the literature *Int. J. Learning and Intellectual Capital*, *14*(2), 154–172. doi:10.1504/IJLIC.2017.084071

Amin, S. (1972). Underdevelopment and Dependency in Black Africa Origins and Contemporary Form. *The Journal of Modern African Studies*, *10*(4), 503–524. doi:10.1017/S0022278X00022801

Åström, S., Roth, S., Wranne, J., Jelse, K., & Lindblad, M. (2013). *Food consumption choices and climate change*. Swedish Environmental Research Institute.

Bakan, J. (2004). *The Corporation: The Pathological Pursuit of Profit & Power*. Toronto: Penguin Group.

Bello, W. (2003). *De-globalization:Ideas for a New World Economy*. Fernwood.

Berne Declaration. (2013). *Agropolicy. A handful of corporations control world food production*. Available at http://www.econexus.info/sites/econexus/files/Agropoly_Econexus_BerneDeclaration.pdf

Brenkert, G. (1998). The Marketing Challenge: Towards Being Profitable and Socially Responsible. *Journal of Business Ethics*, *7*(7), 497–507.

Bryant, R., & Bailey, S. (1997). *Third world political ecology: An introduction*. London: Routledge.

Carlsson-Kanyama, A. (1997). Weighted average source points and distances for consumption origin- tools for environmental impact analysis? *Ecological Economics*, *23*(1), 15–23. doi:10.1016/S0921-8009(97)00566-1

Chang, H. (2003). Kicking Away the Ladder: Development Strategy in Historical Perspective. In Rethinking Development Economics. Anthem Press.

Chase-Dunn, C. (1989). *Global Formation*. Cambridge, MA: Blackwell.

Chen, S., & Ravallion, M. (2001). How did the World's Poor Fare in the 1990s? *Review of Income and Wealth*, *4*(3).

Chew, S. (1995). Environmental Transformations: accumulation,Ecological Crisis and Social Movements. In D. Smith & J. Borocz (Eds.), *A New World Order?Global Transformations in the Late Twentieth Century*. Westport, CT: Praeger Publishers.

Conway, D., & Hynen, N. (2006). *Globalization's Contradictions: Geographies of Discipline,destruction and Transformation*. New York: Routledge.

Davidson, D., & Hatt, K. (2005). *Consuming Sustainability Critical Social Analysis of Ecological Change*. Fernwood Publishing.

Eastwood, L. (2006). Contesting the Economic Order: Resisting the Media Construction of Reality. In S. Best & A. J. Nocella (Eds.), *Igniting a Revolution: Voices in Defense of Mother Earth*. Edinburgh, UK: AK Press.

Eberle, U., Hayn, D., Rehaag, R., & Simshäuser, U. (Eds.). (2006). *Ernährungswende. Eine Herausforderung für Politik, Unternehmen und Gesellschaft* [Nutrition Change: A Challenge for Politics, Business, and Society]. Munich: Oekom. (in German)

Ekekwe, E. (1986). *Class and State in Nigeria*. London: Longman.

Estavo, G., & Prakash, M. (1998). *Grassroots Post-Modernism: Remaking the soil of cultures*. Zed Books.

FAO. (2002). The State of Food Insecurity in the World 2001. Rome: FAO.

FAO. (2010). *Biodiversity for a World without Hunger*. Retrieved from http://www.fao.org/biodiversity/biodiversity-home/en/

FAO. (2012). *Food Outlook Global Market Analysis*. FAO.

FAO. (2015). *The State of Food and Agriculture. Social protection and agriculture: Breaking the cycle of rural poverty*. FAO.

FAO. (2017). *The State of Food Security and Nutrition in the World*. FAO.

Frison, E. (2009). Director General calls for investment in true food security. *Biodiversity News*. Retrieved from http://www.bioversityinternational.org/news_and_events/news/news/article

Fuchs, D., & Lorek, S. (2001). Sustainable Consumption Governance in a Globalizing World. *Global Environmental Politics*, *2*(1), 19–45. doi:10.1162/152638002317261454

Georges, M. (2001). Recent progress in livestock genomics and potential impact on breeding programs. *Theriogenology*, *55*(1), 15–21. doi:10.1016/S0093-691X(00)00442-8 PMID:11198079

Gepts, P. (2006). Plant genetic resources conservation and utilization: The accomplishments and future of a societal insurance policy. *Crop Science*, *46*(5), 2278–2292. doi:10.2135/cropsci2006.03.0169gas

Giddens, A. (1990). *The consequences of Modernity*. Stanford, CA: Stanford University Press.

Gonzalez, C. (2004). Trade Liberalization, Food Security and the Environment: The Neoliberal Threat to Sustainable Rural Development. *Transnat'l L. & Contemp. Probs.*, *14*(4), 419.

Gupta, P., & Tsuchiya, T. (1991). *Chromosome Engineering in Plants: Genetics, Breeding Evolution*. Amsterdam: Elsevier.

Harvey, D. (2005). *Brief History of Neoliberalism*. New York: Oxford University Press.

Haws, K., Winterich, K., & Naylor, R. (2014). Seeing the world through GREEN-tinted glasses: Green consumption values and responses to environmentally friendly products. *Journal of Consumer Psychology*, *24*(3), 336–354. doi:10.1016/j.jcps.2013.11.002

Heyman, Y., Vignon, X., Chesne, P., Le Bourhis, D., Marchal, J., & Renard, J. P. (1998). Cloning in cattle: From embryo splitting to somatic nuclear transfer. *Reproduction, Nutrition, Development*, *38*(6), 595–603. doi:10.1051/rnd:19980602 PMID:9932293

Hinrichsen, D., Salem, R., & Blackburn, R. (2002) Meeting the Urban Challenge. Population Reports, Series M, No. 16. Baltimore, MD: The Johns Hopkins Bloomberg School of Public Health, Population Information Program.

Hobson, K. (2003). Consumption, Environmental Sustainability and Human Geography in Australia: A Missing Research Agenda? *Australian Geographical Studies*, *41*(2), 148–155. doi:10.1111/1467-8470.00201

Holsti, O. (1969). *Content Analysis for the Social Sciences and Humanities.* Addison-Wesley.

IOM (Institute of Medicine). (2015). *A framework for assessing effects of the food system.* Washington, DC: The National Academies Press.

Jacobsen, S., Sorensen, M., Soren, M., & Weiner, J. (2013). Feeding the world: Genetically modified crops versus agricultural biodiversity. *Agronomy for Sustainable Development, 0138-9.* doi:10.1007/s13593-013

Jansson, J., Marell, A., & Nordlund, A. (2010). Green consumer behavior: Determinants of curtailment and eco-innovation adoption. *Journal of Consumer Marketing, 27*(4), 358–370. doi:10.1108/07363761011052396

Janvry, L. (1981). *The Agrarian Question and Reformism in Latin America.* Johns Hopkins University Press.

Jongen, W., & Meerdink, G. (1998). *Food Product Innovation: How to Link Sustainability and the Market.* Wageningen Agricultural University.

Klak, T. (2013). World System theory core,semi-peripheral and peripheral regions. In The companion to Development Studies. Routledge.

Klein, N. (2008). *The Shock Doctrine The Rise of Disaster Capitalism.* Metropolitan Books Henry Holt and Company.

Korten, D. (1995). *When Corporations Rule the World.* West Hartford, CT: Kumarian Press Inc.

Lawrence, G., Lyons, K., & Tabatha, W. (2010). *Food Security, Nutrition and Sustainability.* New York: Earthscan.

Lock, I., & Ikeda, S. (2005). *Clothes Encounters: Consumption, Culture, Ecology and Economy. In Consuming Sustainability Critical Social Analysis of Ecological Change.* Fernwood Publishing.

Lönnroth, M. (2010). The Organization of Environmental Policy in Sweden - A Historical Perspective. *The Swedish Environmental Protection Agency Report, 6404*, 319.

Lukaszewski, A. (2004). Chromosome manipulation and crop improvement. In *Encyclopedia of Plant and Crop Science.* New York: Marcel Dekker. doi:10.1081/E-EPCS-120005593

Martin, A. (2008, May 18). One Country's Table Scraps, Another Country's Meal. *The New York Times.*

Marx, K. (1978). *The Marx-Engels Reader*. New York: Norton.

Masters, B. (2004). *Safety of Genetically Engineered Foods: Approaches to Assessing Unintended Health Effects*. Washington, DC: The National Academies of Sciences.

Milner, D. (2012). *Consumption and Its Consequences*. Cambridge Polity Press.

Moomaw, W., Griffin, T., Kurczak, K., & Lomax, J. (2012). *The Critical Role of Global Food Consumption Patterns in Achieving Sustainable Food Systems and Food for All, A UNEP Discussion Paper*. United Nations Environment Programme, Division of Technology, Industry and Economics.

Moore, J. (2000). Environmetal Crisis and the Methabolic Rift in World –historical perspective. *Organization & Environment*, *13*(2), 2. doi:10.1177/1086026600132001

Muller, J. (2013). Capitalism and Inequality, What the Right and the Left Get Wrong. *Foreign Affairs*, *92*(2), 30–51.

Nesheim, M., Oria, M., & Yih, P. (Eds.). (2015). *A Framework for Assessing Effects of the Food System*. Washington, DC: The National Academies Press.

NRC (National Research Council). (2004). *Safety of Genetically Engineered Foods: Approaches to Assessing Unintended Health Effects*. Washington, DC: The National Academies Press.

NRC (National Research Council) & IOM (Institute of Medicine). (2015). A framework for assessing effects of the food system. Washington, DC: The National Academies Press.

O'Connor, J. (1998). *Natural Causes: Essays in Ecological Marxism*. Guiford Press.

Paalberg, R. (2013). *Food Politics: What Everyone Needs to Know*. Oxford University Press.

Paarlberg, R. (2008). *Starved for Science: How Biotechnology is Being Kept Out of Africa*. Harvard Univ. Press. doi:10.4159/9780674041745

Paehlke, R. (2003). Egalitarian Perspectives on Sustainability. In UNESCO's Encyclopaedia of Life Support Systems (vol. 36). UNESCO.

Panitch, L., & Leys, C. (Eds.). (2007). Coming to Terms with Nature: Socialist Register 2007. Fernwood.

Peet, E., & Watts, W. (1996). Liberation Ecology: Development, Sustainability and Environment in an Age of Market Triumphalism. In R. Peet & M. Watts (Eds.), *Liberation Ecologies:Environment,Development,Social movements* (pp. 1–45). London: Routledge. doi:10.4324/9780203286784

Pinstrup-Andersen, P. (2010). *The advantages of genetic engineering in agriculture include increased food production and reduced hunger - benefits for hungry and malnourished in developing countries outweigh disadvantages*. Retrieved from http://www.monsanto.com/biotech GM/asp/experts.asp?id=P

Ravallion, M. (2001). *On the urbanization of poverty*. Washington, DC: World Bank. doi:10.1596/1813-9450-2586

Rees, T. (1998). *Mainstreaming Equality in the European Union: Education, Training and Labour Market Policies*. Routledge.

Reisch, L., Eberle, U., & Lorek, S. (2013). Sustainable food consumption: an overview of contemporary issues and policies. *Sustainability: Science, Practice, & Policy*, *9*(2), 7–25.

Riley, L., Kohlbacher, F., & Hofmeister, A. (2012). A cross-cultural analysis of pro-environmental consumer behavior among seniors. *Journal of Marketing Management*, *28*(3-4), 290–312. doi:10.1080/0267257X.2012.658841

Robbins, P. (2004). *What is political ecology? In Political ecology: A critical introduction* (pp. 5–16). Malden, MA: Blackwell Publishing.

Roberts, J., & Grimes, P. (2002). World system Theory and the environment: Towards a new Synthesis. In Sociological Theory and the Environment: Classical Foundations, Contemporary Insights. New York: Rowman and Littlefield.

Rogers, K. (1997). *Ecological Security and Multinational Corporations*. Kluwer Academic Publishers.

Rosin, C., Stock, P., & Campbell, H. (Eds.). (2011). Food Systems Failure The Global Food Crisis and the Future of Agriculture. London: Routledge.

Schnaiberg, D., & Gould, K. (1994). *Environment and Society The Enduring conflict*. New York: St Martin's.

Schor, J. (2005). Prices and quantities: Unsustainable consumption and the global economy. *Ecological Economics*, *55*(3), 309–320. doi:10.1016/j.ecolecon.2005.07.030

Sears, E. (1981). Transfer of alien genetic material to wheat. In L. T. Evans & W. J. Peacock (Eds.), *Wheat Science, Today and Tomorrow* (pp. 75–89). Cambridge, UK: Cambridge University Press.

Shapouri, H., Gallagher, P., Nefstead, W., Schwartz, R., Noe, R., & Conway, R. (2010). *Energy balance for the corn–ethanol industry. Agricultural Economic Report 846. Office of the Chief Economist.* Washington, DC: U.S. Department of Agriculture.

Shopouri, S., Rosen, M., Peters, S., Tandon, F., Mancino, G. L., & Bai, J. (2011). *International Food Security Assessment, 2011-12. GFA-22.* United States Department of Agriculture-Economic Research Service. Retrieved from http://www.ersusda.gov/Publications/GFA22/GFA22.pdf

Spaargaren, G. (1997). *The ecological modernisation of production and consumption: in environmental. sociology.* Wageningen, The Netherlands: Landbouw University Wageningen.

Steger, M. (2002). *Globalism: The new Market ideology.* New York: Rowman and Littlefield Publishers.

Stein, H. (2003). Rethinking African Development. In Rethinking Development Economics. Anthem Press.

Stern, P. (1997). *Environmentally Significant Consumption.* National Academy Press.

Stiglitz, J. (2010). *Free Fall: America, Free Markets and the Sinking of the World Economy.* New York: W.W. Norton.

Straughan, R., & Roberts, J. (1999). Environmental segmentation alternatives: A look at green consumer behavior in the new millennium. *Journal of Consumer Marketing*, *16*(6), 558–575. doi:10.1108/07363769910297506

Swedish Environmental Protection Agency (SEPA). (2010). *The Climate Impact of Swedish Consumption.* Author.

Tiu, J. (2017). *Nestlé, Unilever, P&G Among Worst Offenders for Plastic Pollution in Philippines Beach Audit.* Available at https://www.ecowatch.com/plastic-pollution-philippines-beach-audit-2488280848.html

UN. (2015). *The Zero Hunger Challenge – Advisory notes for action.* The UN Secretary General's High level Task Force on Global Food and Nutrition Security. Retrieved from http://www.un.org/en/issues/food/taskforce/pdf/HLTF%20 %20 ZHC%20Advisory%20Notes.pdf

UNEP. (2012). *The critical role of global food consumption patterns in achieving Sustainable Consumption*. Nairobi, Kenya: UNEP.

United Nations Commission for Environment and Development (UNCED). (1992). *Agenda 21*. Oxford Press.

Vlaemincka, P., Jiangb, T., & Vrankena, L. (2014). Food labeling and eco-friendly consumption: Experimental evidence from a Belgian supermarket. *Ecological Economics, 108*.

Wallerstein, I. (1976). World System Analysis:the second Phase'. *RE:view*, *13*(2).

Wang, S., Just, D., & Pinstrup-Andersen, P. (2006). *Tarnishing Silver Bullets: Bt technology adoption, bounded rationality and the outbreak of secondary pest infestations in China*. Paper presented at American Agricultural Economics Association annual meeting, Long Beach, CA.

Wapner, P. (1996). *Environmental Activism and World Civic Politics*. State University of New York Press.

Weinstein, B. (2008). Developing Inequality. *The American Historical Review*, *113*(1), 1–18. doi:10.1086/ahr.113.1.1

Wise, T. (2015). Two roads diverged in the food crisis: Global policy takes the one more travelled. *Canadian Food Studies*, *2*(2), 9–16.

Wolf, E. (1997). *Europe and the People Without History*. Berkeley, CA: University of California Press.

World Bank. (1991). *Urban policy and economic development: An agenda for the 1990s*. Washington, DC: World Bank. Retrieved from http://wwwds.worldbank.org/servlet/WDSContentServer/WDSP/IB/1999/0/10/000178830_9810190142135/Rendered/PDF/multi_page.pdf

World Bank. (2006). *The World Bank Annual Report*. Retrieved from http:// siteresources.worldbank.org/INTANNREP2K6/ Resources/2838485 1158333614345/AR06_final_LO_RES.pdf

KEY TERMS AND DEFINITIONS

Capitalist Consumption: A materialistic pattern of consumption informed primarily by profit motives and value augmentation at the expense of the natural environment or ecosystem.

Eco-Efficiency: The less wasteful use of the ecosystem's natural resources.

Green Consumer: A natural resource friendly consumption pattern that does not taint the environment.

Political Ecology: A Marxian theory which underscores the primacy of resource equality as a critique of the neo Malthusian debate.

Sustainable Food Consumption (SFC): A development term linked to socially just and ecologically friendly consumption pattern.

Chapter 6
Comparing the Effects of Unsustainable Production and Consumption of Food on Health and Policy Across Developed and Less Developed Countries

Josue Mbonigaba
University of KwaZulu-Natal, South Africa

ABSTRACT

The unsustainable food consumption across high-income countries (HICs) and low-income countries (LICs) is expected to differ in nature and extent, although no formal evidence in this respect has been documented. Documenting this evidence is the aim of this chapter. Specifically, the chapter seeks to answer the following questions: 1) Do the contexts in less developed countries (LDCs) and developed countries (DCs) make the nature and extent of unsustainability in food consumption different? 2) Do the mechanisms of the linkage between unsustainability of food consumption and health outcomes independent of countries' contexts? 3) Are current policies against unsustainable food consumption equally effective in DCs and LDCs? These questions are answered by means of a systematic review of the literature for the period 2000-2017. The findings are that the nature and extent of unsustainability is quite different across contexts of LICs and HICs.

DOI: 10.4018/978-1-5225-3631-4.ch006

INTRODUCTION

Unsustainable activity refers to actions that result in outcomes now that are due to worsen over the next period (Hobson, 2002). Consequently, in terms of production processes, unsustainable development has been referred to as processes resulting in the satisfaction of the current generation's needs without a guarantee of the same benefits for the next generation (Johnson et al., 2014:422). Unsustainable food consumption is defined in a similar fashion. It refers to food consumption/production levels and patterns that are likely to result in a continual decrease in welfare across generations. This welfare is expected to be lost in three domains as reported in the literature, these being environmental degradation, economic inefficiency, and negative social outcomes (Lefin, 2010). In the environmental domain, for example, food consumption/production becomes unsustainable when biodiversity that is crucial to the welfare of the next generation is reduced through food consumption/production processes. In the economic domain, the technical and allocative inefficiencies that arise from current consumption/production processes deprive the next generation of resources through wastage (Clapp, 2017;91). The unsustainability of food consumption in the social domain emerges from consumption/production processes that are degrading the environment whilst at the same time resulting in sicknesses, social conflicts and lower human capital formation, among many others social ills (Hobson, 2002, Hertwich and Katzmayr, 2003).

To date, the literature has referred anecdotally to the possibility that the nature and extent of unsustainability in food consumption might differ in both nature and extent depending on the livelihood and lifestyles of communities (Barnidge et al., 2011, Connell, 2010, Vermeir and Verbeke, 2006, Lehota, 2004, Schloster et al., 2012, Capone et al., 2014), on the economic processes (Vermeir and Verbeke, 2006), and on socio cultural-political set ups (Vermeir and Verbeke, 2006, Thøgersen, 2010, National insitute of consumer research, 2010, Nemecek et al., 2016, Lorek and Fuchs, 2013). In this chapter the "nature of unsustainability" refers to specific features of unsustainability. As an example, a type of production process might result in deforestation in one setting, while it might lead to greenhouse gas emission in another setting. Furthermore, unsustainability might arise predominantly from inefficiency in production processes (weak unsustainability) or it might stem from consumption behaviours (strong unsustainability). With reference to the "extent of unsustainability", this chapter refers to quantities of damages arising from unsustainability. The differences in extent across different settings might arise from mitigating or aggravating factors affecting unsustainability that prevail in these different settings. The effects of consumption/production processes and institutions on unsustainable food consumption mean that differences in these processes imply differences in the vulnerability of the various populations and their coping strategies

(Gautam and Andersen, 2016, Alemu, 2010, Charles et al., 2010), which in turn have a bearing on the environmental, economic and social domains of unsustainability.

DCs and LDCs more generally differ in many respects regarding food consumption/production processes and in institutional set-ups (Herrero and Thorntonb, 2010:20880, Thøgersen, 2010). DCs use mechanized and industrial agriculture in their food production processes (Mózner, 2013) and have more established regulations concerning environmental degradation (Charles et al., 2010). Economic and social behaviours are likely to be different to those in LDCs due to established social security systems in DCs that mitigate the impacts of unsustainable food consumption. In these countries over nutrition and consumption of saturated fats from meat products prevail (Herrero and Thorntonb, 2010). Diseases in these countries are predominately non-communicable such as heart diseases, diabetes and obesity-related disorders, which are related to the consumption of high-energy foods and fats (McKenzie and Williams, 2015b). In LDCs, agriculture is less mechanized and largely traditional (Pretty et al., 2013) and the population is increasing rapidly. Hunger and deficiency in micronutrients in diets are characteristic of the majority of the population (Randolph et al., 2007, Seligman et al., 2009, Alinov, 2010, Oni, 2010, Pretty et al., 2003). Furthermore, these countries suffer from very limited or an absence of social security systems and safety nets. Diseases for the majority of people in these countries are largely infectious.

These differences imply different consequences of unsustainability either directly or through the indirect effects of these set-ups, which also mean that standard policies across DCs and LDCs might not be equally effective (Connell, 2010, Randolph et al., 2007, Hobson, 2002). Because set-ups and lifestyles in these countries are expected to impact on the nature and extent of unsustainability in food consumption/production (Mont and plepys, 2005, Kjærga et al., 2013), it follows that the patterns and extent of unsustainability are likely to be related to specific settings (Hobson, 2002, Mackay and Wolbring, 2013a, Hawkes, 2006, Mackay and Wolbring, 2013b). In spite of this possibility and its relevance for policy making, whether or not the nature and extent of unsustainability follows these contexts has not been investigated across DCs and LDCs. Therefore, through a systematic review of the literature, this chapter analyses the evidence to answer the question as to whether patterns of unsustainability and mechanisms causing ill health follow consumption/production processes and institutional set-ups in DCs and LDCs. Also investigated in this chapter is whether policies have been equitable in the context of differences in institutions, consumption behaviours and production processes across DCs and LDCs. The chapter ends with suggestions for new ways in which policies can be managed to impact effectively on unsustainability at a global level, by addressing factors in specific countries.

The rest of the chapter is structured such that the next section deals with the methodology, section 3 presents the findings and the chapter ends with a discussion of the findings and the conclusions.

MATERIALS AND METHODS

A systematic review of the literature was conducted in line with the standard practice of literature review (Littell et al., 2008, Khalid et al., 2003). Different databases were searched using search strategies to yield the maximum number of materials to be screened. The following terms or their combinations were used: "unsustainable food consumption" and "health" OR "policy", OR "high-income countries" OR "low-income countries", OR "developing countries", OR "developed countries" or "low-resource counties", OR agriculture OR farming, OR sustainable, OR context OR livelihood OR production OR institutional. Databases searched included EBSCOhost, Econlit, Cochrane Library, Medline, Social Sciences Citation Index, and Web of Science databases. For the search to yield sufficient material to screen, the period covered by the review was 2000-2017. References in the retrieved literature were searched to identify and search for additional studies. Finally, a search of gray literature was conducted to identify documents that are non-academic in nature to provide information especially on policies.

Overall, this search process resulted in identifying 1565 documents. These included journal articles, books, book chapters, working papers and conference proceedings. The number of the final papers reviewed was arrived at after a screening of titles and the abstracts for relevance to the focus of this chapter. The screening was done independently by research assistants and then by the main investigator. The main investigator, based on the screenings of his own and assistants', decided on the documents to include in the review. The screened document was included in the review if a source analysed (un)sustainability of food consumption/production or its relation to health, if it reported evidence for DCs or LDCs or both; or if it was peer reviewed or published by a reliable source such as policy making bodies; and if it reported on evidence relevant to the chapter's research question(s). With regard to DCs and LDCs, a country was classified as DC or a LDC if at the time of a source's publication, a country covered was classified as such by the World Bank. The screening process resulted in 75 documents being included in the final review.

The study classified the screened documents under three themes, each related to a research question under investigation. These themes were 1) "unsustainability of food consumption", 2) "unsustainability of food consumption and health"; and 3) "unsustainability of food consumption and policies". Because some papers covered more than one research question under investigation, these appeared under more than one theme. In the end, 44 documents focused or had sections focusing on the first theme, 40 focused or had sections focusing on the second theme, while 34 focused or had sections focusing on the third theme. The critical review of the evidence of studies classified under each theme resulted in answers to the three research questions. The extent of unsustainability under theme 1 was reviewed in terms of mitigating or aggravating factors within DCs or LDCs. The evidence arising from this methodology is reported next.

RESULTS AND DISCUSSION

(Un)sustainability of Food Consumption: DCs vs. LDCs

There have been many studies focusing on the unsustainability of food consumption/production in DCs and LDCs (Table 1). These studies used a variety of methods to characterize and explain the unsustainability of food consumption/production. The most common methods used by the studies were qualitative, consisting of analysing information through discussion with people or document reviews (see for example Reisch et al., 2013, Thøgersen, 2010). Some studies reported on evidence from specific countries whilst most of the studies reported results on DCs and LDCs as groups. A few studies used quantitative methods, notably descriptive statistics, inferential methods or their combination (Amendah et al., 2014, Grunert et al., 2014 for instance). Data sources included food balance surveys (FBS) constructed by the Food and Agriculture Organisation of the United Nations (FAO) based on national accounts, households or individual budgetary and dietary surveys. Other sources of data included the United Nations Environment programme (UNEP), Biodiversity International, Water Footprint Network, IFAD, and the World Food Programme (WFP). Studies included micro- (individuals or households) and macro-level (cross-country) analyses and covered different geographical regions of the world.

The evidence about unsustainability in food consumption from these studies can be analysed according to three pillars of unsustainable consumption, these being environmental, economic and social unsustainability (Table 1). Indicators of environmental unsustainability included loss of biodiversity, greenhouse gas emissions, water shortage, non-responsiveness of soil to chemicals, soil erosion, extreme weather conditions, ecosystem loss, ocean acidification and sea level rise. Social unsustainability was dominated by illnesses that arise from unsustainable consumption such as heart diseases and stunting. In very few studies in LDCs, social unsustainability included also social conflicts over land, unsustainable strategies to cope with food shortages such as contracting debts, selling assets, and dropping children from school (Alemu, 2010, Alinov, 2010). Economic inefficiencies consisted

Table 1. Nature and extent of unsustainability across DCs and LDCs: summary of studies

	Studies	Environmental Degradation		Economic Inefficiency		Social Unsustainability		Data Sources Methodology	Contextual, Institutional Factors ant Extent
		LDCs	DCs	LDCs	DCs	LDCs	DCs		
1	(Alinov, 2010)	Environment degradation through traditional survival methods	NA	Insufficiency of of food, limited micronutrient	NA	Selling productive assets, dropping kids out of school	NA	Secondary survey data analysed with statistical methods	Social ill more prevalent in LDCs than in DCs, aggravated by absence of infrastructure in LDCs
2	(Amendah et al., 2014)	Environment degradation through traditional survival methods	NA	Insufficiency of food, limited micronutrient	NA	Selling productive assets, dropping kids out of school	NA	Household survey, use desctiptive statistics and logistic regression analysis	Absence of food more acute in LDCs than in DCs, aggravated by lack social safety nets
3	(Brown and Jacobson, 2005)	Destruction of rainforest to plant palm tree	NA	Costly to treat illness from saturated oil from palm tree	NA	Ill health from consumption of palm oil	NA	Source of data is documentary analysed and argue in relation to research question	Deforestation in LDCs, oil processing in DCs, extent: trade aggravate situation for LDCs
4	(Thøgersen, 2010)	NA	Green gas house emission, loss of biodiversity	NA	Pricing distortions	NA	NA	Data on published research analysed using qualitative methods	LDCs benefit less of subsidies as buffer to price shocks than Dcs
5	(Reisch et al., 2013)	NA	Mechanized agriculture, use of pesticides	NA	Process of globalization resulting in inequalities	NA	Heart related diseases	Source of data is documentary analysed and argue in relation to research question	Mechanized agriculture more in DCs than in LDCs, LDCs, food in high energy increasing in LDCs
6	(Kearney, 2010)	Degradation of through limited usage input	Excessive water and land use in dunstrial sector	NA	NA	NA	Saturated consumption fat leading to heart related diseases	Source of data is documentary analysed and argue in relation to research question	Governments in LDCs and policies in these countries are less responsive to these issues than in DCs

continued on following page

Table 1. Continued

	Studies	Environmental Degradation		Economic Inefficiency		Social Unsustainability		Data Sources Methodology	Contextual, Institutional Factors ant Extent
		LDCs	DCs	LDCs	DCs	LDCs	DCs		
7	(Han and Hansen, 2012)	NA	NA	NA	NA	Health effect of food consumption and behaviour	High energy meat consumption leading to cardio vascular diseases	Literature Meta-analyses	Increasing consumption of unhealthy food in urban area of LDCs is due to aggravate health outcomes
8	(vonMeyer-Höfer et al., 2015)	Depletion of resources, unequal access, loss of biodiversity in LDCs	NA	Depletion of resources, unequal access, loss of biodiversity in LDCs	NA	Unequal access to resources In LDCs	NA	Online consumers survey analysed within the context theory of planned behaviours and structural equation modelling in China and India	LDCs suffer more unequal access to resources than in DCs
9	(Vermeir and Verbeke, 2006)	NA	NA	NA	Consumption processes that result in higher prices	NA	Consumption behaviour in high energy food	Survey data Documents reviews	Policy intervention on more prevalent in DCs than in LDCs
	(Hart et al., 2013)	NA	Degraded environment as a result of degradation	NA	Not halting the loss of biodiversity in the production process	NA	Exposure to pollution resulting in respiratory diseases	Source of data is documentary analysed and argue in relation to research question	DCs institutions own promoting fresh food non existing in LDCs
10	(Vogit, 2014)	NA	Process where prod and consumer relate in DCs	NA	More power from producers when they are not connected to food	NA	Increase in consuming unhealthy food with no influence of consumer	Source of data is documentary analysed and argue in relation to research question	DCs institutions own promoting fresh food non existing in LDCs
11	(Grunert et al., 2014)	NA	Consumption of environment unfriendly food and unhealthy	NA	NA	NA	Consumption related social unsustainability in terms caaaaardio vascular diseases	Online survey data Data analysed were hierarchal regression	Prevalent in DCs and in LDCs but LDCs have less mitigating set ups
12	(McKenzie and Williams, 2015b)	System that lead to water and sold degradation, loss of biodiverse	System that lead to water and sold degradation, loss of biodiverse	Productivity growth due to decreased owing soil degradation	Productivity growth due to decreased owing soil degradation	Absence of balanced nutrients	Over consumption of meat products	Source of data is documentary analysed and argue in relation to research question	Absence of governance and management of unsustainability more prevalent in LDCs

continued on following page

Table 1. Continued

	Studies	Environmental Degradation		Economic Inefficiency		Social Unsustainability		Data Sources Methodology	Contextual, Institutional Factors ant Extent
		LDCs	DCs	LDCs	DCs	LDCs	DCs		
13	(Bryceson, 2004)	Livelihood that do not cater for environmental shock	NA	Inefficient in food production system	NA	Livelihood results in insufficiency of food	NA	Source of data is documentary analysed and argue in relation to research question	Policy based on contextual vulnerabity less prevalent in LDCs than in DCs
14	(Hobson, 2002)	NA	Environmental impacts of consumption pattersn	NA	Inefficient use of land in terms of production	NA	Ignorance Toxic materials and waste emission	Documentary, conceptual, argumentative	Politically dominant approach to prevent unsustainable actions more in DCs than in LDCs
15	(Horrigan et al., 2002)	NA	Consumption process that degrade environment in DCs	NA	Externality of unstainable production not included in pricing	NA	Production process that results in health outcome	Source of data is documentary analysed and argue in relation to research question	More institutions to deal with these issues in DCs than in LDCs
16	(Johnson et al., 2014)	Production that results in Poverty and nutrition	Loss of biodiversity Large scale acquisition of land	Allocative inefficiency as some community do not get basic food	NA	Production systems that result in hunger in LDCs	Increased consumption rich in energy and meat in LDCs and DCs	Nutrition survey Uses a causal model	Institutional in DCs are more socially inclusive that those in LDCs
17	(Mózner, 2013)	Production that do not allow calories intake	Production and consumption that damage environment	Production inefficient	Demand for land based resources in other countries	Hunder and food poverty	High beef meat consumption	Biophysical methodology	Institutions that treat environmental issues and health simultaneously, less prevalent in LDCs than in DCs
18	(Nemecek et al., 2016)	NA	Production system that degrades environment	NA	Poor production methods. unfair market through international trade	Health and other social dimension		Source of data is documentary analysed and argue in relation to research question	LDCs suffer more effects of international trade in food than DCs
19	(Clapp, 2017)	Crop that are associated with deforestation	Crop that are associated with a lot of water usage	High food price		Hunger, famine, aggravated	NA	Documentary, reviewed with argumentative approach	Food output and modern agriculture more limited in LDCs than in DCs

continued on following page

Table 1. Continued

	Studies	Environmental Degradation		Economic Inefficiency		Social Unsustainability		Data Sources Methodology	Contextual, Institutional Factors ant Extent
		LDCs	DCs	LDCs	DCs	LDCs	DCs		
20	(Randolph et al., 2007)	Process that produce livestock are beneficial	NA	Not maximizing productivity per animal	NA	Insufficient production of animal stock	NA	Documentary, analytical	Livestock production, limited in LDCS than DCs
21	(Garnett, 2014)	NA	Environment degradation from the supply chain	Unequal distributing of the product in LDCs and DCs	Pricing and subsidies distortions	System that produce food that not culturally acceptable	NA	Documentary, argumentative	Institutions to minimize environment degradation less prevalent in LDCs compared to DCs
22	(Cullet, 2004)	NA	NA	Production process that do not assure fairness in food distribution		Technological development not aimed at providing food security	NA	Documentary, qualitative analysis	Legal regime needed to take account of the food needs is less established in LDCs
23	(International Institute of Social Studies, 2015)	NA	NA	NA	NA	NA	Consumption of unhealthy processed product	Online surveys from consumers Descriptive statistics and hierarchical regressions	Limited institutions to respond to consequences in LDCs
24	(Pretty et al., 2003)	Technologies that sustain environment	NA	Limited agricultural productivity	NA	NA	NA	Survey data Analysis of based on the questionnaire	Insufficient means to acquire inputs, lack of institutional
25	(Freibauer et al., 2011)	NA	biofuel and industrial materials compete with food for biomass	NA	Less variety of food,		Absence of non-sustainable diets	Documentary, argumentative analysis	Instituting that balance diets
26	(Godfray et al., 2010)	Competition for land and water resources	Completion of land and water aggravated by climate change		Feeding grain to live stock rather than feeding them directly to human is inefficient	High growth of population, income inequality	Change in consumption patterns	Literature Reviews of the evidence	Adopting new technologies Is a challenge
27	(Capone et al., 2014)	Food systems that degrade environment	Food systems that degrade environment			Food systems that result in hunger		Documentary argumentative Research	Institutions that moderate good production system more prevalent in DCs

continued on following page

Table 1. Continued

	Studies	Environmental Degradation		Economic Inefficiency		Social Unsustainability		Data Sources Methodology	Contextual, Institutional Factors ant Extent
		LDCs	DCs	LDCs	DCs	LDCs	DCs		
28	(Alemu, 2010)	Production process that result in poverty	NA	Farming activities generating less income	NA	Strategies that might affect some other aspects of well-being	NA	Data from general households survey analysed with descriptive stats	Institutions aimed at understanding socioeconomic fabric of the population more available in DCs
29	(Alinov, 2010)	Production process affected by climate shock	NA	Suffer price instability of the food system	NA	Production process result in food insecurity	NA	Used a survey, analysed using descriptive statistics	Institutions that cater for the most vulnerable household to price and climate shock
30	(Risku-Norja, 2011)	NA	Production process that degrade environment	NA	NA	NA	Consumption of animal food	Documentary, analysis	
31	(Stewart et al., 2013)	NA	Production system that do not provide enough food to urban population in LDCs	NA	Inefficiency in urban food market	Production system that do not provide enough food to urban population in LDCs	NA	Documentary, literature analysis	Urban population more growing in LDCs than in DCs
32	(Dobermann and Nelson, 2013)	Absence of agro-ecological intensification	Unsustainable use of waters and land soil nutrients		Absence of agro ecological intensification process	absence of economic, social and Eco local principles of farmers need	Process that don not consider social context of famers	Documentary review, argumentative	Inequity in terms of accessing inputs
33	(Carlsson-Kanyama and Gonzalez, 2009)	NA	Agricultural practice that are damaging in DCs	NA	Losses in production due wastage in energy	NA	Agricultural practice leading to unhealthy diets in DCs	Documentary, argumentative analysis	Institutions that control the greenhouse emissions
34	(Pretty and Noble, 2006)	Production lacking technology that are not harmful in LDCs	NA	Urbanisation is causing shift in consumption and pricing		Processes that result in insufficiency of food in LDCs	NA	Interview with farmers analyse using descriptive statistics	Political involvement is the way forwards
35	Bert et al. (2017)	NA	Food consumption that lead to unsustainable production n DCs	NA	Process do not use efficiently resources	NA	Process not producing health food	Documentary, argumentative analysis	Institution on heath information and better nutrition more available in DCs

continued on following page

Table 1. Continued

	Studies	Environmental Degradation		Economic Inefficiency		Social Unsustainability		Data Sources Methodology	Contextual, Institutional Factors ant Extent
		LDCs	DCs	LDCs	DCs	LDCs	DCs		
36	(Lorek and Fuchs, 2013)	Governance that does not take care of sustainability	Governance that does not take care of sustainability	NA	Policy that focus on growth only without considering environment	Governance that does not take account of social burden of consumption	NA	Argumentative analysis	Incorporate external costs into prices, technological innovation
37	(Kasa, 2008)	NA	Process resulting in overconsumption	Process that result in diversified foods	Beef consumption process might result in shortage of global		Process allowing social issue to arise from beef consumption	Documentary, argumentative methods	Information's and discouragement
38	(Seyfang and Smith, 2007)	NA	Technologies that do not safeguard environment	NA	Markets that are not transformed for sustainable choice	NA	The lack of a social and active citizenship	Documentary, argumentative community level factor	Institution that incorporate social culture into consumption
39	(Temme et al., 2015)	NA	Process that do not replace plant-based product by high energy food	NA	NA	NA	Absence of sustaible diets	Survey,, energy and nutrient intake observed and analysed	Public health institution combine sustaible message
40	(Friedl et al., 2014)	NA	Increase in consumption that degree environment	NA	Influence of these on energy use	NA	Create obesity, disability and chronic diseases	Survey, household analysis of sustainable consumption and regression analysis	Public canteen to cater for fresh food more prevalent in DCs
41	(Amendah et al., 2014)	Consumption system that degraded envir in LDCs	NA	Buying good on credit	NA	Removing children from school	NA	Secondary data analysis from another study using descriptive methods	Public insition that cushion social ills
42	(Herrera-Estrella, 2000)	Process that make water and soil dwindling	NA	Limited use of efficient methods	NA	Crop yields are significantly lower	NA	Literature Critical analysis	Limited capacity to store food that are produced
43	(Khan et al., 2014)	Production system not taking account of climate change in LDCs	NA	Productions are inefficiency	NA	Food insecurity in sub-sharan Africa	NA	Literature Document reviews	Use push pull technology to allow the dealing
44	(Herrero and Thorntonb, 2010)	Livestock Taking land	livestock	NA	Overconsumption	Some part	All of the developed world	Literature Documents review	Developed and developing

NA means that there is no evidence in the study referring to that types of unsustainability of wastes in food production, food markets distortions, unequal distribution of food, price distortions and externalities (Horrigan et al., 2002 for example).

In DCs, unsustainability in the form of environmental degradation stemmed mainly from industrial agriculture. This type of agriculture resulted in losses of biodiversity in the food consumption/production process, non-response of soil microorganisms due to overuse of chemicals, and greenhouse gas emissions of which DCs contributed 14% of global output due to meat production (Thøgersen, 2010, vonMeyer-Höfer et al., 2015, Hart et al., 2013 for instance). In the form of social unsustainability, the evidence revealed heart related diseases, diabetes, respiratory diseases, and hypertension as the most salient feature of social unsustainability related to food consumption/production in DCs (Kearney, 2010, Reisch et al., 2013). Economic inefficiency consisted of waste of cereals, which were processed industrially to get output used to feed livestock rather than feeding these cereals directly to the livestock.

Different patterns of unsustainability were observed in LDCs. Although there was some sort of environmental unsustainability in the form of carbon emissions from industrial agriculture, this was not the predominant one. The environment in these countries is degraded rather through traditional methods of food consumption/production, which result in deforestation, soil erosion from rain, and soil deterioration because of limited technological inputs. Regarding social unsustainability, the evidence revealed hunger and limited variety in food, deficiency in micronutrients (Amendah et al., 2014), and conflicts over land tenure as the most common features of social unsustainability. Furthermore, LDCs suffer limited food output which leads to strategies that are likely to perpetuate and complicate social unsustainability. These strategies consist of contracting debts to get food (Alinov, 2010, Gautam and Andersen, 2016) and removing children from school to help in work related to food search. Economic inefficiency is more abundant in LDCs and included inappropriate storing of food, and misuse of land by planting inappropriate crops (Pretty et al., 2003).

These patterns of unsustainability across DCs and LDCs stem from their food consumption/production processes and their institutional makeup. In DCs, most of the evidence (about 40% of reviewed studies) focus on environmental degradation in terms of loss of biodiversity, water resources, and the non-response of soil to microorganisms. This degradation stem from mechanized agriculture linked to industrial production of processed high-energy food in these countries. Mass processing of high-energy food is mainly driven by increasing consumption in these products in urban areas of LDCS due to globalization and less regulated trade of these products (see for example Brown and Jacobson, 2005). Consumption of high-energy food is in contrast stable and likely to decrease in DCs. It is also worth to note that declining population in DCs which takes place jointly with increasing incomes per capita result in more high-energy meat products per capita. The resulting outcome is the development of cardiovascular diseases. Such outcomes clearly follow the production process and other contexts (declining population and high income). Similarly, the patterns observed in LDCs stem from their contexts of production and social conditions. Even if some features of industrialized agriculture as well as rising incomes in the developed world are observed in some parts of LDCs (Connell, 2010, Bryceson, 2004, Bulte and Soest, 2001), there are notable differences. The evidence shows that soil degradation through erosion, traditional agricultural methods, limited use of technology and excessive degradation of soil productivity, characterize these countries. These outcomes stem from traditional agricultural practices involving deforestation as a way to gain agricultural land and firewood (Pretty et al., 2003). These traditional methods result further in very low crop yields amidst growing populations and inefficiencies in production processes, which result in hunger and malnutrion. The absence of or very limited safety nets, the illiteracy rate, and the lack of strong institutions complicate matters, resulting in conflicts over land in the quest for survival, which lead to injuries and deaths. Furthermore, hardship in these countries leads to the adoption of survival strategies such as dropping children from school, contracting debts and selling assists (Gautam and Andersen, 2016) which are likely to worsen the well-being of future generations.

The evidence revealed further that, even in case of the unsustainability being the same in nature across LDCs and DCs, the extent of the damages is higher in LDCs. In these countries' contexts, insufficiency of food and food insecurity prevail side- by -side with urban situations similar to those in DCs. While some urban population of DCs are being exposed to high energy food and suffer the same illness as in DCs, they suffer additional inconvenience specific to their predicaments such as limited social infrastructures. DCs' production process established in LDCs through multinationals add to the vulnerability of the environment and people in these countries, a situation that is aggravated by the interdependence and trade through globalization. The evidence also indicate that production processes in DCs influence environment degradation in LDCs when food processing in DCs require the use raw food commodities in LDCs (Brown and Jacobson, 2005 for example). The extent of damages in LDCs is further inflated by absence or limited safety and social systems, environmental management policies, and health systems to cater for these impacts. LDCs suffer most of food price shocks and unequal distribution of the food products, and benefit less from subsidies by their governments than DCs. Low literacy rates and low health literacy rates amid fast urban migration in LDCs increase unawareness of health risks associated with exposure to processed food than is the case in DCs. Evidence of land competition exists in DCs and LDCs, but might not result in serious injuries due to regulation and the definition of property rights in DCs (Barnidge et al., 2011).

Briefly, the institutions, settings and social contexts in DCs and LDCs shape the features of unsustainability as well as the extent of this unsustainability. The consumption/production processes in DCs results in greenhouse gas emissions, the overproduction/consumption of unhealthy food, long term non-communicable diseases but with a trend and movement towards decreasing these patterns. The institutional setting mitigates these impacts through social safety nets and proper regulation more than institutions in LDCs do.

Unsustainability of Food Consumption and Health: DCs vs. LDCs

The reviewed literature also revealed that consumption/production processes in DCs and LDCs determine mechanisms leading to ill health (Table 2). Starting with the review of diseases arising from unsustainable consumption/production, the diseases included communicable and non-communicable diseases. The review revealed that diseases most common to DCs were predominantly non-communicable such as cardiovascular and respiratory diseases. The review showed also that these non-communicable diseases were prevalent in urban settings of LDCs. Nevertheless, infectious, short term diseases, were the most prevalent diseases in LDCs while DCs suffered mainly from non-communicable diseases of a long term duration. LDCs also suffered diseases arising from injury, emotional stress and fatigue. Although many factors can be responsible for these diseases, income levels, lifestyles systems, and contexts, which are believed to differ across DCs and LDcs, are likely to determine the mechanisms linking food production/consumption to these illness

A close analysis of the evidence about how these diseases occur reveals that non-communicable diseases in DCs occur through a long-time exposure to risk of air pollution or some type of consumption for a long period of time. Over an extended period, the exposure to polluted environment produces the effects that lead to respiratory diseases. Other diseases occur in these countries, as noted earlier, because people eat a lot of processed industrial food. The long-term effects of consuming this type of food is the development of heart diseases, obesity and diabetes, which have serious long-term consequences for the population.

In contrast, the review of the evidence suggests that diseases in LDCs are predominantly infectious diseases. In thc rural parts of these countries, exposure to air pollution is less serious. Rather, a different kind of pollution in the form of debris prevails. This type of pollution arises from soil erosion which in turn happens as a result of deforestation which is such that when it rains, debris gets dumped into rivers whose water is used for drinking and cooking in households. The use of this water culminates in infectious waterborne diseases such as cholera and diarrhea and other infectious diseases. The common crowding and the absence of sanitation in these

Table 2. Mechanisms of linkage to health in DCs and LDCs: summary of the studies.

Study no	Study	Production	Consumption	Mechanism	Type of Ill Health	LDCs	DCs	Contextual and Institutional Factors	Methodology
1	(Kasa, 2008)	Production of beef degrades health and environment	Process focusing on animal product	Long exposure to air pollution	Respiratory diseases	NA	✓	DCs have more food processing in DCs but LDCs also exposed	Source of data is documentary analysed and argue in relation to research question
2	(Amendah et al., 2014)	Food shortage	Insufficient food	Working excessive hours in informal, conflicts over land tenure	Fatigue, emotional stress, injury	✓	NA	Mechanism more unique to LDCs	Source of data is documentary analysed and argue in relation to research question
3	(Barnidge et al., 2011)	Shortage of land, poor distribution of structural factor	Absence of sustainable consumption	Involvements in conflicts to get	Different types of injury	✓		Conflicts over land is a mechanism more unique to LDCs	Qualitative participatory methods to get perception from community
4	(Gauker, 2010). (Reisch et al., 2013)	Chemicals effects consumed in the food	Chemicals effects consumed in the food	Chemical infected food leading to resistance to antibiotics	Heart diseases and respiratory diseases		✓	Some regulations exist to stop mechanism	Source of data is documentary analysed and argue in relation to research question
5	(Godfray et al., 2010)	Production process that tend to safeguard livestock as a capital in LDS	Lifestyles that reduce the consumption of meat	Insufficient calorie intake Drinking contaminated water	Chronic diseases, diseases of poverty	✓	✓	Insufffient food intake, contaminated water mechanism unique to LDCs	Documentary, argumentative method
6	(vonMeyer-Höfer et al., 2015)	NA	unwariness of health motives in choosing consumption in china and india	Consumption of high saturated fat	Herat diseases	✓	NA	DCs have more saturated fat but LDCs also exposed	Online consumer survey analysed with econometric analysis
7	(Reisch et al., 2013)	Social: Food shortage	Deforestation	Injury due to mud sliding as a result of deforest	Various types of diseases	✓	✓	Injury due to mud sliding unique to LDCs	Documentary analysis
8	(Kearney, 2010)	Production of food rich in animal products	High saturated food	Increase in consumptions of saturated fat	Cardio diseases and diabetes	NA	✓	DCs have more food processing in DCs but LDCs also exposed	Systematic reviews
9	(Pretty et al., 2003)	Food shortage	Health saturated fats	Shift towards diversified diets	Heart diseases	NA	✓	DCs have more food processing in DCs but LDCs also exposed	Analysis of survey data and qualitative information

continued on following page

Table 2. Continued

Study no	Study	Production	Consumption	Mechanism	Type of Ill Health	LDCs	DCs	Contextual and Institutional Factors	Methodology
10	(Han and Hansen, 2012)	Degradation of environ related food	Food choices that are unsustainable	Compelled to consume these foods. Pyscho-social	Heart diseases	NA	✓	DCs have more food processing in DCs but LDCs also exposed d	Databases of empirical studies, evidence of which is synthetized
11	(Vermeir and Verbeke, 2006)		Consumption processes that do not take into account social responsibilities	Attitude behaviour and perception	Obesity and	✓	✓	DCs have more food processing in DCs but LDCs also exposed	These more placed in DCs
12	(Freibauer et al., 2011)	Production of waste	Overconsumptions	Waste, overconsumption	Heart diseases	✓	✓	DCs have more food processing in DCs but LDCs also exposed	Documentary evidence
13	(Vogit, 2014)	Production process not allowing consumer	Consumption patterns showing limited consumer	Getting food from supermarkets	Chronic illness	NA	✓	DCs have more of the mechanism but LDCs also exposed	Documentary, argumentative analysis
14	(Temme et al., 2015)	Not based of dietary patterns resulting in sustainable	Not linking public health to consumption	Getting high energy food	Chronic diseases	✓	✓	DCs have more of the mechanism but LDCs also exposed	Survey, energy and calorie s intake observed
15	(Friedl et al., 2014)	Process of consumption that affect prod process	Consumption process that affect heath	Overproduction of food	Chronic diseases	✓	✓	DCs have more of the mechanism but LDCs also exposed	Source of data is documentary analysed and argue in relation to research question
16	(Thøgersen, 2010)	Absence of organic food	Consumption of food that is not organics	Heath problem and consumption of animal fat	Chronic diseases	✓	✓	DCs have more of the mechanism but LDCs also exposed	Documentary, analysis
17	(Risku-Norja, 2011)	Production in animal rich energy food	Production of unhealthy food	Increasing overnatution in animal meat	Chronic	✓	✓	DCs have more of the mechanism but LDCs also exposed	Documentary, analysis
18	(Oni, 2010)	Poverty and food insufficiency	Absence of sufficient food	Food insecurity and hunger	Chronic and infectious	✓	NA	Limited food and related illness unique to LDCs	Source of data is documentary analysed and argue in relation to research question
19	(McMichael et al., 2007)	Livestock production consume more energy	Consumption results in fat	Increase in meat consumption		✓	✓	Policy that prevent both health risk and environment degradation	Documentary

continued on following page

Table 2. Continued

Study no	Study	Production	Consumption	Mechanism	Type of Ill Health	LDCs	DCs	Contextual and Institutional Factors	Methodology
20	(Mont and plepys, 2005)	Consummers influence productive process	Consumes consume unsustainably	Health problems	Short and chronic diseases	✓	✓	Instruction that motivate behaviour	Documentary analysis
21	(Mózner, 2013)	Production options not orientated towards health	Consumption options not orientated toward health	Overocumption of unhealthy animal products	Cardio	NA	✓	Intuitions linking consumption to health	Biophical methods
22	(World Health Organisation, 2003a)	Contribution to greenhouse gas emissions	Consumption of energy dense diet	Over consumption	Chronic diseases	✓	✓	Food policy to deal with health issues	Documentary evidence, descriptive analyses
23	(World Health organisation, 2003b)	Production process that result in ovary affect health	Insufficient consumption	Insufficient calories	All sorts of diseases	✓	NA	Institutions in DCs	Documentary analysis
24	(Godfray and Garnett, 2014)	Deforestation, water use overfishing, pollution and biodiversity loss	Growing populations and increasing demand for food	Insufficient food or over consumption of unhealthy food	All sort of diseases	✓	✓	Adopt intensification that is coherent through governance	Argumentative methods
25	(McKenzie and Williams, 2015a)	Process that results in food insecurity in LDCs, and more unhealthy food in DCs	Process that lead to over consumption and under nutrient	Overconsumption and under consumption	Underweight And overweight in LDCs overweigh in DC	✓	✓	DCs have more saturated fat but LDCs also exposed	Documentary, argumentative research
26	(Horrigan et al., 2002)	Factory style animal agriculture affect health	Process involving consumption of animal fact in LDCs and DCs	Degenerative disease in LDCs and DCs	Chronic heart related diseases	✓	✓	DCs have more saturated fat but LDCs also exposed	Documentary, argumentative
27	(Pretty et al., 2003)	Use of nonrenewable unhealthy inputs	Very limited consumption and food insecurity	Through inhalation, through hunger	Infectious disease	✓	NA	DCs have more saturated fat but LDCs also exposed	Questionnaire data, analysed using descriptive statistics
28	(Herrero and Thorntonb, 2010)	NA	More avaaibility in processed food	Consumption of high-energy processed food		✓	✓	DCs have more saturated fat but LDCs also exposed	Source of data is documentary analysed and argue in relation to research question
29	(DeBon et al., 2010)	Not taking account of urban migration	NA	Persistence inefficiency in agriculture resulting in limited food	Food insecurity and unhealthy food	✓ NA		Persistence inefficiency in agriculture resulting in limited food more in LDCs	Documentary analysis

continued on following page

Table 2. Continued

Study no	Study	Production	Consumption	Mechanism	Type of Ill Health	LDCs	DCs	Contextual and Institutional Factors	Methodology
30	(Johnson et al., 2014)	Biodiversity loss and ecosystem degradation	Obesity, suffering micro-nutrition	Eating patterns, poverty, eating meat product, urbanisation	Heart disease and others		✓ Long	DCs have ways of avoiding but process being initiated in LDCs	Source of data is documentary analysed and argue in relation to research question
31	(Clapp, 2017)	Insufficient production in food of some types	Insufficient type of healthy diet	Overconsumption of unhealthy of available but unhealthy food	Heart diseases	NA	✓	DCs have more saturated fat but LDCs also exposed	Documentary, argumentative analysis
32	(Carlsson-Kanyama and Gonzalez, 2009)	Processes not taking into account healthy diets	Consumptions process not informed by campaign	Consumption of unhealthy diets	Non communicable disease		✓	DCs have more saturated fat but LDCs also exposed	Documentary sources analysed with argumentative methods
33	(Seligman et al., 2009)	Food insecurity	Food insecurity and health outcomes in DCs	Diet in high energy food ases	Heart diseases	✓	✓	DCs have more saturated fat but LDCs also exposed	Survey data analysed quantitatively
34	(Kjærga et al., 2013)	Not including health promotion strategies in production	NA	Production process not taking care of health consideration	Malnutrition	✓	✓	Production process not taking care of health consideration more in LDCs	Documentary analysis
35	(Vermeir and Verbeke, 2006)		Consuming high fat meat	Attitude and behaviour	Obesity, heart diseases			Attitude and behaviour more looked at in DCs	Documentary
36	(Europian public health association, 2017)	Process that follows unhealthy consumption	Consumption process affect environment	Consumption of unhealthy product influencing production	Chronic diseases	✓	✓	Consumption of unhealthy product influencing production more in DCs	Documentary, argumentative analysis
37	(Hawkes, 2006)	Insufficient taking into account agriculture and health	NA	Hunger and obesity happening simultaneous	Hunger related diseases	✓	✓	Hunger and obesity happening simultaneous in LDCs	Source of data is documentary analysed and argue in relation to research question

continued on following page

Table 2. Continued

Study no	Study	Production	Consumption	Mechanism	Type of Ill Health	LDCs	DCs	Contextual and Institutional Factors	Methodology
38	(Mackay and Wolbring, 2013b)	Not linking production and health	Not linking consumption and health	Hunger and obesity happening simultaneous	Hunger related diseases	✓	✓	Hunger and obesity happening simultaneous	Documentary
39	(Ekpenyong, 2015)	Production process orientated towards urban consumption	Food insecurity in urban areas	Hunger and obesity happening simultaneous	Chronic diseases and poverty diseases	✓	✓ NA	Institution that deal with food security more in DCs	Relied on secondary data based on growth analysis
40	(Connell, 2010)	Not linking livelihood and health	Not linking livelihood and health	Injury and other possible outcomes	Chronic and other diseases	✓	NA	Net to operationalize livelihood and health to go hand in hand	

NA means that there is no evidence in study referring to that types of unsustainability

circumstances aggravate the situation by facilitating contamination. Furthermore, deforestation in these countries results in mudslides when it rains, leading to injuries and deaths. The other mechanism leading to diseases arises from survival strategies. People face shortages of food and consequent nutritional deficiencies lead to fatigue, dizziness and poor immune function. It should be noted that on way of responding to the shortage of food in order to survive is to try and get more arable land. This often results in clashes between communities, which lead to injuries and deaths. Working an excessive number of hours a day in an effort to satisfy household needs for food is also a mechanism leading to ill health in these countries. Finally, as in DCs, increasing numbers of people are exposed to air pollution and to the consumption of processed food following a process of urbanisation, implying that this segment of the population is set to suffer from respiratory diseases as well.

A review of the mechanisms through which these diseases occur suggests some patterns in these mechanisms and inherent types of diseases. While diseases patterns can be attributed to a myriad of factors, these factors are in fact embedded in types of unsustainable production/consumption. The extensive production of processed food in fact results in air pollution. Not only air pollution, but also the inhalation of chemicals used in the production of food such as vegetables. The greater availability of processed meat products results in the overconsumption of these products, which results in diseases such as heart diseases. Prolonged exposure to air pollution gives rise to other chronic diseases such as respiratory diseases. The evidence of these linkages explains why the problem of chronic non-communicable diseases is most prevalent in DCs and among the well-off urban populations of LDCs.

In the poor parts of LDCs in contrast, processes that lead to diseases are different. In these countries, traditional agriculture is inefficient with respect to land usage, which results in very limited crop yields. The associated deforestation not only results in loss of biodiversity but also in degradation of land through erosion. The aggravating factors in these countries are weak institutions, the high growth rate of the population and the fact that firewood is the main source of energy. As reported in the literature these factors are related to health outcomes in many ways. First, low crop yields means that people have not enough food, and limited food leads to deaths and disease associated with malnutrition. Second, sections of the population are exposed to contaminated water resulting in waterborne infectious diseases aggravated by prevailing poor sanitation, third, coping strategies include fighting to get better land for crops, and in social and psychological behaviours that result in drug addiction and prostitution. This evidence now makes it clear why we have more hunger-related deaths, diseases of malnutrition and more infectious diseases in these countries.

Briefly, the evidence discussed above suggests that consumption and production processes result in different mechanisms leading to ill health. These processes in addition to existing institutions and ways of life result in patterns of diseases that are different across DCs and LDCs. It was shown in particular that DCs and urbanized areas of LDCs experience similar mechanisms leading to chronic diseases with some exceptions of food insecurity (Ekpenyong, 2015), while the rural parts of the developing world are likely to suffer mainly short-term and infectious diseases. The evidence in particular showed that production processes result in diseases of overconsumption, while mechanisms of production lead to diseases of poverty. Since processes, circumstances and institutions determine the nature and extent of unsustainability and mechanisms leading to ill health, policy making should take into account differences in these processes across DCs and LDCs.

Unsustainability of Food Consumption and Policies: DCs vs. LDCs

In spite of differences across LDCs and DCs, a review of the evidence on the response policies, however, revealed that these policies took into account neither the differences in institutional set-ups nor the production processes that shape mechanisms to ill health and unsustainable food consumption. Most of the policies reviewed consisted of policies to increase output while safeguarding the environment. These were policies such as watershed and irrigation management, integrated ecological demonstrations (Pretty et al., 2003), bans on pesticides, support for organic agriculture, ecological farming to implement Agenda 21, placing value on the natural capital and taxing unsustainable use. Other policies were aimed at sustainable consumption that safeguards the environment, such as restrictions on international trade in processed food while nurturing healthy lifestyles, and labelling processed food to make the consumer aware of the unsustainability of certain food (Table 3).

To be effective and equitable under prevailing circumstances in DCs and LDCs, policies need to incorporate aspects of differences in institutional settings and production processes that impact on unsustainability. To assess whether policies took into account production processes and policies across DCs and LDCs, policies were evaluated with respect to the source of the policy, the proposer of the policy, the issue of focus, and the involvement of local political institutions.

With respect to the context from where policy proposals emanate, a review of the evidence revealed that most policy proposals arise from research and from policy makers in DCs (Table 3). The evidence reviewed showed that most of the policy documents and policy research papers make up about 90% of the reviewed documents. This evidence implies that policies did not take into account, at least sufficiently, the effects of institutions and consumption/production processes in LDCs.

With respect to the issues of focus, policies focused on issues such as reducing unsustainable food consumption through trade, placing a value on natural resources

Table 3. Policy equity across LDCs and DCs.

	Study	Source: Policy Design Context	Degradation	Social Unsustainability Policy	Econo Inefficiencies	Data Methodology	Areas of Focus	Policy	Suitability to LIC ? Suitability to LIC ?
1	(Brown and Jacobson, 2005)	Policy makers		Negative effects of palm oil on health		Reports and document information Descriptive analysis	Developed country	Labels dangers of product. discourage processing	No. No related measure from source Yes. Policy directly targeted at consumption in DCs
2	(Garnett et al., 2016b)	Research in DCs	Completion of land water and energy	Food shortage	NA	Documentary, argumentative analysis	Developed country	Sustainable intensification	Most relevant to both DCs and LDCs
3	(Seyfang and Smith, 2007)	Research	Degraded environment	Social concerns	NA	Reports	Developed country	Combine environment and social policy	Most relevant to DCs
4	(Carlsson-Kanyama and Gonzalez, 2009)	Research	Green gas house emission	NA	NA	Documentary, review and analysis	Developed country	Reduce meat consumption, encourage energy efficient food	Most relevant to DCs
5	(Reisch et al., 2013)	Research in DCs and LDcs	Environment degradation	Reduced meat consumption	Wastages	Documentary, analysis of documentary evidence	Developed country	Avoid product by airplane	Most relevant DCs
6	(Pretty et al., 2003)	Research focusing in LDCs	Negative effects of sustainable agriculture	NA	NA	Survey questionnaire analysed, reports Qualitative analysis and descriptive statistics	Developed country	Political involvements	Most relevant to DCs
7	(Griggs, 2013)	Research general	Negative effect on environment	NA	NA	Online survey	Developed country	Taxing unsustainable use	No. inability to place a sound taxation system Yes. Taxation enforcement possible
8	(Kearney, 2010)	Research paper for DCs and LDCs	NA	Health degradation due to shifts	NA	Documentary review, argumentative	DCs and LDCs	Consider trade, agricultural and health factors	Relevant to both countries but mostly DCs
9	(Pretty et al., 2003)	Research Paper	Soil erosion	Hunger, poverty	Low productivity	Survey data Descriptive statistics	LDCs	National policy reforms	Non conducive, social and political conditions NA

continued on following page

Table 3. Continued

	Study	Source: Policy Design Context	Degradation	Social Unsustainability Policy	Econo Inefficiencies	Data Methodology	Areas of Focus	Policy	Suitability to LIC ? Suitability to LIC ?
10	(vonMeyer-Höfer et al., 2015)	Research in LDCs	NA	Health related are the key issues, food insecurity	NA	Structural equation modelling online survey and theory of planned behaviour	China and india	Change prices of food and improve the safety information	No, pricing information important to significantly reduce commotion No, pricing more important for poor community Bu marginally affected
11	(Vermeir and Verbeke, 2006)	Research in DCs	NA	Unsustainable consumption based on behaviours	NA	Surveys, descriptive and inferential methods	DCs	Change diets, marketing to influence behaviours	More relevant to DCs
12	(Kasa, 2008)	Research in DCs and LDCs	NA	Increase in beef consumption through trade	NA	Documentary, argumentative analysis	DCS LDCs	Regulate international trade in	Most relevant to DCs
13	(Europian public health association, 2017)	Policy documents	Consumption	diets that damage health and associated with		Documentary evidence Critical analysis	Europe	Reduce green gas house. Increase plant based consumption	Most suitable to DCs and LDCs
14	(Thøgersen, 2010)	Research in DCs	NA	Consumption practices	NA	Documentary, analysis of documentary evidence		Deal with institutional and culture practices	Most relevant to LDs
15	(Kesavan and Swaminathan, 2008)	Research in LDCs	Degradation	Food shortage	NA	Documentary, review and analysis	LDCs in Asia	Ecofriendly agriculture and on farm enterprise to intensify crop	Most relevant to LDCs
16	(Godfray and Garnett, 2014)	Research in DCs and LDCs	Degradation	Food shortage Over consumption	Wastages of resources e	Documentary evidence, argumentative analysis	DCs and LDCs	Moderate demand, reducing waste, good governance, sustainable intensification	Most relevant to DCs
17	(Herrero and Thorntonb, 2010)	Research in DCs and LDCs	Degradation	Food consumptions	NA	Documentary, argumentative analysis	DCs and LDCs	Combination of policy proposed	Most relevant to DCs

continued on following page

Table 3. Continued

	Study	Source: Policy Design Context	Degradation	Social Unsustainability Policy	Econo Inefficiencies	Data Methodology	Areas of Focus	Policy	Suitability to LIC ? Suitability to LIC ?
18	(Horrigan et al., 2002)	Research in Dcs	Water wastage, pesticides	Health issues	Externalities of unsustainable consumption	Documentary, argumentative methods	DCs	Pollution policies, consumption policies	Most relevant to DCs
19	(Britsih Dietitic Association, 1013)	Policy document in UK	NA	overconsumption	NA	NA	Developed country	Pricing, of unsustainable food	Most relevant for DCs
20	(Stewart et al., 2013)	Research	NA	NA	NA	NA	Developed country	Increase veg consumption	NA
21	(Lorek and Fuchs, 2013)	Research in DCs	Degradation	Overconsumption	NA	Documentary, review of stats	DCs	Focus on the consumption patterns	Most relevant DCs
22	(Oni, 2010)	Research	NA	NA	NA	NA	Developed country	Government should play a bigger role in address food policy	LDCs less involved in policy make king
23	(Sonigo et al., 2012)	Research	NA	NA	NA	NA	Developed country	NA	NA
24	(Clapp, 2017)	Research in DC	Trade in unsustainable food	Insufficiency of food	NA	Documentary, conceptual and argumenta tie methods	Europe	Selective self f-sufficient policies	Most relevant for DCs .
25	(Pretty and Noble, 2006)	Research in DCs and LDCs	Traditional methods	Low crop yield hunger	NA	Commentary evidence Critical analysis	LDCs	Institutional, international and local, and institutional	Most relevant to DCs
26	(UNEP, 2012)	Research in DCs and LDCs		Unsustainable Diets	NA	Documentary evidence	LDCs and DCs	Reduce subsidies unsustainability	Most relevant
27	(Hobson, 2002)	Research paper in DCs	Consummation induced environment tal degradation	NA	NA	Interview with consumers	DCs	Reduce unsustainable consumption	More relevant for DCs

continued on following page

Table 3. Continued

	Study	Source: Policy Design Context	Degradation	Social Unsustainability Policy	Econo Inefficiencies	Data Methodology	Areas of Focus	Policy	Suitability to LIC ? Suitability to LIC ?
28	(Ekpenyong, 2015)	Reach paper in LDCs	NA	Urban poverty, food insecurity	NA	Secondary data, descriptive research design	Nigeria	Nutrition projects needed in cities	More relevant to LDCs
29	(Charles et al., 2010)	Research paper in Dcs and LDs	Water mismanagement in DCs	Food insecurity in DCs	Economic inefficiencies in DCs and LDCs	Documentary reviews	DCs and LDCs	Increase yield without expanding land	More relevant for both DCs a and LDCs
30	(Dobermann and Nelson, 2013)	Policy document in DCs and LDCs	All sorts of degradation	NA	NA	NA	DCs and LDCs	All environment policies aimed	More oriented to DCs due to health
31	(Freibauer et al., 2011)	Policy in DCs and LDCs	Resource scarcity due increasing demand	Unhealthy eating patterns	NA	Documentary, meta-review of documents	DCs and LDcs	Control food supply chain, change mindset, new technologies	Most relevant to DCs
32	(Johnson et al., 2014)	Research paper	NA	Unsustainable diets	NA	Survey, descriptive analysis and causal model	DCs	Find and respond factors of unsustable consumption	Most relevant to DCs
33	(vonBraun, 2007)	Policy document	NA	Food shortage and prices increase	NA	NA	DCs and LDCs	increase market access, impact food prices	Most relevant to DCs than to LDCs
34	(Europian commisison, 2015)	Policy documents	Environment degradation	Food shortage	Inefficiency	NA	LDCs	Variety of policies proposed	Most relevant

NA means that there is no evidence in study referring to that types of unsustainability

and taxing unsustainable use (Griggs, 2013), self-sufficiency and limiting international trade in processed food (Kearney, 2010, Clapp, 2017), labeling of processed food, emphasizing nutritional consideration in agricultural practices, and reducing the production of saturated animal products (Pretty et al., 2003). Policies that refer to the use of biological fertilizer, soil conservation, and biotechnological meant to meet food shortages in LDCs, did not specify how these are to be practical given the current predicament in these countries. The evidence indicated that these countries are characterized by poor soil quality as a result of inefficient usage practices aggravated by climatic conditions, lack of economic resources as well as issues relating to keeping stock, but how these would be effectively overcome were not addressed.

An analysis of these issues shows that, in spite of the fact that contexts in LDCs countries lead to more impact of unsustainability, policy has been disproportionally directed towards the DCs issues. For instance, reducing unsustainable food consumption by restricting international trade in some of the processed foods traded is likely to work for DCs. For the majority of the population in DCs, this policy appears to be irrelevant since the production processes in these countries result in shortages of food. They would, therefore, be better off consuming processed food via international trade than consuming no food at all. Furthermore, valuing and taxing the usage of natural resources is less relevant in LDCs due to limited expertise to value different types of natural resources in these countries. Given that the issues focused on are more likely to reflect the institutions and realities in DCs, the evidence suggest inequities within these policies.

With respect to political involvement, most policy proposals involve political stakeholders in DCs than they do in DCs ones. Political governance has been found to be important in enforcing policies. Given that the role of political governance in countries is not taken into account by policy makers, it follows that policies that seek to deal with unsustainability would not be effective. Briefly, the above evidence and the implied equity for DCs and LDCs suggest that policies on unsustainable food production/consumption have not been guided by production processes and characteristics. Policies have been formulated more on the basis of insights on issues as they are in DCs rather than those in the developing world. New ways of dealing with unsustainable food consumption that are equitable for both DCs and LDCs need to be sought.

Discussion of the Evidence and the Way Forward

Institutional set-ups, people's lifestyles and their interactions have been found to be instrumental in determining food consumption/production processes. These processes are in turn crucial in explaining the unsustainability in food consumption/production as well as mechanisms leading to ill health. A reviewed literature indicated that these institutions and consumption/production processes in DCs and LDCs influence unsustainability of food consumption to a different extent. In spite of these differences and their implications for the nature and extent of unsustainability in food consumption/production across these countries, limited research has delved into assessing the possible role of institutions and consumption/production processes in these countries in determining the patterns of the nature, extent and mechanisms of unsustainability in food consumption/production. The questions in this study that have been asked specifically in this respect are 1) whether there is evidence for institutional set-ups and consumption/production processes influencing the nature and extent of unsustainability across DCs and LDCs; 2) whether mechanisms arising from these production processes link to ill health in different ways across these countries; and 3) whether policies dealing with unsustainable consumption/production have been equitable in taking into account different institutional and production processes across DCs and LDCs. These questions were answered by means of a systematic review of the literature.

The evidence revealed that declining populations, increasing income per capita and industrial-orientated food consumption/production processes in DCs result in the overproduction of processed food, which degrades the environment while hampering the health of people through overconsumption in high-energy diets. In contrast, DCs, which have a higher population growth rate, suffer from limited technology in food production processes, and from hunger and inadequate micronutrients alongside a degraded environment. The extent of unsustainability is deeper in LDCs as it is associated with the absence of safety nets, resulting in increased impoverishment and limited human capital formation for the next generation. This situation is aggravated by the movement of rural-to-urban migration (Ekpenyong, 2015:31), which exposes the urbanized part of the population in these countries to issues similar to those suffered by the population in DCs. The evidence revealed further that mechanisms leading to ill heath are different. LDCs suffer ill health that is both short- and long-term in nature due to direct exposure to degraded environments, but also due to malnutrition and hunger. Most of the contexts in LDCs give rise to burdens of diseases that are essentially infectious. The unsustainable situation in DCs culminates instead in long-term and non-communicable diseases.

In spite of these differences, policies have not been equitable. Policies have been developed from within the context of DCs and have largely ignored issues in LDCs due to limited involvement in policy making by stakeholders in LDCs. This oversight has been happening in spite of the recognition that policies tailored to issues in LDCs work (Pretty et al., 2003:217). The important role of paying attention to institutions, political and socio-cultural outcomes has also been highlighted as instrumental in policy equity (McKenzie and Williams, 2015a, Garnett et al., 2016a). Given that these differences in institutions across LDCs and DCs has not been contemplated, it follows that policies have not addressed equitably the issues of low food productivity and hunger and their ramifications in LDCs. The proposals to date, aimed at feeding the population projected to be 9 billion people in 2050 in a sustainable manner do not address many of the issues in LDCs, such as decreasing farm productivity and limited crop diversity. Most of the additional growth in population is, however, projected to come from these countries. Without localized sustainability policies, some LDCs will depend on the more generalized policy making, which is doomed to be ineffective in addressing prevailing issues equitably.

This situation calls for a change in the policy response to the challenges of sustainable food consumption brought about by different consumption and production processes across DCs and LDCs. Policies that focus on specific aspects in these countries, would involve local political bodies and country-specific polices tackling both underfeeding and overfeeding in LDCs. While geopolitical issues do not allow to take needs and possibilities for poorest countries (Freibauer et al., 2011:16, UNCTAD, 2011) change is needed for these countries to be considered in a more integrated policy frameword. As a way forward, a unified policy framework is needed that incorporates aspects such that these policies work for both DCs and LDCs. According to the status quo, these policies are inclined towards DCs and need to change to include focus on issues affecting most parts of LDCs, such as the shortage of food, inclusion of localized policy proposals and reforms, and ensuring a buy-in from political stakeholders in these countries.

REFERENCES

Alemu, Z. G. (2010). *Livelihood strategies in rural South Africa: Implications for poverty reduction. International Association of Agricultural Economists (IAAE).* Foz do Iguacu, Brazil: Triennial Conference.

Alinov, L. (2010). Livelihoods strategies and household resilience to food insecurity: An empirical analysis to Kenya. In Promoting Resilience through Social Protection in Sub-Saharan Africa. Dakar, Senegal: Academic Press.

Amendah, D. D., Buigut, S., & Mohamed, S. (2014). Coping strategies among urban poor: Evidence from Nairobi, Kenya. *PLoS One*, *9*(1), 1–8. doi:10.1371/journal.pone.0083428 PMID:24427272

Barnidge, E., Baker, E. A., Motton, F., Fitzgerald, T., & Rose, F. (2011). Exploring community health through the Sustainable Livelihoods Framework. *Health Education & Behavior*, *38*(1), 80–90. doi:10.1177/1090198110376349 PMID:21169478

Birt, C., Buzeti, T., Grosso, G., & Justesen, L. (2017). Healthy and Sustainable Diets for European Countries. European Public Health Association – EUPHA. https://biblio.ugent.be/publication/8521128/file/8521129.pdf

British Dietetic Association. (2013). *Policy on sustainable food*. Author.

Brown, E., & Jacobson, M. F. (2005). *Cruel Oil: How palm oil Harms health, rainforest & wildlife*. Center of Science in the Public Interest.

Bryceson, D. F. (2004). Agrarian vista or vortex: African rural livelihood policies. *Review of African Political Economy*, *31*(102), 617–629. doi:10.1080/0305624042000327831

Bulte, E. H., & Soest, D. P. V. (2001). Environmental degradation in developing ountries: Households and the reverse environmental Kuznets Curve. *Journal of Development Economics*, *65*(1), 225–235. doi:10.1016/S0304-3878(01)00135-3

Capone, R., Bilali, H. E., Debs, P., Cardone, G. & Driouech, N. (2014). Food system sustainability and food Security: Connecting the Dots. *Journal of Food Security, 2*, 13-22.

Carlsson-Kanyama, A., & Gonzalez, A. D. (2009). Potential contributions of food consumption patterns to climate change. *The American Journal of Clinical Nutrition*, *89*(5), 1704S–1709S. doi:10.3945/ajcn.2009.26736AA PMID:19339402

Charles, H., Godfray, J., Crute, I. R., Haddad, L., Lawrence, D., Mui, J. F., & Whiteley, R. et al. (2010). The future of the global food system. *Philosphical Transactions of the Royal Socieity B*, *365*(1554), 2769–2777. doi:10.1098/rstb.2010.0180 PMID:20713383

Clapp, J. (2017). Food self-sufficiency: Making sense of it, and when it makes sense. *Food Policy*, *66*, 88–96. doi:10.1016/j.foodpol.2016.12.001

Connell, D. J. (2010). Sustainable livelihoods and ecosystem health: Exploring methodological relations as a source of synergy. *EcoHealth*, *7*(3), 351–360. doi:10.1007/s10393-010-0353-7 PMID:21104294

Cullet, P. (2004). Intellectual property rights and food security in the South. *The Journal of World Intellectual Property*, *7*(3), 261–286. doi:10.1111/j.1747-1796.2004.tb00209.x

Debon, H., Parrot, L., & Moustier, P. (2010). Sustainable urban agriculture in developing countries. A review. *Agronomy for Sustainable Development*, *30*(1), 21–32. doi:10.1051/agro:2008062

Dobermann, A. & Nelson, R. (2013). *Opportunities and solutions for sustainable food production*. Academic Press.

Ekpenyong, A. S. (2015). Urbanization: Its implication for sustainable food Security, health and nutritional nexus in developing economies - A case study of Nigeria. *Journal of Studies in Social Sciences*, *11*, 29–49.

European Commission. (2015). *Sustainable food consumption and production in a resource-constrained world*. https://ec.europa.eu/research/scar/pdf/scar_3rd-foresight_2011.pdf.

Freibauer, Mathijs, Brunori, Damianova, Faroult, Gomis, O'Brien, & Treyer. (2011). Sustainable food consumption and production in a resource-constrained world. Belgium: European Commission-Standing Committee on Agricultural Research (SCAR).

Friedl, B., Omann, I., & Pack, A. (2014). *Socio-economic drivers of (non-) sustainable food consumption. An analysis for Austria*. Academic Press.

Garnett, M. C., Appleby, A., Balmford, I. J., Bateman, T. G., Benton, P., Bloomer, B., & Vermeulen, H. C. J. et al. (2016a). Sustainable intensification in agriculture: Premises and policies. *Scinece*, *30*, 1–2.

Garnett, T. (2014). *What is a sustaiable healthy diet Discussion paper*. Food Climate Research Network.

Garnett, T., Appleby, M., Balmford, A., & Bateman, I. (2016b). Sustainable Intensification in Agriculture: Premises and Policies. *Science*, *30*, 1–2. PMID:23828927

Gauker, C. (2010). *The impacts of sustainable and industrial agriculture on human health*. Academic Press.

Gautam, Y., & Andersen, P. (2016). Rural livelihood diversification and household well-being: Insights from Humla, Nepal. *Journal of Rural Studies*, *44*, 239–249. doi:10.1016/j.jrurstud.2016.02.001

Godfray, H. C. J., Crute, I. R., Haddad, L., Lawrence, D., Muir, J. F., Nisbett, N., … Whiteley, R. (2010). The future of the global food system. *Philosophical Transaction of the Royal Society B*, *365*, 2769-2777.

Godfray, H. C. J. & Garnett, T. (2014). Food security and sustainable intensification. *Philosophical Transaction of the Royal Society B*, *369*, 1-12.

Griggs, D. (2013). *Sustainable development goals for people and planet*. Academic Press.

Grunert, K. G., Hieke, S., & Wills, J. (2014). Sustainability labels on food products: Consumer motivation, understanding and use. *Food Policy*, *44*, 177–189. doi:10.1016/j.foodpol.2013.12.001

Han, Y., & Hansen, H. (2012). Determinants of Sustainable Food Consumption:A Meta-Analysis Using a Traditional and a Structura Equation Modelling Approach. *International Journal of Psychological Studies*, *4*(1), 22–44. doi:10.5539/ijps.v4n1p22

Hart, K., Allen, B., & Karsten, J. (2013). *EU policy options to encourage more sustainable food choices*. Academic Press.

Hawkes, C. (2006). The links between agriculture and health: An intersectoral opportunity to improve the health and livelihoods of the poor. *Bulletin of the World Health Organization*, *84*(12), 984–990. doi:10.2471/BLT.05.025650 PMID:17242835

Herrera-Estrella, L. R., & Alvarez-Morales, A. (2000). Genetically Modified Crops and Developing Countries. *EMBO Reports*, *2*(4), 256–258. doi:10.1093/embo-reports/kve075 PMID:11306538

Herrero, M., & Thorntonb, P. K. 2010. Livestock and global change: Emerging issues for sustainable food systems. *Proceedings of the National Academy of Sciences of the United States of America,* 110.

Hertwich, E., & Katzmayr, M. (2003). *Examples of sustainable consumption*. Academic Press.

Hobson, K. (2002). Competing discourses of sustainable consumption: Does the rationalisation of lifestyles' make sense? *Environmental Politics*, *11*(2), 95–120. doi:10.1080/714000601

Horrigan, L., Lawrence, R. S., & Walker, P. (2002). How sustainable agriculture can address the environmental and human health harms of industrial agriculture. *Environmental Health Perspectives*, *110*(5), 445–456. doi:10.1289/ehp.02110445 PMID:12003747

International Institute Of Social Studies. (2015). *Global governance/politics, climate justice & agrarian/social justice: linkages and challenges*. Author.

Johnson, J. L., Fanzo, J., & Cogill, B. (2014). Understanding sustainable Diets: A descriptive analysis of the determinants and processes that influence diets and their impact on health, food security, and environmental sustainability. *Advances in Nutrition*, *5*(4), 418–429. doi:10.3945/an.113.005553 PMID:25022991

Kasa, S. (2008). Globalizing unsustainable food consumption:Trade policies, producer lobbies, consumer preferences, and beef consumption in Northeast Asia. *Globalizations*, *5*(2), 151–163. doi:10.1080/14747730802057480

Kearney, J. (2010). Food consumption trends and drivers. *Philosophical Transaction of the Royal Society B*, *365*(1554), 2793–2807. doi:10.1098/rstb.2010.0149 PMID:20713385

Kesavan, P. C. & Swaminathan, M. S. (2008). Strategies and models for agricultural sustainability in developing Asian countries. *Philosophical Transaction of the Royal Society B, 363*, 877-891.

Khalid, S., Khan, M. B., Regina, K., Jos, K., & Gerd, A. (2003). Five steps to conducting a systematic review. *Journal of the Royal Society of Medicine*, 96. PMID:12612111

Khan, Z. R., Midega, C. A. O., Pittchar, J. O., Murage, A. W., Birkett, M. A., Bruce, T. J. A., & Pickett, J. A. (2014). Achieving food security for one million sub-Saharan African poor through push-pull innovation by 2020. *Philosophical Transactions of the Royal Society of London. Series B, Biological Sciences*, 369. PMID:24535391

Kjærga, B., Land, B., & Pedersen, K. B. (2013). Health and sustainability. *Health Promotion International*, *29*, 10–15. PMID:23300191

Lefin, A. L. (2010). *Food consumption and sustainable development: An introduction*. Institut pour un Développement Durable.

Lehota, A. L. (2004). National and international trend of food consumption and behaviour. *Marketing*, *1*, 7–14.

Littell, J. H., Corcoran, J., & Pillai, V. (2008). *Systematic reviews and meta-analysis*. New York, NY: Oxford University Press. doi:10.1093/acprof:oso/9780195326543.001.0001

Lorek, S., & Fuchs, D. (2013). Strong sustainable consumption governance – precondition for a degrowth path? *Journal of Cleaner Production*, *38*, 36–43. doi:10.1016/j.jclepro.2011.08.008

Mackay, R., & Wolbring, G. (2013a). *Sustainable Consumption of Healthcare: Linking Sustainable Consumption with Sustainable Healthcare and Health Consumer Discourses*. The 3rd World Sustainability Forum.

Mackay, R., & Wolbring, G. (2013b). *Sustainable Consumption of Healthcare: Linking Sustainable Consumption with Sustainable Healthcare and Health Consumer Discourses*. Academic Press.

Mckenzie, F. C., & Williams, J. (2015). Sustainable food production: Constraints, challenges and choices by 2050. *Food Security*, *7*(2), 221–233. doi:10.1007/s12571-015-0441-1

Mcmichael, A. J., Powles, J. W., Butler, C. D., & Uauy, R. (2007). Food, livestock production, energy, climate change, and health. *Lancet*, *370*(9594), 1253–1263. doi:10.1016/S0140-6736(07)61256-2 PMID:17868818

Mont, O., & Plepys, A. (2005). Motivating sustainable consumption. *Sustainable Development Research Network*, *29*, 30.

Mózner, Z. V. (2013). *Towards Sustainable food consumption? An Ecological footprint in Hungary* (PhD thesis). Corvinus University of Budapest.

National Institute Of Consumer Research. (2010). *Sustainable Consumption and some contemporary European issues*. Author.

Nemecek, T., Jungbluth, N., Canals, L. M., & Schenck, R. (2016). Environmental impacts of food consumption and nutrition: where are we and what is next? *Int J Life Cycle Assess, 21*, 607-620.

Oni, O. A. (2010). *Food Poverty and Livelihoods Issues in Rural Nigeria*. Academic Press.

Pretty, J., Noble, A., Bossio, D., Dixon, J., Hine, R. E., & Morison, I. L. (2013). Resource-conserving agriculture increases yields in Developing Countries. *Environmental Science & Technology*, *40*(4), 1114–1119. doi:10.1021/es051670d PMID:16572763

Pretty, J. N., Morison, J. I., & Hine, R. E. (2003). Reducing food poverty by increasing agricultural sustainability in developing countries. *Agriculture, Ecosystems & Environment*, *95*(1), 217–234. doi:10.1016/S0167-8809(02)00087-7

Pretty, J. N., & Noble, D. (2006). Resource-Conserving Agriculture Increases Yields in Developing Countries. *Environment Science and Technology.*

Randolph, T., Schelling, E., Grace, D., Nicholson, C. F., Leroy, J., Cole, D., & Ruel, M. et al. (2007). Role of livestock in human nutrition and health for poverty reduction in developing countries. *Journal of Animal Science*, *85*(11), 2788–2800. doi:10.2527/jas.2007-0467 PMID:17911229

Reisch, L., Eberle, U., & Lorek, S. (2013). Sustainable food consumption: an overview of contemporary issues and policies. *Sustainability: Science, Practice, & Policy*, *9*, 7–24.

Risku-Norja, H. (2011). *From environmental concerns towards sustainable food provisioning. Material flow and suitable food consumption scenario studies on sustainability of agri-food systems* (Doctoral dissertation). Faculty of agriculture and Forestry, Faculty of Agriculture and Forestry, University of Helsinki.

Schloster, H., Deboer, J., & Boersema, J. J. (2012). Can we cut on the meat of the dish? Constructing consumer-oriented pathway towards meat substitution. *Appetite*, *158*, 39–49.

Seligman, H. K., Laraia, B. A., & Kushel, M. B. (2009). Food Insecurity Is Associated with Chronic Disease among Low-Income NHANES Participants. *Journal of Nutrition and Diseases, 5*, 304-310.

Seyfang, G., & Smith, A. (2007). Grassroots innovations for sustainable development: Towards a new research and policy agenda. *Environmental Politics*, *16*(4), 584–603. doi:10.1080/09644010701419121

Sonigo, D. P., Bain, M. J., Kong, M. M. A., Fedrigo, M. D., Withana, M. S., Watkin, M. E., & Dresner, D. S. et al. (2012). *Policies to encourage sustainable consumption Full.* France Europian Commsion.

Stewart, R., Korth, M., Langer, L., Rafferty, S., & Silva, N. R. D. (2013). What are the impacts of urban agriculture programs on food security in low and middle-income countries? *Environmental Evidence*, *2*(1), 7. doi:10.1186/2047-2382-2-7

Temme, E. H., Bakker, H. M., Seves, M., Verkaik-Kloosterman, J., Dekkers, A. L., Raaij, J. M. V., & Ocké, M. C. (2015). How may a shift towards a more sustainable food consumption pattern affect nutrient intakes of Dutch children? *Public Health Nutrition*, *18*(13), 2468–2478. doi:10.1017/S1368980015002426 PMID:26344035

Thøgersen, J. (2010). Country differences in sustainable consumption: The case of organic food. *Journal of Macromarketing*, *30*(2), 171–185. doi:10.1177/0276146710361926

UNCTAD. (2011). Sustainable agriculture and food security in LDCs. *Least Developed Countries Series, 20*, 1-2.

UNEP. (2012). *The Critical Role Of Global Food Consumption Patterns In Achieving Sustainable Food Systems And Food For All.* A UNEP Discussion Paper.

Vermeir, I., & Verbeke, W. (2006). Sustainable food consumption: Exploring the consumer "attitude–behavioral intention" gap. *Journal of Agricultural & Environmental Ethics*, *19*(2), 169–194. doi:10.1007/s10806-005-5485-3

Vogit, M. (2014). *Meanings attached to food and sustainable food consumption* (Master's thesis). Uppsala University, Department of Earth Sciences.

Vonbraun, J. (2007). The world food situation: New riving forces and required actions. *IFPRI's Bi-Annual Overview of the World Food Situation presented to the CGIAR Annual General Meeting.*

Vonmeyer-Höfer, M., Juarez Tijerino, A. & Spiller, A. (2015). *Sustainable food consumption in China and India.* Global Food Discussion Papers, No. 60.

World Health Organisation. (2003a). Diet, nutrition and the prevention of chronic diseases. *World Health Organization Technical Report Series*, 916. PMID:12768890

World Health Organisation. (2003b). *Poverty and Health.* Paris: OECD.

Section 4

System Innovations and the Transition to Sustainable Food Systems in Developing and Emerging Economies

This section discusses the roles of innovations to sustainable food systems in developing and emerging economies.

Chapter 7
Harmonising Roles of Agricultural Extension With Other Agencies Involved in Agricultural Development Towards Sustainable Food Systems in Nigeria

Oluwole Matthew Akinnagbe
Federal University of Technology, Nigeria

ABSTRACT

The success of sustainable agriculture depends not just on the motivations, skills, and knowledge of individual farmers, but on action taken by the agricultural extension agency in harmonising the activities of other agencies involved in agricultural development. This chapter looks into the coordinating roles of agricultural extension with other agencies involved in agricultural development towards sustainable food systems in Nigeria. The agricultural extension agency needs to coordinate other agencies in rural development to ensure unity of purpose and avoid duplication of efforts. Agricultural extension is considered the best institution to coordinate other agencies involved in agricultural development towards sustainable food system in Nigeria because of the edge agricultural extension has at grassroots level in light of community organizing and empowerment role. Government in all tiers should support the agricultural extension agency by ensuring that all development partners liaise with the agency before carrying out their work to enhance sustainable rural development.

DOI: 10.4018/978-1-5225-3631-4.ch007

INTRODUCTION

Agricultural extension and advisory services plays a critical role in facilitating linkages with farmer-based organizations and other relevant accleartors including government agencies, private sector and non-governmental organizations, research institutes and education centres (Davis and Heemskerk, 2012). According to Feder *et al.* (2010), Swanson and Rajalahti (2010), agricultural extension has a tremendous potential to improve agricultural productivity and increase incomes through transfer and facilitation of knowledge, skills, and technologies. Extension is an activities that provide information and services needed and demanded by farmers and other actors in rural setting to assist them in developing their own technical, organizational, and management skills and practices so as to improve their livelihood and well-being. It could also be seen as systems and mechanisms designed to build and strengthen the capacity of rural farmers and other stakeholders.

Changes and a range of other pressures are forcing a re-examination of public extension services. Re-examination also shaped by a climate of today's perception of poor performance of past investments in extension. The monopoly public services model for extension is obsolete in the more competitive, market-oriented agriculture (Alex, *et.al.*, 2004). In Nigeria, there are numerous extension service providers, including public-funded institutions, non-governmental organizations (NGOs), commodity processors, farmers' associations, and private agrochemical input suppliers. Smallholder farmers rely heavily on public extension systems such as the agricultural development programme (ADP), which are experiencing increasing operational difficulties as a result of dwindling resources (Madukwe, 2008).

Extension systems and delivery methods in many developing countries have been constantly viewed as ineffective in responding to the demands and technological challenges of various types of clients and in reaching the rural poor (Birner *et al.* 2009). According to Davis et.al. (2010), the number farmers per extension agent in Nigeria is 3330. This sometimes was as a result of financial challenges. On the other hand, NGOs have better financial resources, but service only small clienteles and deal with only limited numbers of commodities.

Various grassroots'-level extension providers operate with specific objectives and outputs that are of little significance to productivity and sustainable agricultural development. Pluralism in agricultural extension allows farmers to choose among alternatives because the various extension providers offer different services. However, the levels of pluralism and uncoordinated extension services at the grassroots level are resulting in lower outputs and confusion, at farmers' expense (Hanyani-Mlambo, 2002).

There are so many international organisations and developmental partners working in the rural areas in Nigeria: Food and Agricultural Organization of the United Nations (FAO), International Fund for Agricultural Development (IFAD), World Bank (WB), International Institute of Tropical Agriculture (IITA), Department for International Development (DFID), European Commission, and United States Agency for International Development (USAID) among others and other national agencies with the aim of improving farmers' life and increase productivity. Some of the activities of these organisations calls for the facilitation of coordination and collaboration among both public and private extension providers in order to ensure a unified service and to avoid duplication and wastage of scarce resources. The purpose of this paper therefore is to x-ray the roles of agricultural extension services in harmonising the activities of the various organisations/development partners working with different clientele in the rural areas towards sustaining food system in Nigeria.

COORDINATION AND ITS CHALLENGES

Coordination is the extent to which organizations attempt to ensure that their activities take into consideration those of other organizations in the pursuance of their goals. It ensures that harmony of individual efforts towards the accomplishment of a group goal is achieved. It is a way of integrating different parts of an organization to accomplish a collective set of tasks or integrating or linking together different resources to accomplish a collective set of tasks. According to Malone and Crowston (1994), it is the act of managing interdependencies between activities. It is the interrelation of functions, structures, and resources in an organizational context, which can take place at different levels (Mangham, 1986). The more efficient coordination is in all levels of administration, the common outcome, cohesion, will be reached in a more efficient manner; because coordination is a tool of cohesion (Viinamäki 2004). Every activity in an organization requires coordination of a variety of functions within and between firms in order to avoid complexities and unintended loses.

Co-ordination is the process of synchronizing the effort of a group to achieve a desired goal. This can be achieved by making the different agencies work together in an efficient and organized way. This entails generating and managing information effectively and efficiently. Extension has proven itself to be a cost-effective means of bringing about greater economic returns for farmers with significant and positive effects on knowledge, adoption, and productivity. Extension is thus a cost-effective tool that can play an important role in dealing with climate change while at the same time helping to increase productivity and reduce poverty (Davies, 2009).

Coordination produces performance, because it produces the necessary trust needed for achieving performance through networking. It tells what to include in a good plan and how to execute it, specifies who will be a staff and the rational placement, gives a clear focus, makes things realistic, and gives a good appraisal. It is therefore part of planning, organizing, staffing, directing reporting and budgeting. For there to be an effective coordination, a clear determination of role and responsibility must be emphasized (Viinamäki 2004). According to Akinnagbe and Olaolu (2016), it is essential to develop and establish a habit of doing casual evaluation to determine if proper coordination has been done.

Four kinds of elements can be coordinated: resources, programmes, clients and information. Coordination of all elements may not be possible or even desirable. Once the elements for coordination have been determined, decisions should be reached regarding how they will be coordinated. Certain elements generally can best be coordinated at specific levels. Coordination of resources that is joint decisions as to which organizations will get which resources, is best conducted by representative of the Board. However, coordination with regard to programmes, programme development and the avoidance of duplicate programmes may best be done or guided by relevant agency decision makers when they are motivated to do so. Those who work closely with the clients or people who receive the benefits from the organization, can do the best job of coordination that directly involves the clients. Information should be coordinated at all levels. For example, information relevant to the coordination projects should flow freely between all participants.

If the coordination is for programmes and programme development, only relevant decision makers may need to be involved. If the coordination involvers clients' services, line staff should do it. Sometimes administration have personality differences that make coordination difficult, but effective coordination may be possible with other elements. Coordination and leadership are closely bound together as each affects the other. Coordination cannot be achieved without effective leadership. They jointly ensure that all efforts are channeled effectively towards the desired goal. A feature of the large complex organization is that it is not possible to rely on a common superior to coordinate all aspects of two or more managers' work. As a result of this, a number of other coordinating devices must be introduced.

For coordination to be effective, certain principles are required.

- Professionally competent staff must be provided.
- Objectives of an organization must be clearly defined and understood.
- Two-way channels of communication must be kept open.

- A feeling must be created that field staff are important members of the organization. This means that they should be paid regularly given incentive and their effects well commended.
- Administrative staff in key positions must be cooperative and enjoy mutual respect and confidence among staff is built on:
  - Foundation of comparable professional training and ability.
  - Willingness to examine a given problem and facts associated with it in an objective manner.
  - Recognition and acceptance of honest differences in opinions.
  - Give and take philosophy.
- The location of responsibility for coordination should be placed within the administrative structure so that the:
  - Person responsible for coordination have the authority to deal with administrative problems and report directly to the chief administrative officer.
  - Officers at the zonal or divisional headquarters have status commensurate with their responsibility.
  - Persons doing the coordination should not spend most of their time manipulating budget.

There are several factors that act as barriers to coordination. These include:

- Threat to autonomy: One of the biggest barriers is the fear of losing organizational autonomy. Coordination almost always involves a loss of some autonomy for the organizations. That is, coordinating organizations have to "give and take" with regard to goals, services and clients and hence compromises are reached, which reduce the freedom to make decisions and run programmes.
- Professional staff fear of loss of freedom. They may be committed to different ways of working with clients.
- Fear that clients representatives will try to dominate the organizations that serve the clients.
- Disagreement among resource providers as to client needs and services to be provided.
- Coordination is complicated if too many organizations are involved.
- Lack of domain consensus: disagreements among the organization regarding the right of one or more organizations to be involved as well as disagreement about which organizations should function on which geographical areas, provide which services and to which clients.

- Different expectations from different quarters e.g. from federal, state and local levels, as to which clients are to be served and which services should be provided.
- Low priority accorded to coordination by participating organizations.
- Availability of financial resources.

OVERVIEW OF AGRICULTURAL EXTENSION ROLES TOWARDS SUSTAINABLE FOOD SYSTEM

Agricultural Extension agency has been performing different roles at the grassroots level towards sustainable food system in Nigeria and these roles enable extension agency has an edge over other agencies and hence consider as institution to coordinate other agencies involved in agricultural development. Some of these roles performed by extension agency include among others:

1. **Community-Organizing Role:** Extension has a unique working relationship with local government officials that can serve as a strong link between development agencies technical assistance and research and the community. It's this uniqueness that provides the basis for Extension's comparative advantage in rural development activities. In most states in Nigeria, extension has a reservoir of local government-based and rural development experience. Extension as an institution is known and respected among traditional local government groups and has the experience to develop a close collaborative relationship with other local government entities that haven't traditionally had a working relationship with development agencies (Schutjer, 1991).
2. **Empowerment Role:** The term empower means to enable, to allow, or to permit and can be viewed as both self-initiated and initiated by others. Telling adults what to do provokes reaction, but showing them triggers the imagination, involving them gives understanding, and empowering them leads to commitment and action (Chamala, 1990). Extension personnel have developed a philosophy where their role is to help farmers and rural communities organize themselves and take charge (empowerment) of their growth and development. The development that is envisaged is not merely a handout of benefits to people in need, but a process of empowerment where rural communities can acquire mastery over their own destiny through the realization that they, individually and collectively can do something to improve their circumstances. This is a refocusing strategy from waiting on government, oil companies or other organizations/donors to solving their own problems themselves (Ozor and Nwankwo, 2008).

3. **Human Resource Development Role:** The human resource development approach empowers people and gives new meaning to all other roles. Development of technical capabilities must be combined with management capability. The entire philosophy of human capacity building is to encourage rural communities to understand their personal and group styles of managing themselves and to improve their planning, implementation, and monitoring skills. According to Orapin (1996), one approach in creating sustainable rural development is through giving the main actors (villagers living in the community) an equal opportunity to think and plan their own future. This underpins the need for effective leadership at the local community levels in order to harness the efforts of the rural people towards their own development.
4. **Problem-Solving and Education Role:** Problem-solving is one of the key objectives of extension. This is achieved by helping farmers to identify their problems and seek the right solutions by combining their indigenous knowledge with improved knowledge and by using their resources properly. Extension aims at changing the outlook and attitude of the farming community in general and it seeks means to improve the farm operations and farmer's family life in totality on their own initiative. As the farmers are mostly small and marginal, they lack direct access to developing agricultural technology. Educating such a group of farmers has to be, therefore, a sustained process to keep pace with rapidly changing agricultural technology.
5. **Programme Continuity:** Most rural development projects are created by different development agencies that engage the services of the consultants, external provisions of personnel who are often present only for a short period, in opposition to long time development. Extension workers been part of the day to day affairs of the rural people can be incorporated effectively into a regular and institutionalized extension programme which has a longer time span.
6. **Technology and Management Information:** Extension traditionally has played a role in providing information and promoting new technologies or new ways of managing crops and farms. Extension also links farmers to researchers and other actors in the innovation system. Farmers, extension agents, and researchers must work together on farmers' fields to prioritize, test, and promote new crop varieties and management techniques. While extension must now go beyond such methods, there is still a need for simple technology transfer in order to increase resilience to recent activities in agricultural development such as climate change, flooding (Davies, 2009). Extension agents can introduce locally appropriate technologies and management techniques that enable farmers to adapt to climate change by, for example, developing and disseminating local cultivars of drought-resistant crop varieties with information about the

crops' advantages and disadvantages. Additionally, extension staff can share with farmers their knowledge of cropping and management systems that are resilient to changing climate conditions such as agro forestry, intercropping, sequential cropping, and no-till agriculture. Some of these practices have the added advantage of improved natural resource management. Tree planting can also help to improve soil, prevent soil erosion, and increase biodiversity. It is important to provide farmers with information about how the various options will potentially increase income and yields, protect household food security, improve soils, enhance sustainability, and generally help to alleviate the effects of climate change. At the same time, extension staff can play an important role in transferring indigenous technical knowledge to help farmers worldwide.

7. **Facilitating, Brokering, and Implementing Policies and Programmes:** Another role of extension, which will be critical for agricultural development in today's world, is that of acting as an honest broker, bringing together different actors within the rural sector. Traditionally this has meant linking farmers to transport agents, markets, and inputs suppliers, among others. With coordinating rural development, it will be increasingly important for the extension system to link farmers and other people in rural communities directly with rural development partners, be they private and public institutions or non-governmental organisations.

REASONS WHY AGRICULTURAL EXTENSION AGENCY NEED TO COORDINATE OTHER AGENCIES TO ACHIEVE SUSTAINABLE FOOD SYSTEMS IN NIGERIA

The following are the reasons why agricultural extension agency needs to coordinate other agencies in rural development:

1. **Ensures Unity of Purpose:** Every unit, level, agricultural agencies or non-agricultural agencies involve in rural agricultural development operates in line with the overall goal of agricultural extension. There is therefore need for unity of purpose between these agencies working in the rural areas. Extension as an agricultural agency taking the lead in ensuring that all other agricultural agencies are embracing the need to create the conditions for the small farmers especially those in the rural area who are responsible for the highest provider of workforce required to produce for in the country, to be productive as possible so they can feed their families and the country Nigeria.

2. **Ensures Integration of the Efforts of Different Units and Levels:** The Research Extension Farmer Input Linkages (REFIL), involves a feedback mechanism between the farmers and the research institutes with extension acting as an intermediary between the farmers and the research institutes. The differences in objectives and motivations between public and private research and extension organizations mean that one or the other may be more appropriate for certain kinds of extension activities and clients. With the focus now shifted to demand-driven, location-specific, customized and tailor-made technologies and services to serve different categories of farmers, a vital link in the integration process, missing is the "farmer", needs to be included in the integration process.
3. **Enhances Supervision of These Agencies:** The agricultural extension agency need to monitor the activities of these agencies working in the rural area to ensure they are working in accordance with the overall goal of people.
4. **Ensures Efficient Utilization of Resources:** There is need for efficient utilisation of resources supplied by the agricultural and non-agricultural agencies, thus achieving optimum result in farmers overall output. The extension service providers within the system need coordination to optimize the use of limited resources: personnel, funds and logistics to increase coverage within the system. This implies sharing of information and expertise among the agencies involved, and participation where appropriate in each other's extension activities and jointly organized ones. Extension helps agencies to complement each other for effectiveness of the system and enhanced service delivery to farmers.
5. **Ensures Cooperation and Avoid Unnecessary Rivalry and Conflicts:** Extension develops close cooperation among agencies and formal research institutes, inputs, credits and marketing to provide farmers with efficient services. Extension agency coordinating other agencies could help avoiding unnecessary rivalry and conflicts among the agricultural and non-agricultural agencies.
6. **Avoid Duplication of Efforts:** Since there are many agricultural and non-agricultural agencies, agricultural extension as an agency has records of areas already supplied with inputs and directs the agencies to areas yet to be supplied.

CONCLUSION

Agricultural extension is crucial to development in the agricultural sector and overall national development. As an agency, it has played a major role in the supervision, coordination and ensuring cooperation among the agricultural and non-agricultural agencies involved in rural development. Extension has proven itself to be a cost-effective means of bringing about greater economic returns for farmers with significant and positive effects on knowledge, adoption, and productivity. As extension takes on a broader coordination role, this must not be sacrificed for what might appear to be short-term gains. Among many activities in which extension must be involved include technology transfer, offering economical advice (including book-keeping), developing agricultural markets and informational system, developing small enterprises and discovering new alternatives for obtaining profits, participating to rural development implementation, advising in legal and social fields. Most rural development efforts need the coordinating role of extension agency to last and meet up to its stated objectives. The issue of coordination by agricultural extension agency will help guide against duplication of rural development programmes by the government and other agencies who aim at carrying out development programmes at the grassroots as well as making sure that approved programmes will be those that will meet the farmers' felt needs while following approved standards. Government in all tiers should support the agricultural extension agency in coordinating rural development programmes by ensuring that all development partners liaise with agricultural extension agency before carrying out their work in the rural areas to enhance sustainable rural development.

REFERENCES

Akinnagbe, O.M., & Olaolu, M.O. (2016). Policy issues for improving monitoring and evaluation of agricultural extension programmes in Nigeria. *African Evaluation Journal*, *3*(2), a122. doi: .v3i2.12210.4102/aej

Alex, G., Byerlee, D., Helene-Collion, M., & Rivera, W. (2004). *Extension and rural development: Converging views on institutional approaches?* Agriculture & Rural Development Department of the World Bank.

Birner, R., Davis, K., Pender, J., Nkonya, E., Anandajayasekeram, P., Ekboir, J., & Cohen, M. et al. (2009). From Best Practice to Best Fit: A Framework for Designing and Analyzing Pluralistic Agricultural Advisory Services Worldwide. *Journal of Agricultural Education and Extension*, *15*(4), 341–355. doi:10.1080/13892240903309595

Chamala, S. (1990). Establishing a group: A participative action model. In S. Chamala & P. D. Mortiss (Eds.), *Working together for Landcare: Group management skills and strategies* (pp. 13–38). Brisbane: Australian Academic Press.

Davies, K. E. (2009). Agriculture and Climate Change: An Agenda for Negotiation in Copenhagen. The Important Role of Extension Systems. International Food Policy Research Institute. Focus 16. *Brief, 11*, 1–2.

Davis, K., & Heemskerk, W. (2012). *Investment in extension and advisory services as part of agricultural innovation systems. Module 3 of agricultural innovation systems: an investment sourcebook.* Washington, DC: The World Bank.

Davis, K., Swanson, B., Amudavi, D., Mekonnen, D. A., Flohrs, A., Riese, J., . . . Zerfu, E. (2010). *In-Depth Assessment of the Public Agricultural Extension System of Ethiopia and Recommendations for Improvement.* IFPRI Discussion Paper 01041. Washington, DC: IFPRI.

Feder, G., Anderson, J., Birner, R., & Deininger, K. (2010). *Promises and Realities of Community-based Agricultural Extension.* IFPRI Discussion Paper 00959. Washington, DC: IFPRI.

Gullick, L., & Urwich, C. L. (Eds.). (1957). *Papers on the Science of Administration.* New York: Institute of Public Administration.

Hanyani-Mlambo, B. T. (2002). *Strengthening the pluralistic agricultural extension system: a Zimbabwean case study. Integrated Support to Sustainable Development and Food Security Programme (IP).* Rome, Italy: FAO.

Madukwe, M.C. (2008). Practices without Policy. In *The Nigeria Agricultural Extension Service 30th inaugural lecture of the University of Nigeria.* University of Nigeria Press Ltd.

Malone, T. W., & Crowston, K. (1994). Toward an Interdisciplinary Theory of Coordination. *Computer Survey, 26*(1), 87–119. doi:10.1145/174666.174668

Mangham, I. L. (1986). *Power and Performance in Organization.* New York: Basil Blackwell Inc.

Orapin, S. (1996). People's participation in community development. *TDRI Quarterly Review, 11*(3), 19–25.

Ozor, N., & Nwankwo, N. (2008). The role of local leaders in community development programmes in Ideato Local Government Area of Imo State: Implication for extension policy. *Journal of Agricultural Extension, 12*(2), 2.

Schutjer, W. A. (1991). Rural Development and Extension. *Journal of Extension, 29*(1), 1–2.

Sinkaye, T. (2005). Agricultural Extension Participatory Methodologies and approaches in Agricultural Extension in Nigeria. AESON.

Swanson, B., & Rajalahti, R. (2010). *Strengthening Agricultural Extension and Advisory Systems: Procedures for Assessing, Transforming, and Evaluating Extension Systems*. ARD Discussion Paper 20. Washington, DC: World Bank

Viinamäki. (2004). A theory of coordination and its implication on EU structural policy: A comparative study of the challenges for coordination in structural funds in Finland, Ireland and Sweden. Acta Wasaensia 132. Administrative Science 9. Vaasa: University of Vaasa/Publication Unit.

Chapter 8
Making Agricultural Input Subsidies More Effective and Profitable in Africa:
The Role of Complementary Interventions

Joseph S. Kanyamuka
Lilongwe University of Agriculture and Natural Resources, Malawi

Charles B. L. Jumbe
Lilongwe University of Agriculture and Natural Resources, Malawi

Jacob Ricker-Gilbert
Purdue University, USA

ABSTRACT

The combined effects of declining soil fertility, continuous mono-cropping, poor crop residues management, and limited resources are considered the major constraints to increased crop productivity in Sub-Saharan Africa. It is for this reason that most African governments in Sub-Saharan Africa have been implementing farm input support programmes to boost smallholder production. While substantial amounts of resources are committed to support such programmes, evidence suggests that the increased use of modern inputs such as inorganic fertilizers on the main staple food crops appear to be only marginally profitable or even unprofitable. There is a renewed realization that the use of fertilizer input alone to raise farm productivity is likely to be impeded, if sufficient attention is not given to complementary interventions such as integrated soil fertility management technologies and extension services. This chapter provides evidence from several African countries on the role of complementary interventions in enhancing profitability, effectiveness, and efficiency with which farm inputs such as inorganic fertilizer and improved seed are applied.

DOI: 10.4018/978-1-5225-3631-4.ch008

INTRODUCTION

The combined effects of declining soil fertility due to low and inappropriate fertilizer use and inappropriate crop residues management, coupled with limited resources are considered major constraints to increased crop productivity in Sub-Saharan Africa (SSA) (Ngwira et al., 2012 Mhango et al., 2012 and Vanlauwe, 2015). This has been exacerbated by soil nutrient mining through continuous cultivation of crops (mono-cropping), especially maize thereby leading to subsequent reduction in the productivity of soils. To address the declining soil fertility, most African governments (such as Malawi, Kenya, Tanzania, Zambia Mali, Nigeria, Ghana, Senegal, Ethiopia and others) have embarked on farm input support programs to provide farmers with subsidized inorganic fertilizers and improved seed with the goal of helping the farmers break out of a low input/low output trap (Denning et al. 2009; Sachs 2012 & Ricker-Gilbrt & Jayne, 2014).

However, one of the major ironies emanating from applied research is that the greater use of modern inputs such as inorganic fertilizer appears to be only marginally profitable or even unprofitable to use on the main staple food crops (such as maize) in the region. There is an increasing realization that fertilizer input intensification alone to raise farm productivity growth is likely to be impeded if adequate attention is not paid to complementary interventions (e.g. Liverpool-Tasie et al., 2016). In this regard, Dorward and Chirwa (2013) noted that increased production is crucial to the achievement of programme objectives that results from incremental use of inputs (which mainly comprises inorganic fertilizer and improved seeds) that leads to increased yields due to high yield responses to the inputs which is dependent upon weather and the efficiency of input use and of crop production.

Malawi has been a leader in pioneering agricultural input subsidy programmes (ISPs) after well-publicized initial reports about the success of the first targeted input voucher program in 2004/05 that saw at least nine other African governments following suit. Kenya joined the ranks of sub-Saharan African (SSA) countries implementing targeted input subsidy programmes for inorganic fertiliser and improved seed in 2007/08 through the establishment of the National Accelerated Agricultural Inputs Access Programme (NAAIAP).

While agricultural input subsidy programmes have been implemented to enable farmers access chemical fertilizers and improved seeds needed to boost crop production and enhance household and national food security (Holden and Lunduka, 2010, Mason, et al., 2017), such programmes are largely implemented using government resources that take up largest share of total agricultural national budgets. Jayne and Rashid (2013) observes that ten African governments spend roughly US$1 billion every on input subsidy programs (ISPs), amounting to 28.6% of their public expenditures on agriculture. For instance, the Malawi Farm Input Subsidy Programme (FISP) dominated the overall agricultural budget, absorbing approximately 75 percent of the total value of the Ministry of Agriculture, Irrigation and Water Development (MoAIWD)'s financial resources in 2014/15 (World Bank, 2017). In 2011, the Government of the Republic of Zambia (GRZ) spent approximately US$184 million, equivalent to 0.8% of gross domestic product, to provide nearly 182,500 MT of fertilizer and 9,000 MT of hybrid maize seed to participating farmers through its FISP and between 2004 and 2011, Zambia's FISP accounted for an average of 30% of total GRZ agricultural sector spending (Mason et al. 2013).

This poses serious challenges regarding the sustainability of such agricultural input subsidy programmes in terms of cost of maintaining the programme every year (Dorward & Chirwa, 2011). Besides, reliance on chemical fertilizers alone, without paying attention to complementary interventions that are necessary to improve crop response to applied nutrients has adverse implications for achieving sustainable agricultural intensification. This is also compounded by the fact that majority of small-scale farmers that form the bulk of the farming population are resource-constrained and thus cannot afford inorganic fertilizer at market prices. Chirwa et al., (2012) noted that the benefits of fertilizer input subsidy programmes can be strengthened by complementary investments that promote higher responses to fertilisers, among others.

Therefore, there is an urgent need for integrating complementary investments and sustainable intensification strategies that can make agricultural input subsidies that have prominently featured in many African governments' agricultural development and food security strategies, more effective and profitable among the farming population. There is ample evidence, for example, that fertilizer use efficiency is greatly enhanced if it is complemented by other sustainable intensification strategies (e.g. Snapp et al. 2014; Vanlauwe et al, 2014 and Mather et al., *2016*). According to (Ollenburger, 2012 Sommer et al., 2013 & Vanlauwe, 2015), one of the sustainable intensification strategies that that have shown potential to raise the efficiency of inorganic fertilizer use and improve smallholder farm productivity growth while preserving the natural resource base is Integrated Soil Fertility Management (ISFM). While we acknowledge that there is myriad of factors that do affect profitability, effectiveness and efficiency of agricultural input programmes such as efficient input supply systems, this paper focusses on enhancing the profitability of such programmes through the achievement of greater nutrient use efficiency i.e. raising the grain output: N ratio. This is guided by the notion that the efficiency with which fertilizer nutrients are utilized by crops is strongly reduced by soil degradation.

The Concept of Integrated Soil Fertility Management (ISFM)

Fairhurst (2012) defined integrated soil fertility management (ISFM) as a set of soil fertility management practices that essentially includes the use of inorganic fertilizer, organic inputs and improved germplasm, combined with the knowledge on how to adapt these practices to local conditions aimed at optimizing agronomic use efficiency of the applied nutrients and improving crop productivity. The concept of ISFM embraces the combined use of appropriate interventions on soil management, fertilizer use and crop agronomy to drive the main outputs of increased yield and productivity.

Mhango et al. (2012) and Ollenburger (2012) & Vanlauwe (2015), note that integrating legumes and appropriate crop rotation in food production system is key to implementation of ISFM. Figure 1 depicts the concept of ISFM implementation aimed at maximizing the agronomic efficiency of applied inputs through the proper deployment of improved varieties, fertilizer, and organic inputs, adapted to local conditions.

OPPORTUNITIES AND CONSTRAINTS TO INCREASING FERTILIZER USE EFFICIENCY

In this chapter, we argue that any successful fertilizer policy should aim at promoting effectiveness (achieving full agronomic purpose), efficiency (improving the returns

Figure 1. Conceptual depiction of Integrated Soil Fertility Management implementation in practice. Source: Vanlauwe et al. (2014)

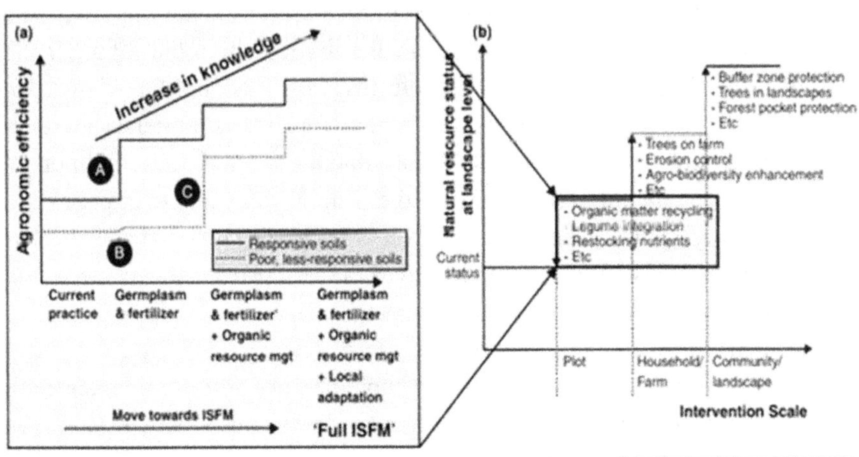

to fertilizer relative to costs) and sustainability (improving agricultural production and productivity while at the same time sustaining soil quality for future generations) (Dittoh et al., 2012). Achieving fertilizer use efficiency will involve improving crop response rates to applied nutrients, the price ratios of fertilizers to crop outputs, availability of transport and distribution as well as credit access among smallholder farmers.

There is now increasing realisation that fertilizer subsidies should not only address demand and supply constraints but also function in a transparent and cost-effective manner. In this regard, the development of commercial input markets coupled with efficient fertilizer distribution systems are critical. For instance, almost all the agricultural input subsidy programmes implemented in Africa have not been supported with effective crop-specific and location-specific fertilizer recommendations (Mutegi et al., 2015) and complementary extension to enable farmers' appropriate use of the inputs. As a result, the agronomic efficiency of Nitrogen (defined as unit of grain produced per unit of fertilizer N applied) is poor (e.g. 14 kg grain/kg N applied in Malawi and 11-20 kg grain/ kg N in Kenya, 7.5 kg grain/kg N in Nigeria) (Liverpool-Tasie et al., 2016; Malawi Government, 2008) and less than half the efficiency that can readily be achieved with good management. Mutegi et al. (2015) identified a number of gaps for increasing fertilizer use efficiency and the subsidy impacts on crop productivity. These include improving input distribution logistics to ensure timely delivery of the inputs, improving targeting of beneficiary households and supplying the correct fertilizers for different soil fertility conditions.

In a quest to improving the effectiveness of agricultural input subsidy programmes and their impact on agricultural productivity and production in SSA region, there has been an outcry for reforms, especially from the donor community. This has been partly necessitated by the notion that input subsidy programmes are crowding out the private sector. As such, one of the reforms implemented by the Malawi government has been increasing the participation of the private sector during the 2015/16 and 2016/17 season in the FISP especially in the distribution and retailing of fertilizer. According to World Bank (2017), the key reforms to the FISP in Malawi in 2016/17 include (i) increasing the share of fertilizer retailed by private suppliers; (ii) reducing the level of subsidy to beneficiaries by introducing a fixed value coupon with beneficiaries to pay the difference between the coupon value and market price; and (iii) implementing a pilot scheme that targets productive farmers in two districts. On the other hand, Kenya's case has been touted as the best private sector-led fertilizer market development programme in Sub-Saharan Africa.

One dimension to increasing fertilizer use efficiency in Africa will involve combining inorganic fertilizer with other complementary agronomic management practices for efficient nutrient utilization (Wallace and Knausenberger, 1997). Such practices include integration of nitrogen fixing legumes in crop rotations, residue retention in the soil, organic matter through the application of manure, water harvesting, and controlling soil erosion in site-specific integrated soil fertility management strategies. Other equally important agronomic management practices include timely fertilizer application following recommended instructions and quantities, timely weeding and its frequency, plant spacing, pests and diseases control and others. These requires that a prudent and sound extension support is in place to ensure that farmers are furnished with such knowledge and skills for appropriate implementation of the ISFM practices. However, most of the input subsidy programmes have not been fully supported with complementary extension services to ensure optimum implementation of the site-specific research-based ISFM recommendations among the farmers. This is the case with the FISP in Malawi, for example. As such, this has compromised the maximum benefits that can be derived from input subsidy programmes in most of the African Countries.

ISFM involves the combined use of appropriate interventions on soil management, fertilizer use and crop agronomy to drive the main outputs of increased yield and productivity. The introduction of interventions is affected by market economics and government policy. When introduced successfully, productivity is increased and less land is required to achieve a given level of production. The impact is the sustainable improvement of food security, increased farm incomes and lower food prices, which benefit the urban population (Fairhurst, 2012)

Figure 2. ISFM involves the combined use of appropriate interventions on soil management, fertilizer use and crop agronomy to drive the main outputs of increased yield and productivity. The introduction of interventions is affected by market economics and govern.
Source: Fairhurst (2012)

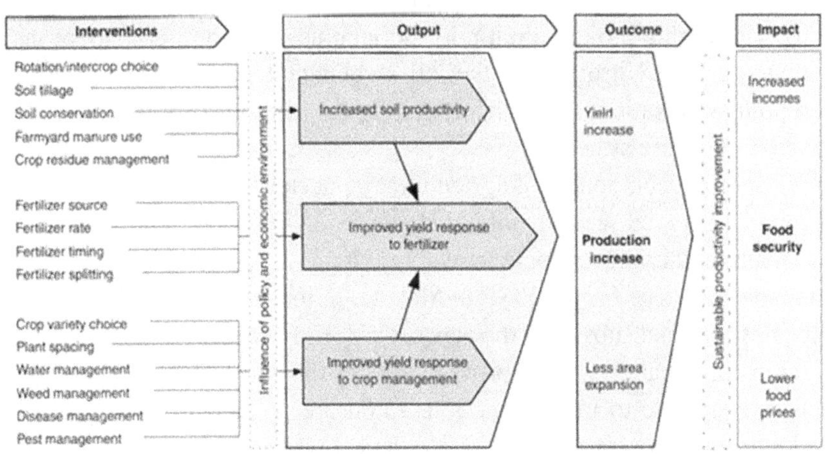

Scaling Up ISFM Technologies and Practices in Africa

While the benefits of ISFM practices in raising fertilizer use efficiency and enhancing crop productivity are widely acknowledged in literature (Marenya and Barrett, 2007; Fairhurst, 2012 and Lambrecht et al., 2014), their adoption among smallholder farmers remains fairly low (Kamau et al., 2013 and Menale et al., 2012). This is despite several promotional efforts by both governments and non-governmental organisations as well as international research institutions. This Chapter also draws farm-level study (Kanyamuka (2017) that explored the potential role that complementary interventions like ISFM practices can play in building sustainable food production systems. The study was aimed at addressing the following important research question: "to what extent does adoption of ISFM practices such as maize-legume intercropping and rotation enhance maize productivity among smallholder farmers?". Addressing this question is vital for understanding the basis for promoting ISFM practices that are deemed viable complementary interventions in raising the efficiency of inorganic fertilizer use and thereby improving crop productivity among small-scale farmers in Sub Saharan Africa (SSA) and beyond.

Understanding ISFM technology adoption process is vital for improving the design and implementation of any input intensification programme in the face of changing climate and population growth. One key aspect that has been widely recognised in increasing fertilizer use efficiency has been the integration of legumes in crop-based farming systems either in intercropping or rotation (Denning et al., 2009; Dorward and Chirwa, 2011; Fairhurst, 2012 and Vanlauwe et al., 2014). That is why the Lilongwe University of Agriculture and Natural Resources (LUANAR), formerly known as Bunda College of Agriculture with financial support from the McKnight Foundation Collaborative Crops Research Programme has been implementing a project titled "Legume Best Bets to Acquire Phosphorous and Nitrogen and Improve Family Nutrition and Soil Quality in the Northern and Central part of Malawi 2006/07. Under this project, multipurpose legumes (pigeon peas, soybeans, and groundnuts) that fix Nitrogen into the soil are being integrated with maize either in intercropping or rotation, while incorporating crop residues within the traditional maize-based farming systems.

One important component of agricultural input subsidies in Africa has been the inclusion of improved seed, mainly of staple food crops particularly cereal grains such as maize and improved legume seed (groundnuts, soy beans and pigeon peas). This is especially true for the FISP in Malawi. In improving the efficiency of and effectiveness of input subsidies, the concept of ISFM encourages the use of mineral fertilizer in combination with improved varieties that are pest and disease resistant as well as adaptable to local conditions to ensure greater responsiveness to the applied nutrients. Improved varieties also have higher harvest index (HI) because more of the total biomass production is converted into the harvested product than in unimproved varieties.

ECONOMIC AND PRODUCTIVITY BENEFITS OF COMPLEMENTARY ISFM PRACTICES

Economic analysis of combining inorganic fertilizer and improved seed with other complementary interventions has proved to be profitable. For instance, Lunduka and Kelly (2012) found that addition of the organic and inorganic fertilizer and intercropping with legumes increases net profit per hectare. Intercropping maize with pigeon peas also provides high net returns when combined with inorganic fertilizers. For example, fertilizing maize under a pigeon peas intercrop had the highest profit of US$ 305/ha which was about US$ 100 more than sole maize crop in in Tanzania, and intercropping maize with half the recommended fertilizer rate gave almost similar profit margins to sole maize with full fertilization (US$ 232/ha and US$ 231/ha respectively). In Kenya, trials indicated combining manure with inorganic fertilizer on maize gave an additional profit of US$ 180/hectare relative to applying sole fertilizer.

A study by Tchale and Saur (2007) in Malawi found remarkable gains in efficiency from using ISFM, with an average technical efficiency score of 91 percent among farmers who practiced ISFM options compared to only 14 percent among the farmers who used chemical fertilizers only. Further, a study by Amare et al. (2012)

Figure 3. Comparison of costs of inputs and profit among ISFM practices
Source: Lunduka and Kelly (2012)

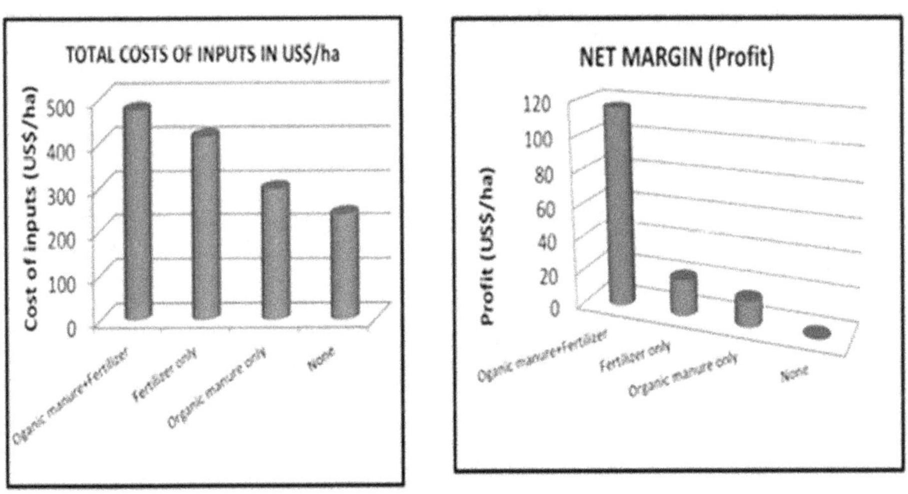

found that application ISFM practices had significant impact on household income and consumption expenditure. The study revealed an average income of maize and pigeon pea adopters of about 150% and 145% higher than had they not been adopting improved maize and pigeon pea, respectively, while for non-adopters, the mean total income per capita would have been increased by 36% and 100% had they been adopted improved maize and pigeon pea, respectively.

Still in Kenya, The Kenya Agricultural Research Institute (KARI) established 136 maize-legume intercrop demos in western Kenya in 2010 through support of AGRA that show-cased intercrops of maize with common beans, soybeans and groundnuts in four districts (Emuhaya, Kakamega, Mumias and Gem). Phosphorus fertilizer was applied at a rate of 20 kg P ha^{-1} at planting while nitrogen fertilizer was applied as top-dress at a rate of 60 kg N ha^{-1}. The control practice was establishment of intercrop of maize and legumes with no fertilizer. Across the four districts, improved cereal-legume intercrop technologies increased maize yield by between 2.8 and 3.3 tons/ha (representing 300%) as shown in Figure 4 (Mutegi and Shamie Zingore (2015).

Kanyamuka (2017) also found that ISFM technology adoption had a positive and significant effect on maize yields in Malawi with treatment effect for ISFM adopters of 0.1 translating to 10.52% increase in yield level for adopters while non-adopters would have increased their maize yield levels by 16.2% from the average had they adopted the ISFM technologies especially integration of legumes and maize-legume rotation in their maize-based cropping systems. Kanyamuka (2017) also found that adoption of ISFM practices such as inorganic fertilizer and improved seed as well maize-legume intercropping and inorganic fertilizer is complementary.

Figure 4. Effect of maize-legume intercropping on maize yield over three consecutive seasons in the farmers' fields
Source: Mutegi and Shamie Zingore (2015)

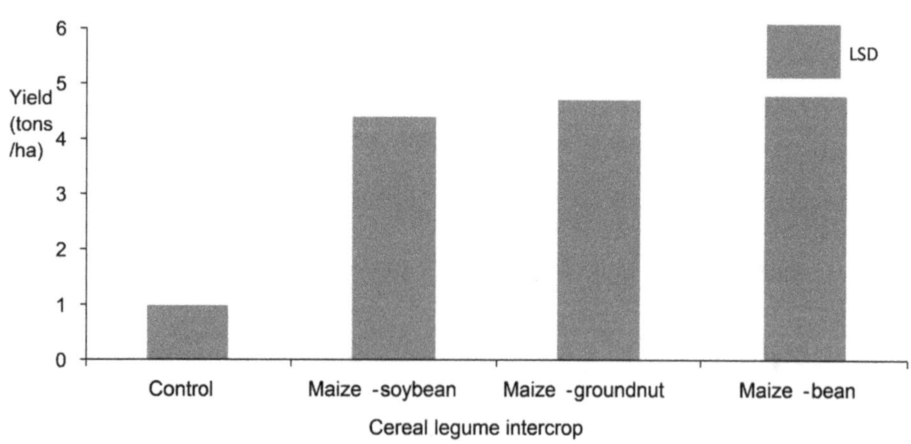

Such relationship between the ISFM practices raises important policy implications, suggesting that inorganic fertilizer use (as in agricultural input subsidy programmes) must not be promoted as a stand-alone practice for the management of soil fertility because the recommended fertilizer application rates are often beyond the reach of resource-constrained smallholder farmers. Therefore, combining the ISP package with complementary interventions such as legume seed with the aim of promoting maize-legume intercropping and rotations will help in raising the efficiency of fertilizer use by accelerating the release of nutrients into the soil while producing long-term benefits for sustainable crop production. Consistent with previous studies (such as Liverpool-Tasie et al., 2016, Marenya and Barret, 2007).

Further economic analysis carried out on data from 10 ISFM projects from across eastern, southern and western Africa yielded benefit-cost ratio values of more than 2 signalling a viable and financially attractive investment (Mutegi and Shamie Zingore, 2015).

While the benefits of ISFM practices in increasing fertilizer use efficiency are widely recognised in literature, the adoption of such practices needed to restore soil properties and improving crop productivity among small-scale farmers in Africa remains fairly low and incomplete. While the number of farmers practicing such legume technologies in the initial stages of the project looked promising, the trend of such numbers over time has generally been on the decline. The study by Kanyamuka (2017) explored several constraints to low uptake of ISFM technologies specially to do with legume integration in farming systems among small-scale farmers. As shown in Figure 5, the major constraint to legume integration has been limited availability and lack of access to improved legume seed. Given the fact that the majority of small-scale farmers are resource-constrained, they cannot afford to purchase the legume seed at market prices. Further, as opposed to maize seed, legume seed is not readily available for farmers use.

Table 1. Effect of ISFM on performance of different crops and financial attractiveness

Country	ISFM Intervention	Crop	Yield Change (tons/ha)	*Benefit-Cost Ratio
Kenya (Western)	Maize-Legume Intercrop	Maize	+4 (300%)	1.8 2.2
Uganda (Isingiro)	Improved seeds + fertilizer + crop rotation	Soybean	+1 (100%)	2.0-2.3
Tanzania (SHT)	Improved seeds + fertilizer + Maizelegume rotation	Maize	+4.5 (300%)	2.1-2.5
Ghana	Maize -legume rotations + improved seeds + fertilizer	Soybean	+1.5 (150%)	2.3-2.7

Source: Mutegi and Shamie Zingore (2015)

Figure 5. Reasons for dis-adoption of legume technologies among farmers
Source: Kanyamuka (2017)

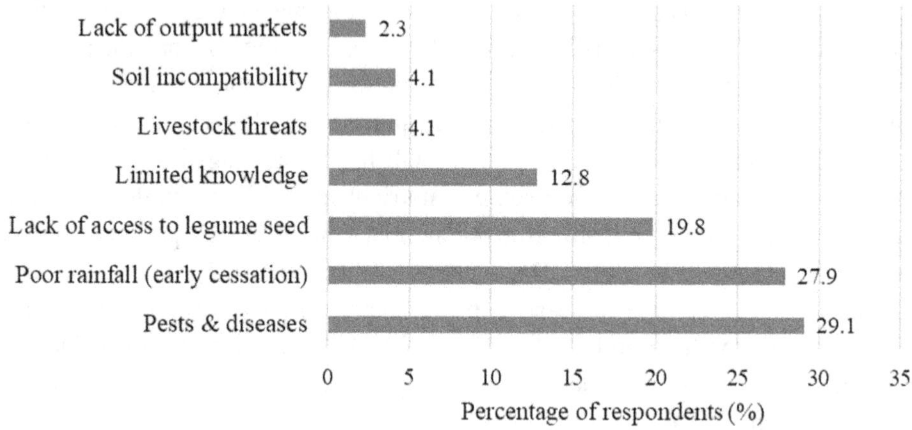

In addition, lack of access to both input and out market has been one of the major hindrances to increasing uptake of legume technologies among the farmers. Despite soil fertility improving benefits attributed to legumes integration, farmers are more driven by immediate cash incentives. This underscores the importance of developing market linkages among the farmers in incentivising adoption of legume technologies such as groundnuts, pigeon peas and soy beans. Vanlauwe et al. (2014) confirms that increasing the adoption of sustainable intensification strategies (such ISFM in which legumes are a key component) will among others, require access to profitable output markets for enhanced productivity growth. According to Sanginga and Woomer, 2009), maximum benefits from ISFM technologies can only be obtained within an enabling environment, where factors such as input and output markets are in place, coupled with functional service delivery institutions and progressive policies. Wiggins and Brooks (2010) confirmed that If market failures persist, farmers could become trapped into low levels of productivity despite existence of the technology and economic opportunity, since they cannot access and afford the seeds and other inputs to take advantage; and thus they remain trapped in poverty.

Other constraints included pests and diseases that attack legumes, limited landholding sizes that constrain crop rotation, tenure insecurity of land that limits long-term investments in soil fertility improving measures and limited extension support services. Denning et al. (2009) noted that delayed access to seed and fertilizer is a recurring complaint of farmers in Malawi and elsewhere in Africa that can be improved by: early tendering and contract signing, early coupons distribution and early stock replenishment to ensure timely application of fertilizer.

SMART INPUT SUBSIDIES AND COMPLEMENTARY INTERVENTIONS

Proponents of smart input subsidies have advanced that input subsidy programmes should be designed with clear exit strategies that put time limit on the support and ensure long-term sustainability (Baltzer and Hansen, 2011). Chirwa et al. (2011) identified potentially sustainable graduation pathways from heavy reliance on fertilizer subsidies. Among others, these include: (1) reduced requirements for purchase of previously subsidised inputs due to increased efficiency in use, (2) reduced requirements for purchase of previously subsidised inputs due to substitution by cheaper inputs, (3) access to low cost credit by poor beneficiary households for purchase of previously subsidised inputs. The first two directly relate to ISFM while third is more of support services needed to enable small-scale farmers' increase uptake of complementary ISFM practices.

There are strong complementarities between the main alternative ways of increasing agricultural productivity, in that high yield responses to inputs require research and extension not only into external input use but also into complementary soil fertility management methods. Rates of return on different investments are affected by these complementarities, and also by increasing and decreasing returns to scale and by the efficiency with which they are implemented. It may be best therefore not to look for the most efficient alternative investments, but for the most efficient alternative combinations or packages of investments.

Among the proposed processes, changes and requirements needed in enhancing the profitability of ISPs, reduced input prices will require efficient and competitive importers, supplier(s), transporters; and improved transport infrastructure while improved agronomy, complementary seed, inorganic and organic fertilisers and improved soil management coupled with investment in agricultural research and extension will translate to increased efficiency in input use. In addition, increased legume integration with rotational offers substitution by cheaper inputs coupled with robust legume seed supply, produce demand and markets; stable and reliable low maize prices and high maize productivity for transition before subsidy removal are all critical.

The empirical evidence presented in this paper suggest a number of factors that need to be considered in enhancing the efficiency and profitability of ISPs in Africa. In the first place, it is apparent that complementary investments in research and extension are needed to promote better farmers' implementation of ISFM practices (of which inorganic fertilizer, improved seed and improved management and agronomy constitute major components) for optimum benefits among farmers. In this regard, improved management practices such as timely weeding, timely planting, timely fertilizer application and appropriate rates are all necessary in enhancing yield responses to fertilizer nutrients. Also, soil testing is vital in informing site-specific fertilizer formulations and/or blends to address area specific soil nutrient deficiencies (Dorward and Chirwa, 2013).

CONCLUSION AND RECOMMENDATIONS

In conclusion, tying agricultural input subsidies to sustainable complementary intervention is crucial in improving their effectiveness and profitability. This can be implemented within the concept of Integrated Soil Fertility Management. Such intervention may include government's increased support to extension services to train farmers about ISFM, integration of legume technologies by scaling up access to legume seed in intercropping and rotation cropping systems, crop residue incorporation, increasing soil organic matter through the application of manure. This could contribute to substantial reductions in national budgets for input subsidies in the long run while at the same time maintaining high crop output with reduced amounts of inorganic fertilizer. A key policy recommendation from this chapter is that for input subsidies that include use of inorganic fertilizer and improved seed to sustainably increase food production, it should go along with complementary interventions such as ISFM practices.

REFERENCES

Asfaw, S., Shiferaw, B., Simtowe, F., & Lipper, L. (2012). Impact of Modern Agricultural Technologies on Smallholder Welfare: Evidence from Tanzania and Ethiopia. *Food Policy*, *37*(3), 283–295. doi:10.1016/j.foodpol.2012.02.013

Baltzer & Hansen. (2011). *Agricultural input subsidies in Sub-Saharan Africa. Evaluation study*. DANIDA.

Chirwa, E. W. (2011). *Dorward, and Mirriam Matita*. Conceptualising Graduation.

Chirwa, Dorward, & Matita. (2012). *Thinking 'Graduation' from the Farm Input Subsidy Programme in Malawi*. Policy brief number 3.

Denning, G., Kabambe, P., Sanchez, P., Malik, A., Flor, R., Harawa, R., & Sachs, J. et al. (2009). Input subsidies to improve smallholder maize productivity in Malawi: Toward an African Green Revolution. *PLoS Biology*, *7*(1), e1000023. doi:10.1371/journal.pbio.1000023 PMID:19175292

Dorward, A., & Chirwa, E. W. (2011). The Malawi Agricultural Input Subsidy Programme: 2005-6 to 2008-9. *International Journal of Agricultural Sustainability*, *9*(1).

Fairhurst, T. (Ed.). (2012). *Handbook for Integrated Soil Fertility Management*. Nairobi, Kenya: Africa Soil Health Consortium.

Holden, S., & Lunduka, R. (2010). *Too poor to be Efficient? Impacts of the Targeted Fertilizer Subsidy Programme in Malawi on Farm Plot Level Input Use, Crop Choice and Land Productivity*. Noragric Report No. 55.

Jayne, T. S., & Rashid, S. (2013). Input subsidy programs in sub-Saharan Africa: A synthesis of recent evidence. *Agricultural Economics*, *44*(6), 1–16. doi:10.1111/agec.12073

Kamau, M., Smale, M., & Mutua, M. (2013). *Farmer Demand for Soil Fertility Management Practices in Kenya's Grain Basket*. Paper Prepared for Presentation at the *Agricultural and Applied Economics Association (AAEA) and CAES joint meeting*. Washington, DC.

Kanyamuka, J. (2017). *Adoption of Integrated Soil Fertility Management Technologies and Its Effect On Maize Productivity: A Case of the Legume Best Bets Project in Mkanakhoti Extension Planning Area of Kasungu District in Central Malawi* (MSc. Thesis). Lilongwe University of Agricultural and Natural Resources, Lilongwe, Malawi.

Lambrecht, I., Vanlauwe, B., & Maertens, M. (2014). *Integrated Soil Fertility Management: From Concept to Practice in Eastern DR Congo*. Bioeconomics Working Paper Series. KU Leuven.

Lenis Saweda, O. (2017). Is increasing inorganic fertilizer use for maize production in SSA a profitable proposition? Evidence from Nigeria. Elsevier. *Food Policy*, *67*, 41–51. doi:10.1016/j.foodpol.2016.09.011 PMID:28413245

Lunduka, R., & Kelly, V. (2012). *Economic Benefits of Integrated Soil Fertility Management (ISFM)*. ASHC policy brief 1.

Marenya, P. P., & Barrett, C. B. (2007). Household-level Determinants of Adoption of Improved Natural Resources Management Practices among Smallholder Farmers in Western Kenya. *Food Policy*, *32*(4), 515–536. doi:10.1016/j.foodpol.2006.10.002

Mason, N. M., Jayne, T. S., Burke, W. J., & Ariga, J. (2016). *Agricultural Input Subsidy Programs in Africa: An Assessment of Recent Evidence*. MSU International Development Working Paper. Department of Agricultural, Food, and Resource Economics and the Department of Economics, Michigan State University.

Mason, N. M., Jayne, T. S., & Mofya-Mukuka, R. (2013). Zambia's input subsidy programs. *Agricultural Economics*, *44*(6), 613–628. doi:10.1111/agec.12077

Mason, N. M., Wineman, A., Kirimi, L., & Mather, D. (2017). The Effects of Kenya's 'Smarter' Input Subsidy Programme on Smallholder Behaviour and Incomes: Do Different Quasi-experimental Approaches Lead to the Same Conclusions? *Journal of Agricultural Economics*, *68*(1), 45–69. doi:10.1111/1477-9552.12159

Mather, D., Waized, B., Ndyetabula, D., Anna, T., & Minde, I. (2016). *The Profitability of Inorganic Fertilizer Use in Smallholder Maize Production in Tanzania: Implications for Alternative Strategies to Improve Smallholder Maize Productivity*. Guiding I investments in Sustainable Agricultural Intensification in Africa/Tanzania Working Paper #4.

Mhango, W. G., Snapp, S. S., & Phiri, G. Y. K. (2012). Opportunities and Constraints to Legume Diversification for Sustainable Maize Production on Smallholder Farms in Malawi. *Renewable Agriculture and Food Systems*, *28*(3), 234–244. doi:10.1017/S1742170512000178

Mutegi, J., & Zingore, S. (2015). *Closing Crop Yield Gaps in sub-Saharan Africa through Integrated Soil Fertility Management*. IPNI-Sub-Saharan Africa Program.

Nakhumwa, T. O. (2004). *Dynamic Costs of Soil Degradation and Determinants of Adoption of Soil Conservation Technologies by Smallholder Farmers in Malawi* (PhD Thesis). University of Pretoria, South Africa.

Ngwira, A. R., Kabambe, V. H., Kambauwa, G., Mhango, W. G., Mwale, C. D., Chimphero, L., & Mapfumo, P et al.. (2012). Scaling Out Best Fit Legume Technologies for Soil Fertility Enhancement among Smallholder Farmers in Malawi. *African Journal of Agricultural Research*, *7*(6), 918–928. doi:10.5897/AJAR11.760

Ollenburger, M. H. (2012). *Modelling Integrated Soil Fertility Management Options in Malawi under Variable Soil and Climate Conditions* (MSc Thesis). Michigan State University.

Snapp, S., Jayne, T. S., Mhango, W., Benson, T., & Ricker-Gilber, J. (2014). *Maize Yield Response to Nitrogen in Malawi's Smallholder Production Systems*. IFPRI Working Paper 09. Malawi Strategy Support Programme.

Sommer, R., Bossio, D., Dester, L., Dimes, J., Kihara, J., Koala, S., & Winowiecki, L. et al. (2013). *Profitable and Sustainable Nutrient Management Systems for East and Southern African Smallholder Farming Systems-Challenges and Opportunities: A Synthesis of the Southern and Eastern African Situation in Terms of the Past Experiences, Present and Future Opportunities in Promoting Nutrient Use in Africa*. CYMMYT.

Tchale, H., & Saur. (2007). Soil Fertility Management and Maize Productivity In Malawi: *Curvature Correct Efficiency Modeling And Simulation. AAAE Conference Proceedings*, 151-163.

Vanlauwe, B. (2015). *Sustainable Agricultural Resources Management: Unlocking Land Potential for Productivity and Resilience. Background Paper*. Nairobi, Kenya: United Nations Economic Commission for Africa.

Vanlauwe, B., Coyne, D., Gockowski, J., Hauser, S., Huising, J., Masso, C., & Asten, P. V. et al. (2014). Sustainable Intensification and the African Smallholder Farmer. *Current Opinion in Environmental Sustainability*, *8*, 15–22. doi:10.1016/j.cosust.2014.06.001

Wallace, M. B., & Knausenberger, W. I. (1997). *Inorganic Fertilizer Use in Africa: Environmental and Economic Dimensions. Winrock International Institute for Agricultural Development and Division of Productive Sector Growth and the Environment Office of Sustainable Development Bureau for Africa U.S. Agency for International Development with the Environmental and Natural Resources Policy and Training (EPAT)* Project.

Chapter 9
Rice Production and Processing in Ogun State, Nigeria:
Qualitative Insights From Farmers' Association

Evans S. Osabuohien
Covenant University, Nigeria

Uchechukwu E. Okorie
Covenant University, Nigeria

Romanus A. Osabohien
Covenant University, Nigeria

ABSTRACT

This chapter examines the importance of indigenous institutional arrangements in rice production and processing activities in Ogun State, Nigeria through the use of key informant interviews (KIIs). Analyses from the study show that agricultural financing constitutes the greatest challenge that affects rice production and processing. Other findings from the discourse reveal that in some rice producing areas where there is the existence of rice farmers' clusters, there is access to modern rice processing machines such as winnowers, threshers, and destoners. The operations of Rice Growers Association of Nigeria (RGAN) in Ogun State are coordinated by the executive committee, which constitute the indigenous institutional arrangement. This chapter recommends that sincere and concerted efforts on the part of the government in implementing the goals of agricultural transformation agenda be

DOI: 10.4018/978-1-5225-3631-4.ch009

made to engender the welfare of rice farmers through the development of the rice value chain. The need for actively involving the rice farmers through the RGAN is also germane. The opportunities identified in the study include: having "pool of land" RGAN that enables the rice farmers to form clusters and increase production, creating platform through which the rice growers could leverage to facilitate access to inputs and technical support, and gaining of visibility and market access to enhance returns on their farming endeavours.

INTRODUCTION

Given Nigeria's discovery of crude oil in commercial quantity, agriculture that was once the prime mover of Nigeria's economy suffered great neglect. The resultant effects include: sharp decline in agricultural production, increasing reliance on food import to meet the domestic demand, and so on. In effect, Nigeria is currently one of the largest food importers in the world with annual food import bill of about $10 billion (Obayelu, 2015). In 2016, the quantity of local rice production in Nigeria was estimated at 4.8 million tonnes (Food and Agricultural Organization-FAO, 2016). This further conforms to an earlier assertion by Ogunsumi, Ajayi, Amire and Williams (2013) on the gap between local demand and supply of rice in Nigeria. The author stressed that the level of rice consumption in Nigeria increases with about 10 percent per annum as a result of changing consumer preferences amidst other factors. The low productivity of rice farmers is occasioned by the use of low technologically empowered agricultural equipment which do not support large scale production. For instance, Fasoyiro and Yaiwo (2012)[1] observed that in Nigeria, rice is mainly produced by small-scale farmers whose production are characterised by low output resulting from production inefficiency, aging farming population, low technological know-how, and so on. In the same vein Uduma, Samson and Mure (2016) noted that the inability of local supply to meet up with rice demand (consumption) has given rise to the high import of rice in Nigeria. According there has been a phenomenal rise in imports of 300 thousand tons annually in recent times which on the average with an estimated cost of 300 million naira annually in foreign reserves. They further stressed that aside from the huge cost to the Nigerian economy, rice imports exposes the country to international market shocks with its associated risk implications on food security.

This is contrary to what obtained a foretime where Nigeria used to be the largest producer of a number of crops including rice in West Africa (Okoruwa, Jabbar & Akinwumi, 1996; Obayelu, 2015). However, in 2010 there was somewhat re-awakening in the agricultural sector with the launch of the Agricultural Transformation Agenda (ATA). The ATA was part of the Federal Government of Nigeria's (FGN) effort to revamp the sector in order to enhance food security, job creation and diversification of the economy (Osabuohien, 2014; Obayelu, 2015; Osabuohien, 2016). The transformation agenda as enshrined in ATA is set to create over 3.5 million jobs from rice, cassava, sorghum and cotton value chains, with more jobs to come from other value chains when fully implemented. Generally, ATA, among others, has the goal of re-defining agriculture as a business by promoting the involvement of private sectors, encouraging the expansion of private sector driven marketing organisations, and promoting incentive-based risk sharing for agricultural credits (Osabuohien, 2016; Okodua, 2017; Osabohien, Osabuohien & Urhie, 2017).

Rice is one of the crops being considered under the FGN's ATA given its growing importance and prominent role among staple food crops in Nigeria. The country has a history of indigenous rice production and high demand (Johnson, Takeshima, & Gyimah-Brempong, 2013).Thus, it is not surprising that rice has emerged as a major staple food crop in Nigeria, given its demand in all the six geopolitical zones, 36 States, all the Local Governments, and across all socio-demographic groups (Gyimah-Brempong, Johnson & Takeshima, 2016). The increasing domestic demand for rice in Nigeria has been attributed to consumer preferences, increasing incomes, rising urban population, among others (Nwanze *et al.*, 2006).

In response to the above challenges, the FGN's policy on agriculture detailed in ATA offers some new incentives with a view to encouraging domestic import substitution in agriculture especially rice production. Some of the efforts include: encouraging private sector investment in agriculture through the removal of restrictions on areas of investment and maximum equity ownership in investment by foreign investors; free transfer of capital, profits and dividends (that is, no currency exchange controls); guarantees against investment expropriation; total removal of import duty on agricultural machineries and tax holiday for agricultural investments (Adesina, 2012; Okodua, 2017)[2]. The above highlighted incentives have resulted in private investors coming into the production of rice in Nigeria. Some of them include: Flour Mills of Nigeria, Bidda-Badeggi, Niger State; Ebony Rice, Ikwo, Ebonyi State; and Dominion Farms, Gassol, Taraba State (Okodua, 2017). A closely related development is the Memorandum of Understanding (MoU) between Dangote Industries Limited and the FGN worth US$1 billion investment in commercial rice farming and modern integrated rice mills, which was signed in August 2014. It is reported that farmlands have been acquired in Edo, Jigawa, Kebbi, Kwara, and Niger States with acreage of about 150,000 hectares as well as the proposed establishment of two large-scale rice mills with capacity of 240,000 metric tons of rice paddy (Okodua, 2017). These private and government participations are expected to encourage local content and value chain development. This will further improve foreign exchange position by reducing of the importation of finished food products such as rice, which is a significant component food imports.

It is based on the foregoing that this study assessed the rice production and processing in Nigeria using Ogun State which is one of the major rice producing States in the country. The methodological approach adopted in achieving this objective is the qualitative technique using Key Informant Interviews (KIIs). The next section briefly discusses rice development in Africa and Nigeria while the information on Ogun State and the Local Government Areas where rice is produced are contained in the Third Section. The theoretical framework is encapsulated in Section four; while Section five contains the level of rice production, traditional versus modern rice processing as well as the constraints to rice production and processing. The last section contains the conclusion and recommendations made from the study.

BACKGROUND

Rice Development in Africa

The major engine for growth in Africa has been Agriculture. However, since majority of the African farmers practice subsistence or small holders farming characterised by low skilled labour force and family units, there is a high incidence of yield gaps, in addition to poor soils and other obstacles to sustainable farming incomes (Gyimah-Brempong, Johnson & Takeshima, 2016). Harold and Tabo (2015) also noted that rice is the single most important source of dietary energy in West Africa and third most important for Africa as a whole. It is evident from the study that despite the increased in local rice production there is still the persistence of the shortage of local production compared to the excess demand for the commodity (Harold & Tabo, 2015; Gyimah-Brempong, Johnson & Takeshima, 2016).

With the fast growing population and the rising food demand, it is important that African continent graduates from the level of food shortage to food surplus. This could be realised by making the challenges confronting the agricultural sector a major priority which must be vigorously and earnestly resolved through strong determination from the political class, exhibiting the right mind set towards agriculture by the private and government bodies and incorporation of youth and women in agriculture. Through this platform the Africa rice framework for Africa's Agricultural Transformation Agenda could be adequately embraced and implemented in various states, agricultural establishments and agencies (Harold & Tabo, 2015).

In previous years efforts have been made by African countries to reposition rice production. For instance, one of the major outcomes of the 'Abuja Food Security Summit' by the African Union in 2006 was the conferment of "region-wide strategic commodity" to rice status. Whilst this confirms the important position of rice in the agricultural sector, it also creates opportunities for positioning it as an important commodity in the secondary and tertiary agricultural sectors (Harold & Tabo, 2015). Rice is, therefore, a priority crop in the implementation of the New Partnership for Africa's Development-NEPAD and Comprehensive Africa Agriculture Development Programme-CAADP (Harold & Tabo, 2015). Consequently, Senegal had launched a national programme for rice self-sufficiency with the objective of increasing production from 215,000 tonnes in 2007 to 1.5 million tonnes of paddy in 2015. Similarly, Mali has started an extensive rice-promotion programme aimed at increasing production by 50% in a cropping season in 2009[3]. Other African countries (such as Cameroon, Ghana, Kenya, Madagascar, Mali, Mozambique, Nigeria, Senegal, Sierra Leone, Tanzania and Uganda) under the umbrella of the Coalition for African Rice Development (CARD) have introduced National Rice Development strategies (NRDS). This programme is aimed to double Africa rice production by 2018 compared to the level of production as at 2008 (Harold & Tabo, 2015).

Rice Sector Development in Nigeria

The Nigerian government is not left out as it has pursued and implemented various agricultural policies at the State and Federal levels on the rice transformation agenda to boost Nigeria's rice production over the years. Among these is the Agricultural transformation agenda (ATA) with the success recorded in local rice production of 4.8 million tonnes per annum (FAO, 2016). Harold and Tabo (2015) further noted that similar rice-sector promotion programs have been embarked upon in other African countries like Ghana and Côte d'Ivoire. These align with the ECOWAS Agricultural Policy-ECOWAP (Olayiwola *et al.*, 2015). Given the rise in food consumption (rice inclusive), some have argued that the production of rice in large quantities (that is, large-scale) should be considered as one of the major ways of ensuring food security for the teaming population in Nigeria (Herrmann, Jumbe, Bruentrup and Osabuohien, 2017; Osabohien, Osabuohien & Urhie, 2017). Others hold contrary view, stressing the need to empower small-holder farmers. Against that backdrop, Juliano (2016) stated the importance of rice over other crops, in terms of its total production in the developing countries and the number of consumers that are dependent on it as a staple food. This has also been stressed by Gyimah-Brempong, Johnson and Takeshima (2016). While Umeh, Joshi and Ukwungwu (1992) discussed that a holistic, broadly based, multidisciplinary pest-management research approach is required due to the immense benefits that integrated pest management (IPM) can provide.

Another important aspect of rice production that requires attention is the issue of technological advancement in rice processing since it has be observed that most of the processes utilised by the rural rice farmers are mostly traditional that are both labour intensive and time consuming. With higher level of technology, the farmers will be able to achieve a higher volume of yield with the best quality of products that will enhance consumer preference for locally produced rice. Technological advancement in the production and harvesting will promote commercialization and profitability of the rice production. Application of modern technology in the production and processing will further guarantee a better packaging of local rice to make it more appealing to consumers and will attract more buyers of the product. The use of modern harvesters suitable to our own ecology will further enhance the standard of the rice production process by reducing the rate of breaking and eliminating contamination by stone and shafts. Kareem (2016) has pointed out that the major obstacle facing the attainment of the potential benefits of agricultural production in many African countries is inadequate science and technological advancement.

Adewumi, Olayanju and Adewuyi (2007) observed that rice production and processing are profitable ventures in Nigeria and what is required is to encourage investment in rice processing activities. Aside the nutritional value of rice and high inclination of people towards its consumption, the by-product of rice could serve as a source of energy generation for domestic purposes. This could serve as a source of biofuel for cooking especially in rural settings where most of the rural dwellers could not readily afford the cost of kerosene or gas for cooking and heating purposes. Yan, Ngadi and Kok (2016) in their study stressed that rice generates large amount of by-products that could be used to produce energy and reduce the amount of firewood required to meet the daily cooking needs. This is crucial in Nigeria where rural dwellers use local means of cooking such as firewood and charcoal. The connotation of the above is that modern processing of rice at the milling centres could help in preserving the rice hub which servers as firewood to the locality thereby reducing the cost of buying kerosene for cooking.

For optimum output to be attained in rice production, it is necessary that ecological consideration be factored in the production process, especially in the choice of land as well as the typological components of the area. In this respect, the method of land preparation plays a significant role in the rice production process. In relation to this, Amb and Ahluwalia (2016) observed that zero tillage in rice-wheat cropping system could have major benefits, such as: improved water usage efficiency, reduced investment cost, higher yield, reduced weed population and a positive environmental effect. In production system with no-tillage or conservation tillage, the crop residues are buried in the soil and thus, the release of *allelochemicals* from both the growing plants and residue decomposition might act synergistically. This is because rice fields have versatile *ecotones* that comprise aquatic habitats as well as dry lands and a large group of biodiversity. Other significant factors that affect rice production include: weeds and pests and diseases infestations. Thus, the control method employed in the rice farm and the timing of the weeding is of essence to prevent its devastating effects in the rice farm development.

Rice Producing Local Government Areas in Ogun State

The focus of this study is on Ogun State, Nigeria where a number of Local Government Areas (LGAs) engage in rice production. Ogun State, created 3rd January 1976, is one of the 36 States in Nigeria, located in South-West Geopolitical Zone. It is bordered by Lagos State to the south, Oyo and Osun States to the north, Ondo State to the east and the Republic of Benin to the west. The map of Ogun State highlighting the rice producing local government areas (LGAs) is depicted in Figure 1. Abeokuta is the capital and the largest city in the State. Besides Abeokuta, other popular cities and towns in Ogun State are: Ijebu Ode, Sagamu, Ijebu Igbo, Ilaro Ayetoro, and Ota.

With regards to agricultural activities in Ogun State, available statistics indicate that the population of the people living in the rural areas, where most of the agricultural engagements take place is approximately 3 million. The number of farming households is about 360,000 persons, which comprises an average family size of 4.8 persons. The total arable land in Ogun State stands at 1,204,000 hectares representing nearly 74 percent of her total land area. Out of the above total arable land, about 350,000 hectares is presently cultivated, which constitutes 29.07 percent of arable land area. The general vegetation cover comprises: rain forest, swamp forest and derived savannah (Ogun State Government-OSG, 2016). The weather and climatic conditions typically follow the tropical pattern with rainy or wet season starting in March and runs till November, which is immediately followed by dry season from December to February. The annual average rainfall ranges between 105 cm (in the northern part of the State) and 128 cm (in the southern areas of the State). Thus, it is not unexpected that the major crops grown in the State include the following: cashew, cassava, citrus species, cocoa, yam, coco-yam, cotton, kola nut, maize, oil palm, pineapple, rice, rubber, sugar cane, and vegetables. They form the main stable food as well as provide raw materials for food processing industries (OSG, 2016; Osabohien *et al.*, 2017).

As depicted in Figure 1, eight out of the 20 LGAs in Ogun State, representing 40 percent, are involved in rice production. These LGAs include: Abeokuta North, Egbado North, Ewekoro, Ifo, Ijebu-North, Ikenne, Obafemi Owode, and Ogun Waterside. In some of these LGAs, there has been the establishment of rice mill that is possibly expected to make Ogun State a major player in rice sector *revolution* in Nigeria in the nearest future, *ceteris paribus*.

THE THEORETICAL FRAMEWORK: RGAN

This study follows a conceptual framework that is built on the theory of New Institutional Economics (NIE) where the role of institutions is considered essential

Figure 1. Rice Producing LGAs in Ogun State
Source: The Authors'

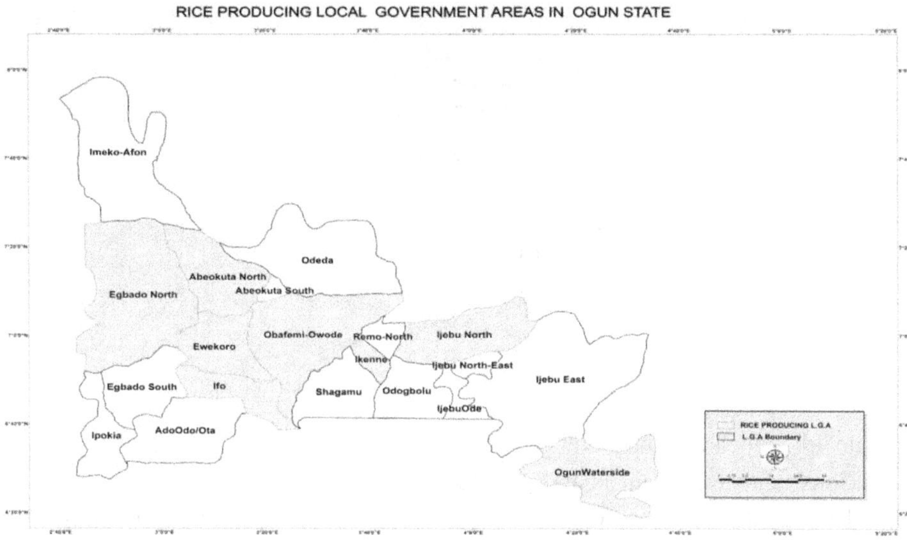

in economic activities and relationships involving economic agents. In general terms, institutions can be formal and informal. While the formal institutions entail well-documented principles that guide and guard the affairs of economic actors in a society, the informal (also called traditional, local or indigenous) institutions comprise moral codes, values, norms and conducts (usually unwritten) that influence the way people or group of persons (for instance a community or an association) are coordinated or organised (North, 2005; Osabuohien & Efobi, 2013; Osabuohien, 2014; Asongu, 2016; Nyarko, 2017). In this context, the indigenous institutions comprise the community leaders who can act on behalf of the members of the communities. The community and/or association leadership is taken as the custodian of the customs, beliefs, norms and values in such communities and association.

Following the approach of local institutional framework as articulated for Nigeria (Osabuohien, 2014) and for Uganda (Osabuohien, Efobi, Gitau & Bruentrup, 2015), this study examines the operations of a typical agricultural association (Rice Growers Association of Nigeria (RGAN) in Ogun State), which makes it quite unique unlike the general concept of non-governmental organisations (NGOs). Osabuohien (2014) enunciates how the existence of NGOs can influence or not the likelihood of Large-Scale Agricultural Land Investments (LALIs) in Nigerian communities. Whereas Osabuohien *et al.* (2015) expounds the case of Uganda to investigate how NGOs can affect the possibility of land investments taking place in a community as well as how such land investments influence the level of community's amenities notably:

Figure 2. The Pyramid of RGAN Membership
Source: The Authors'

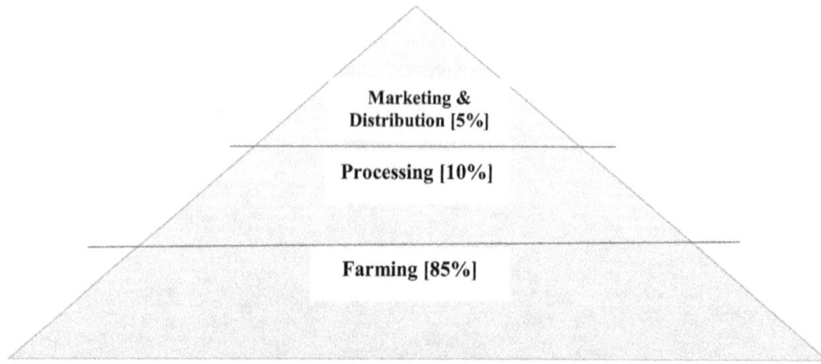

road, health, education and water provision. Thus, this study extends the frontier of institutions-land investment nexus by exploring how RGAN as an informal institution can influence the activities of the members with respect to rice production and processing using the case of Ogun State, Nigeria.

From the foregoing, RGAN can be seen as an informal (local, traditional or indigenous) institutions that is distinct as it agglomerates individuals that are connected directly to rice value chain ranging from production, processing, distribution and marketing of rice (paddy and processed). In effect, the RGAN[4] membership is opened to interested individuals that are involved in: a) rice farming; b) rice processing; and c) rice marketing to distribution as shown in Figure 2. RGAN was established in Ogun State in the year 2000, and as at May, 2016, it has grown to a membership of about 5000. Though membership is voluntary to all of the three aforementioned categories, members usually contribute the sum of Two Thousand Naira (N2, 000) per annum as membership due (levy) to help in running the affairs of the association as well as maintaining their membership status.

The operations of RGAN in Ogun State are overseen by an executive committee made up of nine persons including the Chairperson, Vice Chairperson and Secretary. The Chairperson was very instrumental in establishing RGAN in Ogun State (and is presently the National Deputy Chairperson South of RGAN). He has held the office since its inception and he will join the league of Octogenarian before the end of 2016. He devotes his Wednesdays (at times in conjunction with other executive members) to attend to members who come to see him at his residence located in Obafemi Owode LGA of Ogun State. Apart from having rice farms, few of the members have rice processing machines, such as: thresher, winnower and destoner.

The average farm size of the members is about 2 hectares. Thus, connoting that rice farming Ogun State is dominated by small-scale farmers. Based on needs and discussions with the members, the leadership of RGAN also raises issues of common interest to the government usually through Agricultural Development Programme (ADP) and the Ministry of Agriculture, which is the organ of the government that helps in coordinating policies that are related to food and agriculture (OSG, 2016). As shown in Figure 2, majority of the members of RGAN are into rice cultivation while the least are in marketing and distribution. A number of those in rice cultivation are also into processing, though many use traditional processing methods. In sum, there are far more members in the farming level, followed by processing and then those in marketing and distribution.

RICE PRODUCTION AND PROCESSING IN OGUN STATE

Status of Rice Production

In recent times, rice has been identified as one of the six major cash crops that Ogun State has comparative advantage in producing, and therefore has started drawing attention. The other cash crops apart from rice are: cassava, cocoa, cotton, kola-nut, and oil palm. In spite of the preparation and adoption of Ogun State's document on cash crop policy, there have not been significant improvements in the level of rice production in the State. Current estimates suggest that Ogun State's rice output revolves between 15,000 and 20,000 tons per annum. The area under cultivation is about 12,000 hectares which accounts for a share of 0.7% of national area cultivated to rice (Onabanjo, 2011).

From the interview conducted, it was observed that some of the rice farmers come together to form clusters with a view to creating a kind of self-assistance and boost their production level. Each farmer within these rice farmers' clusters usually cultivates an average of 2 hectares of farmland per planting season. The major reason behind this idea is traceable to the capital intensive nature of cultivating large hectares of rice farm which is beyond the reach of many small-scale farmers. Hence, they make effort to pool resources such as renting tractors together to during land preparation. Some of the respondents were of the opinion that the replication of such rice farmers clusters in different communities within the State will increase the level of rice production. For instance, a farmer needs a total of about 10 workers per hectare to work during the process of rice production, harvesting and processing. According to the Deputy National Chairperson of RGAN, there is very low availability of farming labour resulting to high labour cost and aging farming population. The reasons are not far-fetched, the process is both unattractive to the youth and there is low level of '*tractorisation*'. This is worsened by the fact that the method of rice production by the peasant farmers is labour intensive and characterised by drudgery.

The above necessitates the essence of embarking on a research of this nature, given the vast amount of resources and opportunities Ogun State has for rice production. These opportunities include: huge ready market for rice paddy; growing market for *OFADA* rice, available land for lowland rice cultivation; availability of land for expansion of upland rice cultivation; rich human resource; high yielding rice varieties; and increasing numbers of agricultural service providers. In addition, it has been emphasised that the establishment of the *Ofada/Veetee* rice mill in Itori, Ogun State can strategically position the State as a major player in rice production. With an installed capacity of 75,000 tons per annum, it is expected to expand up to 200,000 tons per annum at full capacity, which will substantially boost the market for paddy rice in Ogun State. This has the potential of increasing food supply, income of the farmers, employment as well as reducing Nigeria's high demand for foreign exchange, strengthening of her currency, and reducing inflation, and so on (Onabanjo, 2011; OSG, 2016).

Rice can be said to be an increasingly important crop in Ogun State (as in other rice producing States such as: Ebonyi, Osun and Niger) as it has become part of the staple food items that people consume as daily diet.

In some areas (e.g. *Ofada*[5] village in Obafemi Owode LGA) of Ogun State, there is a long tradition of rice cultivation. Regarding the rice varieties grown in Ogun State, some of them are considered traditional varieties, others are recently introduced two decades ago usually grown in paddies or on upland fields, depending on the particular variety. It has also been reported that, in the last 20 years, new varieties of rice are produced and disseminated by research institutes, or imported from Asian countries -. Consumption of *Ofada* rice has gained more prominence possibly due to the health awareness of many Nigerians who relate its taste and natural flavour, higher nutritive value to the significance on their health compared to imported rice varieties. The *Ofada* rice variety has been reported to have higher fibre content and better health consideration. Consequently, it now attracts high market price given its relative scarcity compared to other rice varieties including the imported ones.

Traditional vs. Modern Rice Processing[6]

Generally, the paddies of the rice plant are first milled to remove the chaff by subjecting it to steam or parboiling. Raw untreated rice may be ground into flour for many uses such as making of beverages (alcoholic or non-alcoholic), rice flour noodles and food items (*tuwo*). Processed rice seeds may be boiled, steamed or further fried in cooking oil before eating. When combined with milk, sugar, and honey, it is used to make rice desserts. Rice flour and starch are often used in batters and breading to increase crispiness (Ogunsumi *et al.*, 2013). Thus, there are opportunities for value chain in rice production in Ogun State which can lead to job creation and enhanced income for households. There is also the possibility of attracting new industries that will specialise in the processing of rice into different forms suitable for variety of finished products. In addition, through these feeder industries there will be more rice products that will create its own demand and thereby increasing demand for rice.

The traditional methods of processing rice paddy involve: soaking of the paddy in water for 2-3 days; steaming of the soaked paddy for 5–10 minutes and dried in the sun; pounding the dried paddy in a mortar and pestle to remove the husk or use of simple milling machines; then the grain is cleaned using a winnowing basket (Ajala & Gana, 2015). The traditional method of rice processing is simple, but it is tedious, leads to breakage of rice kernels and incomplete removal of husks, and has short storage life. In modern methods, the rice is first cleaned to remove contaminants, and the husks are then removed by a machine called shellers. The shellers are commonly horizontal spaced rotating abrasive stones, but increasing use is being made of rubber roll or rubber belt. The rice and hulls are separated by aspiration and any paddy remaining with the rice is removed in a paddy separator.

The main problem of Nigerian rice (especially those processed through traditional method) is the presence of stones in the rice grains. The responses from the interview conducted with the rice farmers revealed the major sources of the stones were traditional process such as harvesting, sun-drying, and so on. It starts with the parking of paddy rice from the ground after harvesting; then through sun-drying of the paddy after parboiling where the paddy is spread over a mat (or detached sack) on the ground; and parking of the paddy after it has been sun-dried, where some of the paddy that falls off are inadvertently parked together with the pebbles (stones). In some cases, there are perforations in the mat which allows stone penetration and mix-up with the paddy. In the words of the farmers, 'stones do not grow with the rice; it is the weakness of humans in harvesting and processing that introduces the stones'.

In Ogun State, modern stone removal machines (known as destoner as shown in Figure A3 in the Appendix) are located in places like Moloku Asipa in Obafemi Owode LGA, Ijebu, Imuwe, Ago Ijebu, Kobape, Obada-Oko, Ifo, Lafenwa and Iboro, as revealed during KII with the farmers. Whereas the older version of the machine can destone at the capacity of less than 1 ton per day, the modern destoner has the capacity of 2 tons per day equivalent to 250 kg per hour. The modern machines are located where there are cluster of rice farmers in large numbers with a view to having regular patronage. There are slight variations in the machines as it was reported that the type of machine located within a specific area should be compatible with the type of grains produced in that locality.

Another improvement in rice processing in Nigeria is the drying process where the traditional sun-drying can be replaced by mechanical dryer, which can process about 3 tons and remove 50 percent moisture of rice in six hours. However, there is still some demand for the locally processed rice as the farmers interviewed stated that the soaking of paddy rice can be optimal at 4 days, which brings out the aroma (*odour*) that makes the *Ofada* variety unique in the market. One of the RGAN members who is involved in rice farming and processing recalled an event when the paddy was soaked for only 2 days and the resultant effect was shocking: 'one of the distributors returned some bags of 50kg rice on the account that the consumers rejected them because they did not have the usual aroma'.

Major Constraints to Rice Production and Processing

This section presents the summary of the key challenges facing rice production and processing in Ogun State. They are numbered according to the severity as identified by the respondents. Some suggestions on how to ameliorate them are also provided.

Financial Constraint

The challenges encountered by farmers in rice production cuts across the value chain. However, the most outstanding challenge was finance for rice farming. In every segment of the rice production process, it has been observed that finance played a significant role in the development of rice processing in Ogun State. Majority of the rice farmers are aged above 50 years and they serve as the major source of labour as most of the young working population prefer white collar jobs. Consequently, the cost of farm labour has become very expensive making it very difficult for an individual farmer to hire. Hence, before any farmer can carry out rice production beyond the subsistence level, there will be a need for financial support from the government, private investors, individuals, agric-scientist and non-governmental organisations (NGOs). For instance, the government could assist the farmers by procuring group of interested farmers at reduced cost to facilitate farm mechanisation and commercial production. Thus, clusters of rice farmers could afford to hire these tractors to clear large hectares of land

In line with the above insights, our respondents affirm that inadequate finance constitutes the major challenge in rice production in Ogun State. It was further explained thus:

Finance in the sense that if you want to farm on your own, you need a lot of money that is why we are calling on government to get us cluster areas. To develop a large expanse of farmland and distribute to farmers, lets us have clusters, like the value chain now, they are developing a cluster at Onidundu where they will clear about 80 hectares for farmers and distribute to the farmers. All other production processes will be easier.

Land Development Related Challenges

The problem of land development arises because most of the farmlands for rice cultivation are usually located in rainforest, especially areas that have not been used for rice cultivation previously. Given the nature of vegetation, intended lands for rice cultivation will require a great number of labour inputs to clear a large expanse of land particularly when it is not mechanised. However, to effectively clear such a forest zones for rice cultivation, there will be need to hire tractors which could be highly exorbitant; hence, may not be within the reach of an average rice farmer. Apart from the challenge in hiring tractors, the interview revealed that rice seed germinates within the topsoil where the rice roots absorbs soil nutrients; hence, most of the nutrients needed for rice growth are supplied from the surface soil. This further entails that the tractor operation during land preparation should be carried out with some care with a view not to removing the surface soil that houses the essential nutrients, which is crucial for effective rice production. However, it is observed that there are instances whereby the soil nutrients are eroded from the surface soil during tractor operation in the land preparation process which eventually affects the supply of nutrients to the crops. In addition to the constraint of hiring tractors and erosion of the surface soil during land preparation, there are other challenges such as accessibility of land for commercial production.

Input Challenges

The farm inputs refer to availability of farm raw materials required for the rice production. Most important in this aspect is the ability of the farmers to have access to the right variety of rice seed that is suitable for that particular ecology. Secondly the issue of genuine herbicides for effective treatment of weeds is a real challenge at the rice production stage. This owes to the fact that some of the herbicides are now becoming adulterated. This makes it difficult to distinguish between the original and fake herbicides. To access new improved varieties that are suitable for a particular ecology requires a huge sum of money which could highly expensive for an average farmer to afford. However, the government is yet to meet up with the demand of procuring these varieties and making it available to the farmers at subsidised rate as a means assisting the local farmers in boosting rice production in Ogun State. This invariably do not encourage more farmers and the teaming unemployed labour force to go into rice production. This therefore reduces the opportunities for more employment in the real sector economy.

Furthermore, the KIIs conducted show that the adulteration of herbicides constitutes a serious problem in rice production. “It could have been easier, if we had genuine herbicides, the adulteration of herbicides is another problem”, one of the interviewees added. This implicates that when the herbicides are adulterated, the effectiveness of the chemical in killing weeds is reduced such that the farmers after spending their resources in purchasing the herbicides will not even realise the objective of using them. The aftermath is increased cost of production and consequently affects the profitability of the rice farmers adversely coupled with the laborious nature of rice production process.

Other major reasons for insufficient domestic production of rice include: inadequate and untimely availability of necessary inputs, cost of reducing rice production constraints, and cost of adopting proven technologies. Though various governments in Nigeria have tried to improve the domestic production base of rice, there has not been enough incentives for farmers to increase their production. Other challenges revolve around land fragmentation and reliance on crude agricultural equipment. The method of rice cultivation and harvesting mostly relied on labour intensive approach at the subsistence level in addition to low level of Agricultural extension agencies.

Un-Mechanised Rice Production Processes (Low Tractorisation)

Most of the production processes that are connected to rice production in Ogun State still make use of traditional method approach. The process starting from land cultivation to harvesting and processing are mostly done with manual labour, thus making the production process labour intensive. From KII, it was learnt that on the average a farmer needs a total of 10 workers to work on a hectare of land from cultivation to the harvesting stage. Despite the fact that some of these farm workers are family members, the cost of hiring farm labourers poses a significant constraint to rice production in Ogun State. The labour intensive method of rice farming is not only tedious, but time consuming and at the end the farmers only produce at the subsistence levels and in few cases where there is excess it is sold at the local markets. The un-mechanised method (which is also referred to as no or low *tractorisation*) of rice production does not allow for production in commercial quantities that could be harvested, processed and marketed to reach wider consumers.

Furthermore, there is unavailability of farm labour due to competing sectors which employs greater number of the young population such that agricultural business is becoming more and more uninteresting to the teaming young population of the labour force in Ogun State. In a nutshell, the leadership of RGAN are of the opinion that when they have requisite supports from the government agencies such as their counterparts in Indian States, they will not only able to meet with domestic demand, they can have supply for export. This Indian analogue is apt because, the *Ofada* variety of rice in Ogun State is likened to the basmati rice of India in terms of characteristics and nutrient content.

Birds Infestation

Another major challenge associated with the production stage as highlighted in the fieldwork is the issue of bird infestation, which is problematic at the period of maturity of the rice grains in the field. The farmers interviewed explained that rice farmers need to scare the birds for at least 30 days prior to maturity of the rice grains. The challenging aspect of it is that the farmer or the person employed for this job will have to be in the farm early in the morning (about 6.00 am before the birds wake up) to stay on the farm till in the evening (around 7.00 pm after the birds have gone to sleep). In other words, the farmer has to be in the farmer before the waking up of the birds and remain there till evening when the birds go to sleep with routine process of blowing whistle or beating a kind of drum to raise sufficient noise that will drive away the birds. In effect, for a hectare of rice farm, the farmer needs at least two persons to effectively combat the bird infestation challenge, which has huge cost implication. This step is normally taken as a proactive measure to stop the birds from feeding the maturing rice grains while in the field. Without this step, the farmer is bound to incur losses and low output because great proportion of the rice harvest would have been eaten up by the birds while in the field.

Labour Intensive Harvesting and Threshing Process

In this part of the country we were meant to understand that harvesting of the paddy rice is done by pinnacle and by straw. This process basically involves the use of manual labour and is time consuming. The major reason is that *Ofada* rice which is predominantly grown within Ogun State lodges more than any other variety due to its weight. Therefore, harvesting by straw which could be done with machine is not suitable for this variety of rice. When harvesting is done using straw, many of the rice paddies are shattered at the point of harvesting because of the lodging nature of Ofada rice. Hence, the farmers call on agricultural engineers in Nigeria to develop a unique harvester that will be suitable for the type of rice produced within the ecology of the State.

The above lends support to Adewumi, Olayanju and Adewuyi (2007) who worked on rice threshing cluster in Ofada, Abakaliki, Markudi and Kano. Their study revealed that the structure of farm size used for rice production by 80% of the farmers ranges from 1-10 hectares of the rice farm holding, 15% of the farmers cultivated between 11-20 hectares while only 5 percent of the rice farmers have access to 21-30 hectares of land. The average farm size cultivated by the rice famers was 3 hectares. This indicates the low proportion of land currently used by the local farmers for subsistence farming rather than for commercial and export purposes. However, the mechanisation of the rice threshing will to a large extent enhance rice production in Ogun State and enable the farmers to produce high quality rice that is globally acceptable. The traditional method of threshing increases the chances of harbouring dirt and contamination with stones, which adversely affect the perception, acceptability and marketability of the locally processed rice. This scenario does not only affect production and revenue from rice farming but also encourages over reliance on imported rice for local consumption as has been the case in Nigeria for many decades.

Limited Modern Processing Machines

This modern processing machine involves the use of equipment such as modern harvesters, modern milling machines to remove the shaves from the paddy rice, winnowing machines that will facilitate the removal and blowing of the chaffs from the milled rice and destoner machines for removal of stones from the processed rice. Also instead of sun drying the parboiled rice which takes considerable length of time especially during the rainy season farmers can employ the use of modern machine called driers that dry the rice grains within a shorter period of time. Traditionally some farmers make use of their house ceilings made of improved materials such as bamboos, rafters and iron roofing sheets to dry rice due to the heat they generate as a result of the sun's rays. As noted from our interview with the farmers some rice farmers winnow with baskets which does not allow for complete separation of the processed rice from the chaffs. Also the outdated destoner can still be found in some places like *Ifo, Lafenwa, Siun* and *Owode*. The basic source of energy for powering the parboiled rice is basically firewood. Conversely, there are other modern method that could be used such as electricity and solar energy tramping equipment.

Another major processing challenge has to deal with water supply. In some of the areas where rice is processed the nearest source of water supply is about 4 kilometres away from the point of processing. So the farmers will either dig bore holes or install big tanks for water storage and this is capital intensive. The need for water supply arises particularly with the nature of variety of rice produced within Ogun State which *Ofada* rice in which case the farmer need to soak for 4 days to complete a process and another process begins. Before the beginning of another process of soaking it is expected that the farmer replace the old water with a new one to avoid unnecessary odour. The soaking of the *Ofada* rice for 4 days is done to bring out the aroma and taste that make it unique among other varieties.

Weak Linkages for Marketing of Rice Products

The issue of linkages relates with the marketing and distribution networks challenges. There is need to link the rice farmers with the markets for demands of their produce.

In our interview with the farmers we discovered that price of the rice products do not reflect their real value. In this case, the farmers are at a loss. This occurs in instances where the produced and packaged rice is being sold below the real value probably because the farmers are in dire need of cash to solve some other problems and will have no other option than to sell the rice at a price which is below what it would have been sold under normal circumstances.

CONCLUSION AND RECOMMENDATIONS

Government in recent times have realised the need to focus more attention in the production of the cereal crops more especially on rice and cassava production as major plans in the agricultural transformation agenda. Therefore, to make this laudable initiative a reality, it is important that adequate attention is given to rice production and processing at all levels of government particularly in rice producing states like Ogun State. In order to transform the current trends in rice production and processing in Ogun State, it is pertinent that agricultural development agencies in the state need to work closely with the small-scale farmers in the field to understand the real situation of things and the challenges they face. This implies that, strategies on how to improve the working conditions of these farmers should be the central point in the government planning and implementation process.

In Ogun State, rice production and processing have not yet be fully mechanised and to be able to achieve this starting from land development the issue of government financial support plays a significant role. This also entails other financial and technical assistance from agricultural engineers, extension workers, researchers, non-governmental organisations (NGOs), and the private investors. Given the natural endowments of Ogun state which includes favourable weather conditions, fertile land that supports rice production and its favourable ecological composition, the State has competitive advantage in large scale rice production coupled with great economies of scale. This will not only provide more employment opportunities in the state, but will further improve revenue generation in the State while at the same time enhancing the general living standard of the households that are connected to the rice value chain.

Sequel to the issues raised in the course of this study, the following recommendations are proffered:

- To enhance the production of rice, it is therefore germane that additional support for development of efficient rice seed varieties be advocated. It is also essential that the distribution of rice varieties that are resilient to climate changes be encouraged. The promotion of good agricultural practices that will help to compensate for the lapses rice production processes cannot be overemphasised in this regard.

- In terms of enhancing commercialization of rice production, modern means of rice production and processing that will enhance the quality and quantity of production should be adopted.
- In addressing the issues of poor content agro-chemicals, this study is of the opinion that adequate measure on the part of the government agencies such as: agricultural development programme (ADP) and Ministry of Agriculture charged with the responsibility of agricultural development to ensure proper supervision and inspection of the herbicides and other agro-chemical inputs before they sold to the farmers. These agricultural agencies could work in congruence with farmers association to procure these requisite agro-chemicals and supply to the different clusters of rice farmers within the State. This will help in reducing the incidences of adulterated agro-chemical inputs as well as the prices as it will become possible to deal directly with the producers of such agro-chemical inputs.
- It was noted in this study that the method of rice cultivation and harvesting mostly relied on labour intensive approach at the subsistence level in addition to low level of Agricultural extension agencies. An alternative, however, to encouraging farmers to increase their production base is through agricultural extension services. Therefore, for the rice farmers to operate at mechanised level of rice production government support is earnestly advocated for. For instance, information gathered from the interview conducted unveils that there is no machine for rice planting which results to drudgery in the planting process.
- To effectively control for bird's infestation, the farmers need to be financially buoyant in order to employ workers to assist them to do the job. Consequently, this would be incorporated into the production cost which will determine the final price of the processed rice at the point of sales and marketing, which makes the local rice variety uncompetitive compared to the imported varieties. Thus, developing a kind of mechanism that will be deployed to chase or scare away birds during the period when the grains are matured will be a welcome development in ameliorating this problem. Similarly, integrated pest management (IPM) can be developed and implemented for rice insects, which can be narrowed to a given locality as an effective control and management of pest and diseases that affect rice.
- In addressing the challenge of labour intensive harvesting and processing, the provision of modern threshing machine at the farmers' disposal will no doubt boost rice production and processing; reduce drudgery and time involved in the production process. This will further enhance the market potentials for the

finished product. This is because evidences have shown that modern threshing machines significantly reduce the tedious and time consuming nature of the traditional threshing method. The availability of modern processing machines will not only speed up rice production process but will further reduce the risk of exposure to whether due to moulding and shattering especially when the paddy rice is packed in the field to dry prior to the time of threshing. Mostly importantly, other losses arising from grain breakages could be controlled when the threshing duration is minimised. This consequently improves the quality and competitive nature of the rice in the international market.

- In the aspect of weak markets and linkages, it is important to have a market where the rice producers will be linked with their direct customers for supply. In this case, there will be regulatory authorities to regulate and standardise the price of processed rice according to their respective standard measurements which is to be determined by the agency in charge. However, the farmers interviewed explained that currently the demand for Ofada rice is higher than the supply. This could be explained by the reason that *Ofada* rice is like the plasmatic rice of India and contains more nutritional value compared to the polished imported rice.
- With possible multiplier effects for employment and input effects the implementation of the on-going Agricultural Transformation Agenda (ATA) can be re-strategized to reflect on how to improve the welfare of the small-scale farmers who are directly involved in the value chain. Government agricultural policies on rice production should focus more on means of replacing the traditional method of rice processing with modern equipment to enhance the productive capacity of this sub-sector. Provision of processing centres in selected LGAs; common pool of resources that the farmers can coordinate; electrification and provision of sound system to scare away birds especially when the paddy is matured will be right steps in right directions towards boosting rice production and thus, making it more affordable to majority of the citizens.

In conclusion, this study submits that the rice farmers alone cannot record a remarkable mass production because rice production and processing is capital intensive project which requires financial support from government, non-governmental organisation (NGOs), private investors interested in agriculture (*agri-preneurs*), among others. There is need for all stakeholders connected with agricultural productivity to work in synergy in cross fertilisation of ideas on the current challenges and prospects of agricultural production and modern food processing techniques that will enhance food security policy measures. Thus, this study calls on government, non-governmental organisations and private investors and stakeholders in the provision of financial assistance, agricultural grants, subsidy and accessible credit facilities in support of rice production and processing among the rice producing States in Nigeria especially in Ogun State.

ACKNOWLEDGMENT

This study draws from the research project funded by DTT Farms Ltd, Abuja. The respective assistance received from the Officers of Ogun State's Agricultural Development Programme (ADP); the leadership of Ogun State Rice Growers Association of Nigeria (RGAN) led by the Chairman, Pastor Bado Adeneko; Mr Israel Oloruntoba; and Miss Ibukun Beecroft during the fieldwork, are acknowledged. In addition, the authors appreciate Dr Gideon Adewale of Geographic Information System (GIS) Centre, Department of Civil Engineering, Covenant University, Ota, Ogun State, Nigeria for helping to plot the map of Ogun State and its rice producing Local Government Areas (LGAs). Earlier versions of this paper have been presented at: Inaugural Conference of Young African Researchers in Agriculture-YARA (Cape Town, South Africa, 6-8th February, 2017); *3rd Witten Conference on Institutions in Development Research* (Witten/Herdecke University, Witten, Germany, 1st-2nd December, 2016); and *1st Nigeria Agricultural Policy Project (NAPP) Conference* (Abuja, Nigeria, 27-28th September 2016). Thus, the supports to attend the conferences from the organisers as well as helpful comments from the discussants and participants are appreciated. The support from Covenant University Centre for Research, Innovation and Development (CUCRID) during the manuscript preparation is equally acknowledged. The views expressed are those of the authors.

REFERENCES

Abea, S. S., Yamasakib, T., & Wakatsukib, C. (2016). Assessing Silicon Availability in Soils of Rice-Growing Lowlands and Neighbouring Uplands in Benin and Nigeria. *Rice Science*, *23*(4), 196–202. doi:10.1016/j.rsci.2016.06.002

Adesina, A. (2012). *Agricultural Transformation Agenda: Repositioning Agriculture to Drive Nigeria's Economy*. Federal Ministry of Agriculture and Rural Development.

Adewumi, J. K., Olayanju, M. A., & Adewuyi, S. A. (2007). Support for Small Rice Threshers in Nigeria. London: DFID.

Ajala, A. S., & Gana, A. (2015). Analysis of Challenges Facing Rice Processing in Nigeria. *Journal of Food Processing*, *6*. doi:10.1155/2015/893673

Amb, M. K., & Ahluwalia, A. S. (2016). Allelopathy: Potential Role to Achieve New Milestones in Rice Cultivation. *Rice Science*, *23*(4), 165–183. doi:10.1016/j.rsci.2016.06.001

Asongu, S. A. (2016). Law and Investment in Africa. *Institutions, Etc.*, *8*(2), 91–120.

Fasoyiro, S. B., & Taiwo, K. A. (2012). Strategies for Increasing Food Production and Food Security in Nigeria. *Journal of Agricultural & Food Information*, *13*(4), 338–355. doi:10.1080/10496505.2012.715063

Food and Agricultural Organization (FAO). (2016). *Rice Market Monitor*. FAO.

Gyimah-Brempong, K., Johnson, M., & Takeshima, H. (2016). *The Nigerian Rice Economy: Policy Options for Transforming Production, Marketing, and Trade*. University of Pennsylvania Press. doi:10.9783/9780812293753

Harold, M., & Tabo, R. (2015). *Feeding Africa: United Nations Economic Commission for Africa, An Action plan for African Agricultural Transformation*. Abdon Diouf International Conference Center Dakar, Senegal.

Herrmann, R., Jumbe, C., Bruentrup, M., & Osabuohien, E. (2017). Competition between Biofuel Feedstock and Food Production: Empirical Evidence from Sugarcane Outgrower Settings in Malawi. *Biomass and Bioenergy*. doi:10.1016/j.biombioe.2017.09.002

Johnson, M., Takeshima, H., & Gyimah-Brempong, K. (2013). *Assessing the Potential and Policy Alternatives for Achieving Rice Competitiveness and Growth in Nigeria*. The International Food Policy Research Institute (IFPRI). IFPRI Discussion Paper, 01301.

Juliano, B.O. (2016). Rice: Role in Diet. *Reference Module in Food Science Encyclopaedia of Food and Health,* 641–645.

Kareem, O. I. (2016). European Union's Standards and Food Exports from Africa: Implications of the Comprehensive Africa Agricultural Development Programme for Coffee and Fish. *Journal of African Development*, *18*, 83–97.

Ministry of Budget and National Planning. (2017). *Economic Recovery and Growth Plan 2017-2020*. Abuja: Ministry of Budget and National Planning.

National Universities Commission (NUC). (2016). *Universities in Nigeria*. Retrieved 20th May 2016, from http://nuc.edu.ng/

North, D. C. (2005). *Understanding the Process of Economic Change*. Princeton University Press. doi:10.1515/9781400829484

Nwanze, K. F., Mohapatra, S., Kormawa, P., Keya, S., & Bruce-Oliver, S. (2006). Rice Development in Sub-Saharan Africa. *Journal of the Science of Food and Agriculture*, *86*(5), 675–677. doi:10.1002/jsfa.2415

Nyarko, Y. (2017). Transforming Rural Africa— Economics, Technology and Governance. *Journal of African Development*, *19*(1), 160–167.

Obayelu, E. A. (2015). Transformation from Subsistence to Commercial Agriculture in Nigeria: The Effects of Large-Scale Land Acquisition on Smallholder Farmers. In E. Osabuohien (Ed.), *Handbook of Research on In-Country Determinants and Implications of Foreign Land Acquisitions* (pp. 409–431). Hershey, PA: Business Science Reference. doi:10.4018/978-1-4666-7405-9.ch021

Ogun State Government (OSG). (2016). *Ogun State, Nigeria, Investors' Forum*. Retrieved 30th June 2016, from http://ogunstateinvestorsforum.ng/agriculture/

Ogunsumi, L. O., Ajayi, A., Amire, C. M., & Williams, S. (2013). Sustainability of Agricultural Transformation Agenda. The place of Rice Farmers in Ogun State Nigeria. *Research on Humanities and Social Sciences*, *3*(13), 66–78.

Okeh, C., Onwosi, C. O., & Fredrick, J. O. C. (2014). Biogas Production from Rice Husks Generated From Various Rice Mills in Ebonyi State, Nigeria. *Renewable Energy*, *62*, 204–208. doi:10.1016/j.renene.2013.07.006

Okodua, H. (2017). *Assessing the Impact of Rice Sector Policy Reforms on Income Mobility of Rural Households in Nigeria*. AGRODEP Working Paper 0039, February.

Okoruwa, V., Jabbar, M. A., & Akinwumi, J. A. (1996). Crop-Livestock Competition in the West African Derived Savanna: Application of a Multi-objective Programming mModel. *Agricultural Systems*, *52*(4), 439–453. doi:10.1016/S0308-521X(96)00005-4

Olayiwola, K. W., Osabuohien, E. S., Okodua, H., & Ola-David, O. (2015). Economic Integration, Trade Facilitation and Agricultural Exports Performance in ECOWAS Sub-Region. In M. Ncube, I. Faye, & A. Verdier-Chouchane (Eds.), *Regional Integration and Trade in Africa* (pp. 31–46). New York: Palgrave Macmillan. doi:10.1057/9781137462053_3

Onabanjo, A. A. (2011). *Lessons and Experiences of Participatory Policy Making Process: The Case Study of Ogun State Rice Policy. In Abeokuta: Planning, Research and Statistics*. Ogun State Ministry of Agriculture.

Osabohien, R., Osabuohien, E., & Urhie, E. (2017). Food Security, Institutional Framework and Technology: Examining the Nexus in Nigeria Using ARDL Approach. *Journal of Current Nutrition and Food Science*, *13*(4), 1–10.

Osabuohien, E. (2014). Large-scale Agricultural Land Investments and Local Institutions in Africa: The Nigerian Case. *Land Use Policy*, *39*, 155–165. doi:10.1016/j.landusepol.2014.02.019

Osabuohien, E. (2016, June 20). Economy without Buffer must Experience Shocks. *Daily Times,* p. B8.

Osabuohien, E., & Efobi, U. R. (2013). Africa's Money in Africa. *The South African Journal of Economics*, *81*(2), 292–306. doi:10.1111/saje.12012

Osabuohien, E., Efobi, U. R., Gitau, C. M., & Bruentrup, M. (2015). Agents and Implications of Foreign Land Deals in East African Community: The Case of Uganda. In E. Osabuohien (Ed.), *Handbook of Research on In-Country Determinants and Implications of Foreign Land Acquisitions* (pp. 263–286). Hershey, PA: Business Science Reference. doi:10.4018/978-1-4666-7405-9.ch013

Uduma, B. U., Samson, O. A., & Mure, U. A. (2016). Irrigation Potentials and rice self-sufficiency in Nigeria: A review. *African Journal of Agricultural Research*, *11*(5), 298–309. doi:10.5897/AJAR2015.10284

Umeh, E. D. N., Joshi, R. C., & Ukwungwu, M. N. (1992). Biology, Status and Management of Rice Insect Pests in Nigeria. *Crop Protection (Guildford, Surrey)*, *11*(5I5), 408–413. doi:10.1016/0261-2194(92)90022-W

Yan, A., Ngadi, M., & Kok, R. (2016). Physical properties of rice husk and bran briquettes under low pressure densification for rural applications. *Biomass and Bioenergy*, *84*(January), 22–30.

KEY TERMS AND DEFINITIONS

Formal Institutions: This refers to an officially recognised entity with properly documented ethics and established rules that guide and control the activities of economic agents in the society.

Informal Institution: Connotes unofficial arrangements that consist of societal generally accepted principles, moral values, ethics, norms, and conduct that detect the way and manner groups of individuals are controlled through the instrumentality of community leaders.

Institutional Land Investment: This is mainly concerned with land investments such as farming that involves a group of people who are united by a common goal, especially in the area of enlarging agricultural production and maximising profit from sales.

Key Informant Interview (KII): This involves an interactive section with the key players in the rice production and processing particularly under the auspices of Rice Growers Association of Nigeria, Ogun State chapter.

Large-Scale Agricultural Land Investment: Deals with agricultural production in commercial quantities that could guarantee sustainability of the agricultural system and food security.

OFADA Rice: A special breed of rice mostly produced within the South-West geopolitical region with high fibre and nutritional content compared to the imported polished rice.

OSG (Ogun State Government): This describes a geopolitical location within the South-West region of Nigeria that predominantly grows a special variety of paddy rice (OFADA rice); its species origin is traceable to the state.

Respondents: Comprises of the stakeholder and principal personality with the necessary information required to in the process of executing the research.

Rice Growers Association of Nigeria (RGAN): A group bound by the common objective of ensuring sufficient rice production and marketing through the formation of rice farmers clusters and through such arrangements facilitate the process of government intervention.

ENDNOTES

1 The KIIs conducted with officers of RGAN also confirmed this allusion with the additional information that there are no incentives to attract the youth into the agricultural sector.

2 The new government led by President Muhammad Buhari, in 2016 Launched 'The Green Alternative: Agriculture Promotion Policy (GAAPP), 2016-2020'. The GAAPP is said to build upon success of ATA as their goals are very similar with regards to making agriculture a profitable enterprise (Ministry of Budget and National Planning, 2017). This intervention is expected to encourage local production and value chain development with a view to reducing food import.

3 The extent to which these programmes were achieved in the respective countries is outside the scope of this study.

4 In some States, it is also known as Rice Farmers Association of Nigeria. However, this study sticks to the usage of RGAN for consistency.

5 It was learnt during KIIs that *Ofada* variety of rice was named after this *Ofada* village where it was first planted in Nigeria by a military officer who brought it during World War II. In addition, with the role RGAN, Federal University of Agriculture, Abeokuta (FUNAB) has assisted in breeding two varieties of *Ofada* rice.

6 Figures 3 and 4 in the Appendix provide their pictures taken during the fieldwork.

APPENDIX

Figure 3. Traditional rice processing equipment for water supply during paddy rice soaking
Source: Fieldwork, 2016

Figure 4. Modern Rice Shelling Machine
Source: Fieldwork, 2016

Figure 5. Stone Removal Machine (Destoner)
Source: Fieldwork, 2016

Section 5

Food Controversies in Developing and Emerging Countries

This section examines some controversies in food in developing and emerging countries.

Chapter 10
Rights-Based Approach to Food and Nutrition Security in Nigeria

Clementina Oluwafunke Ajayi
Federal University of Technology – Akure, Nigeria

Kemisola O. Adenegan
University of Ibadan, Nigeria

ABSTRACT

This chapter focuses on the need for a rights-based approach to food and nutrition security in Nigeria. The topic is introduced with the definition of basic terms used throughout the chapter. The objectives of this chapter are to create awareness of the need to adopt a rights-based approach to food and nutrition security and help define the context of the right to food in Nigeria. The rest of the chapter sets out the roles of right holders, duty bearers, and accountable agents in food and nutrition security. It highlights their rights, obligations, and responsibilities, as well as voluntary guidelines and implications for a rights-based approach to food. The chapter also reviews policies that have been developed toward ensuring a right to food in Nigeria.

BACKGROUND

Food and nutrition insecurity is increasing in the world where 925 million people are undernourished. Today, over 900 million people are food insecure across the world despite the fact that the world food production has doubled during the past three decades (FAO, 2010). Helen (2002) noted, food security maintains political

DOI: 10.4018/978-1-5225-3631-4.ch010

stability, and ensures peaceful coexistence among people while food insecurity results in poor health and reduced performance of both children and adult. Poor nutrition weakens children's immune systems and renders them more vulnerable to deadly diseases, which further deteriorates their nutritional status. Hence children get stuck in a cycle of undernutrition and infection, which adversely affects their health, development, and increases the risk of mortality.

The first and most important component of social and economic justice is adequate food production (Otaha, 2013). This can therefore improve food and nutrition security if viewed from rights-based perspective with the ultimate goal of development to guarantee all human rights to everyone. A human rights-based approach is a conceptual framework for the process of human development that is normatively based on international human rights standards and operationally directed to encouraging and defending human rights (United Nations, UN, 2017). It seeks to analyze inequalities which lie at the core of development complications and amends discriminatory practices and unfair distributions of power that hinder development progress.

Under a human rights-based approach, the plans, policies and processes of development are anchored in a system of rights and conforming obligations established by international law. This helps to promote the sustainability of development work, empowering people themselves— especially the most marginalized—to participate in policy formulation and hold accountable those who have a duty to act. A rights-based approach to development is both a vision and a set of tools: human rights can be the means, the ends, the mechanism of evaluation and the central focus of sustainable human development (UN, 2002).

According to World Food Programme (WFP, 2017), here are ten things to know about Nigeria food and nutrition security status:

1. Nigeria is a food deficit country and is Africa's largest importer of rice.
2. One third of children under 5 are stunted. That's twice the rate of Thailand and three times that of Tunisia.
3. A child in the remote northwestern region of Nigeria, where stunting rates are around 55 percent, is four times more likely to experience malnutrition than a child in the south.
4. At the same time, child obesity is increasing and 33 percent of adults are obese or overweight. That is roughly on par with Singapore.
5. Nearly half of women of reproductive age (48.5 percent) are anaemic. This ranks Nigeria 172th best out of 185 countries.
6. Conflict with Boko Haram in northeastern Nigeria has left a large part of the population without access to enough food, water and health services.

7. Displacement, lack of access to many locations, high inflation and reduced purchasing power of communities are worsening the food security situation in northeastern Nigeria.
8. Over 3 million people are in urgent need of humanitarian assistance due to conflict.
9. In March 2016, WFP began providing cash transfers via mobile phones to displaced persons and host communities in critical areas. This gives them the opportunity to buy the food they need.
10. In partnership with the Government and other agencies such as UNICEF, WFP is scaling up its assistance in northeastern Nigeria to reach a total of 431,000 people in desperate need, including malnourished children and pregnant and nursing women.

The right to food and nutrition is a holistic concept which empowers people. The right to food and nutrition relates to diverse dimensions of human life, from the most basic access to sun light, seeds, land, water, which allows food production, up to the moment in which food reaches human beings. Food is not just a commodity, it is part of our life, as a social, cultural and economic right and it impacts our lives as communities: security, democracy, health, among others. It is in the light of these that this chapter aim at creating awareness of the need to adopt rights-based approach to food and nutrition security. It will also help define the context of the right to food in Nigeria so as to ensure attainment of food and nutrition security.

DEFINITIONS OF TERMS

Right to Adequate Food

The International Covenant on Economic, Social and Cultural Rights (ICESCR, 1999) defined the right to adequate food as what "every man, woman and child, alone or in community with others, have physical and economic access at all times to adequate food or means for its procurement". This definition is synonymous to that of food security by the World Food Summit (WFS, 1996) which states that 'a situation that exists when all people, at all times, have physical, social and economic access to sufficient, safe and nutritious food that meets their dietary needs and food preferences for an active and healthy life.' Also, the Constitution of the Federal Republic of Nigeria 1999, article 16 (2d) states that 'suitable and adequate shelter, suitable and adequate food, reasonable national living wage, old age care and pensions, and unemployment, sick benefits and welfare of the disabled are provided for all citizens.'

Food Security

Food security concept is not new in research studies. It has been defined in a variety of ways by different authors and organizations. It is a condition related to the supply of food, and individual's access to it. In 2009, the World Summit on Food Security stated that the "four pillars of food security are availability, accessibility, utilization, and stability" (Food and Agriculture Organization, FAO, 2009).

Food Availability

Food availability relates to the supply of food through production, distribution, and exchange (Gregory *et al.,* 2010). According to FAO (1997), food production is determined by a variety of factors including land ownership and use; soil management; crop selection, breeding, and management; livestock breeding and management; and harvesting.

Economic and Physical Access to Food

Food access refers to the affordability and allocation of food, as well as the preferences of individuals and households (Gregory *et al.,* 2010). An adequate supply of food at the national or international level does not in itself guarantee household level food security. Concerns about insufficient food access have resulted in a greater policy focus on incomes, expenditure, markets and prices in achieving food security objectives. The UN Committee on Economic, Social, and Cultural Rights noted that the causes of hunger and malnutrition are often not a scarcity of food but an inability to access available food, usually due to poverty (UN, 1999). Access depends on whether the household has enough income to purchase food at prevailing prices or has sufficient land and other resources to grow its own food (Garret and Ruel, 1999).

Utilization

Utilization refers to the metabolism of food by individuals (Tweeten, 1999). It is commonly understood as the way the body makes the most of various nutrients in the food. Once food is obtained by a household, a variety of factors affect the quantity and quality of food that reaches members of the household. In order to achieve food security, the food ingested must be safe and must be enough to meet the physiological requirements of each individual. Food utilization is also a function of how food is processed and stored, knowledge of good nutrition and care, and adequate health and sanitation services in the environment in which food is consumed (*Ecker and Breisinger 2012).*

Stability

Food stability refers to the ability to obtain food over time (FAO, 1997). Even if your food intake is adequate today, you are still considered to be food insecure if you have inadequate access to food on a periodic basis, risking a deterioration of your nutritional status. Adverse weather conditions, political instability, or economic factors (unemployment, rising food prices) may have an impact on your food security status.

For food security objectives to be realized, all four dimensions must be fulfilled simultaneously (WFS, 2009).

Nutrition

Nutrition is underpinned by intake, absorption and use of nutrients. There are many, often concurring, socio-political, economic and environmental factors that directly and indirectly impact on nutrition (World Health Organisation, WHO, 2013). Nutrition is the science that explains the role of food and nutrients in the human body during growth, development and maintenance of life. In its broader context, nutrition has also been defined as being concerned with "... how food is produced, processed, handled, sold, prepared, shared, and eaten and what happens to food in the body - how it is digested, absorbed, and used" (King and Burgess, 1993). Nutritional status refers to the nutritional state of the body, as expressed according to anthropometric measure such as weight, height, age or combinations of these. Only focusing on food availability or food accessibility is not sufficient and will not result in a well-nourished, healthy and productive population. Additional interventions are needed such as (environmental) health services, care practices, infrastructural facilities, adequate security and focus on nutritious food.

Malnutrition

Malnutrition arises from deficiencies of specific nutrients or from diets based on inappropriate combinations or proportions of foods; for example, goitre, scurvy, anaemia and xerophthalmia are forms of malnutrition caused by inadequate iodine, vitamin C, iron and vitamin A, respectively (Shetty, 2003). Malnutrition can also result from excess nutrient losses or utilization. It relates to imbalances in energy, and specific macro and micro nutrients intake in relation to the requirements. Malnutrition occurs when the intake of essential macro and micro nutrients does not meet or exceeds the metabolic demands for those nutrients.

Undernutrition

Undernutrition stems from the inadequate quantity and /or quality being consumed and/or repeated infection of disease resulting in improper absorption of vital nutrients. It manifests itself through wasting, stunting and micronutrients deficiencies (WHO, 2013). It is caused primarily by an inadequate intake of dietary energy, regardless of whether any other specific nutrient is a limiting factor (Shetty, 2003). It represents a state or condition of people that have to survive on less than 2100 kcal per day, which is estimated from food balance sheets (food production minus export plus import) – based on the FAO Food Balance Sheets, presented at national level and aggregated at global level. Chronic undernutrition is an inadequate nutrition coupled with recurrent or chronic illness over long period of time leading to failure of linear growth. The terms 'malnutrition' and 'undernutrition' are often used loosely and interchangeably. Malnutrition refers to all deviations from adequate and optimal nutritional status, including energy undernutrition and over-nutrition (obesity is a form of malnutrition). The term 'undernutrition' is used to refer to generally poor nutritional status, but also implies underfeeding.

Hunger

Hunger is a condition, in which people lack the basic food intake to provide them with the energy and nutrients for fully productive lives (Hunger Task Force, 2003)

Food and Nutrition Security

Food and Nutrition Security exists when all people at all times have physical, social and economic access to food, which is safe and consumed in sufficient quantity and quality to meet their dietary needs and food preferences, and is supported by an environment of adequate sanitation, health services and care, allowing for a healthy and active life (World Food Population, 2013). Improving food and nutrition security requires knowing where the most vulnerable are located and understanding what makes them vulnerable. Targeting is a key mechanism for reaching vulnerable populations and ensuring efficient and effective use of limited resources.

Human Rights

Human Rights are moral principles or norms which describe certain standards of human behavior, and are regularly protected as legal rights in municipal and international law. They are commonly understood as inalienable fundamental rights to which a person is inherently entitled simply because she or he is a human being and which are inherent in all human beings regardless of their nation, location, language, religion, ethnic origin or any other status. They are rights that every human being enjoys, agreement between states, norm about state-citizen relationships and can be protective devices or emancipatory tools but also political chips/tools or empty promises.

The Right to Food

The right to food is a human right protecting the right for people to feed themselves in dignity. This implies that sufficient food is available, that people have the means to access it, and that it adequately meets the individual dietary needs. The right to food protects the right of all human beings to be free from hunger, food insecurity and malnutrition. The right to food does not imply that governments have an obligation to hand out free food to everyone who wants it, or that everybody has a right to be fed. However, if people are deprived of access to food beyond their control such as in times of war, disaster etc., the right to food requires the government to provide food directly. The right to food originated from the Universal Declaration of Human Rights (1948) and underwent evolution of International Covenant on Civil and Political Rights (1966-1976), International Covenant on Economic, Social and Cultural Rights (1966-1976), General Comment no 12 of United Nation Commission on Economic, Social and Cultural Rights (1999) before metamorphosing into FAO guidelines on the Right to Adequate Food (RtAF), 2004 and FAO Panther principles (2006).

Right Holder

Every human being is inherently a right holder who should enjoy universal human rights that must be guaranteed (Ljungman, 2004). Every citizen of a country is a right holder for their basic food needs to be meet.'

Duty Bearer

According to Ljungman 2004, by endorsing the diverse United Nations human rights treaties, states (Government) automatically take up the principal roles of guaranteeing these rights. They are termed "principal duty bearers". The duty bearer has three obligations which are to respect, protect, and fulfill.

- To *respect* requires the State and all its organs and agents to abstain from carrying out, sponsoring or tolerating any practice, policy or legal measure violating the integrity of individuals or impinging on their freedom to access resources to satisfy their needs. It also requires that legislative and administrative codes take account of guaranteed rights. This means the duty bearer must respect existing access to adequate food and health and must not take any measures preventing or destroying such access.

- To *protect* the right to food is to take necessary measures to ensure that a third party does not deprive people of their access to adequate food. Where violations do occur the State must guarantee access to legal remedies.
- To *fulfill* involves issues of advocacy, public expenditure, governmental regulation of the economy, the provision of basic services and related infrastructure and redistributive measures. The duty of fulfillment comprises those active measures necessary for guaranteeing opportunities to access entitlements. To fulfill the right to food has two components: to facilitate or create social and economic environments that foster human development, and to provide food to people in an emergency or in circumstances when self-provisioning is beyond their control.

Even though states play the role of the principal duty bearer, there are other non-state entities (moral duty bearer) that have obligations to respect, protect and fulfill the rights of people. They are primary duty-bearers e.g. parents for children, teachers for students, police for crime suspects, doctors/nurses for patients, employers for employees; secondary duty-bearers e.g. institutions and organizations with immediate jurisdiction over the primary duty-bearers e.g. school principals, community organizations, hospital administrations, etc.; tertiary duty-bearers e.g. institutions and organizations at a higher level / more remote jurisdiction (Non-Governmental Organizations, aid agencies, private sector organizations); and external duty-bearers *e.g.* countries, institutions, organizations with no direct involvement e.g. United Nations, Security Council, African Union.

The constitutional recognition of the right to food in Nigeria makes it a normal standard that every human being is a right holder and enjoys obligations from the duty bearer. Also, "each state party to the present Covenant undertakes to take steps, individually and through international assistance and co-operation to the maximum of its available resources, with a view to achieving progressively the full realization of the rights recognized in the present Covenant" (Art. 2, ICESCR). To this end, there is need for every citizen to wake up to the call of ensuring their right to food is established.

Human Rights Principles: PANTHER

The PANTHER framework, developed by FAO in 2006 is a human rights–based approach. The framework stipulates the right to food and food security. As shown in Figure 1, it establishes the seven principles that should govern decision making

Figure 1. Human Rights PANTHER Principles
Source: FAO, (2006)

and implementation processes of the right to food. The first letter of each principle formed the acronym PANTHER.

- *P*articipation requires that everyone have the right to subscribe to decisions that affect them.
- *A*ccountability requires that politicians and government officials be held accountable for their actions through elections, judicial procedures and other mechanisms.
- *N*on-Discrimination prohibits arbitrary differences of treatment in decision making.
- *T*ransparency requires that people be able to know processes, decisions and outcomes.
- *H*uman dignity requires that people be treated in a noble way.
- *E*mpowerment requires that they are in a position to exert control over decisions affecting their lives.
- *R*ule of law requires that every member of society, including decision makers, must comply with the law.

REVIEW OF FOOD PRODUCTION EFFORTS IN NIGERIA

Nigeria is challenged with rapid population growth which affects her ability to assure stable supply of, and access to food. Its population projected to rise from 182.3 million in 2015 to 262.3 million by 2030 to be the world's fifth most populous and the third most populous by 2050 (World Population Prospects, 2016). Also, research findings in Nigeria show that the returns for scaling up nutrition specific intervention has a compound rate of returns of >13% and benefit to cost ratio of 16:1 (Global Nutrition Report, GNR (2015)). The growing evidence food and nutrition insecurity in and around Nigeria calls for a critical analysis of potential challenges concerning citizens and how to cope with these challenges.

Nigeria has made several attempts to increase food production in both quantity and quality. Some of these attempts have cumulated into several programmes and projects such as Operation Feed the Nation, Agricultural Development Project, Fadama I, II and III, Vision 2020, Agricultural Transformation Agenda, The Agriculture Promotion Policy among others. These policies and programmes were aimed at boosting agricultural production thereby solving the problem of food insecurity and poverty.

The 1960s featured strong public intervention in agriculture, with development guidelines and plans established at the federal level and implemented in the states. The government's main concern at the time was to increase domestic production, particularly of cash crops. This period pushed Nigeria to the position of the world's top producer of groundnuts, rubber and palm oil, and the world's second-largest cocoa producer.

The 1970-1986 period, which coincided with intensive petroleum exploitation, was marked by a decrease in policies supporting agriculture. The strong drop in domestic agricultural production reduced the country to growing dependency on imported foodstuffs. In the wake of the major food crisis in the country in 1976, programmes such as "Feed the Nation" (1976-1979) and "Green Revolution" (1979-1983) were set up. These programmes focused on strengthening agricultural production, providing subsidized inputs, community development, and access to credit. However, they were implemented without a transparent framework to structure action, and the successive governments did not ensure continuity. The enactment of the Land Use Act in 1978 marked an historic turning point for land use management in Nigeria. The movement was reversed in 1987 with the structural adjustment programmes (SAPs) that sought to reduce the national economy's dependency on oil and promote the private sector as the engine driving growth. In 1998, the Nigerian government once again turned its attention to the agricultural sector. It adopted an agricultural policy that had the objective, among others, of ensuring food security for the population by developing local production.

Since the reference document "Agriculture in Nigeria: The New Policy Thrust" was issued in 2001, the government has assigned the agricultural sector an ambitious role in its strategic planning frameworks. The strategic document for reducing poverty in Nigeria, "National Economic Empowerment and Development Strategy" (NEEDS II 2008-2011) emphasized economic development driven by the private sector, and the "7-point Agenda", a framework guiding economic reform in the country that was adopted in May 2007. These two documents were the medium-term policy documents intended to help the country achieve the Millennium Development Goals of 2015 and its own "2020 Vision" plan. The latter aims to make Nigeria one of the top twenty economies in the world by 2020. For agriculture, this means increasing current domestic production by six folds. *The National Food Security Programme (NFSP)* issued in August 2008 by the Federal Ministry of Agriculture and Water Resources as it was called then was designed to help Nigeria attain food security by ensuring that all Nigerians have access to good-quality food while making Nigeria a major exporter of foodstuffs. The programme designated priority crops (cassava, rice, millet, wheat) for achieving food security and outlined objectives for all stages of these supply chains. The aim is to create more value in production, particularly downstream in the value chain by improving storage, processing, and access to agricultural markets. The programme also had plans for the creation of irrigation schemes (450,000 ha). The strategic frameworks in NEEDS II and the 7-point Agenda were been translated into short-to-medium-term programmes. The Ministry drew up a "5-point Agenda" for agriculture, a detailed roadmap of steps to be implemented to attain the objectives listed for agriculture in the 7-point Agenda.

Olusegun Obasanjo's administration also launched Presidential Initiatives in 1999 for seven agricultural products (cassava, rice, vegetable oil, sugar, livestock, cultivated trees and dry grains). The aim of these initiatives was not only to boost Nigeria's agricultural exports by taking advantage of preferential agreements in the framework of the World Trade Organization (WTO) and the Economic Partnership Agreements between the European Union and the Africa-Caribbean-Pacific countries but also to make the most of the potential regional market made up of neighboring countries. Although these measures have shown that investment in the agricultural sector can have concrete results in terms of increasing domestic production, their overall outcomes have been mixed in that only the "intensification of production" segment has been taken into account, ignoring the downstream segments of the value chain (such as product processing).

Support for agricultural inputs has been a central element of Nigerian agricultural policy since the 1950s. This support consists primarily of providing public subsidies so that farmers can more easily acquire inputs (fertilizers, improved seeds, phytosanitary products). The level of federal subsidies has followed a spiky path, with highs and lows, and methods of implementation have varied. In addition to federal subsidies, each state allocates its own subsidies for fertilizer. These vary greatly from one state to another in both amounts (50 to 150 kg per farmer) and subsidization rates (from 10% to 50%). Even so, many farmers still find it difficult to obtain good-quality inputs at an affordable price and at the time they are needed. The government has not yet managed to set up an effective regulatory and monitoring system to address quality issues and the diversion of subsidized inputs smuggled outside the country. Some states have been testing the distribution of input subsidy vouchers since 2008.

The Agricultural Transformation Agenda was an initiative of the Federal Ministry of Agricultural and Rural Development (previously named Federal Ministry of Agriculture and Water Resources) to support former President Goodluck Jonathan's Transformation Agenda. The goal of the ATA was to build commodity value chains and the institutions required to unlock the country's huge agricultural potentials with the targeted outcomes such as, add 20million tonnes of food to the domestic food supply by 2015, create 3.4 million jobs and ensuring import substitution through the acceleration of production of local staples; reducing dependence on food imports and turning Nigeria into a net food exporter. A major success story of the ATA in 2014 was the news that Nigerian food import bill had dropped by N466 billion thereby adding N780 billion to the economy during the period. Building on the successes and lessons from the ATA, the current Administration under the leadership of President Buhari, the Agricultural Promotion Policy (APP) has the vision to work with stakeholders to build an agribusiness economy capable of delivering sustained prosperity by meeting domestic food security goals, generating exports and supporting sustainable income and job growth between 2016 and 2020. Among the 11 guiding principles of APP are food as a human right focusing the policy instruments for enterprise development on the social responsibility of government with respect to food security, social security and equity in the Nigerian society; and compelling the government to recognize, protect and fulfil the irreducible minimum degree of freedom of the people from hunger and malnutrition.

Despite successive strategies and programmes implemented, food and nutrition insecurity are still rampant. The 2015 Global Food Security Index, which comprises 109 countries appraised by the level of food sector development (food affordability, availability, quality and safety); put *Nigeria* at 91st place with the average of 37.1 indexes as against 80th among 105 countries in 2011. The food security situation of the country thus call for rights based approach interventions. The APP has recognized food as human right, therefore, there is need to create the awareness among the right holders in order to make the duty bearers accountable.

ISSUES TO BE ADDRESSED WHEN ADOPTING THE RIGHTS-BASED APPROACH TO FOOD SECURITY IN NIGERIA

Adopting the human rights framework for addressing food security seems a natural extension of the progress already made in the health and human development movement. This can also be applicable to food security by paying attention to several key elements that are but not limited to equal and public participation, government accountability, an analytical framework that accounts for vulnerability and discrimination and stronger connections between policies and outcomes. These issues include:

Equal and Public Participation

According to ICESCR (Art. 2), each state party to the present covenant undertakes to take steps, individually as well as through international assistance and co-operation to the maximum of its available resources, with a view to achieving progressively the full realization of the rights recognized in the present covenant." The principle of public participation seeks and facilitates the involvement of those potentially affected by or interested in rights based approach to food security in its decision making process. This is in relation to the facilitator of collective intelligence and inclusiveness, shaped by the desire for the participation of the whole country. Participation in both programs and policy making means and goal should be embraced by all the right holders bearing in mind that ability to claim right to food is their right.

Promote Government Accountability

The human rights framework is premised on the concept of accountability of state authorities through social control to all stakeholders. Since evidence of food insecurity is on the increase in the country, there is no apparent linkage of the report's findings to any action plan to reduce rates. Measurement is a key component of tracking the magnitude of food insecurity. This includes ensuring that there are governmental actors charged with establishing these reference goals with clear timeframes for implementation of action plans to achieve them. Strategies that empower the people they are meant to serve should be embraced.

Address Conflict, Vulnerability and Discrimination

Conflict may be defined as a struggle or contest between people with opposing needs, ideas, beliefs, values, or goals. Conflict on teams is inevitable; however, the results of conflict are not predetermined. Conflict is compromising the food security and nutrition of millions of Nigerians. Vulnerability can be defined as the probability of an acute decline in access to food, or consumption, often in reference to some critical value that defines minimal values of human well-being (WFP, 2002). Vulnerability influences a household's behavior and means of coping with its current wealth and livelihood circumstances. It also has an impact on the state of food security at the individual and household levels. Certain groups by nature of socioeconomic conditions or previous discrimination, are more vulnerable to food insecurity than others. A human rights approach entails focusing on those who are most vulnerable, understanding what causes this vulnerability or susceptibility to adverse outcomes, and changing conditions to improve their situation.

According to the United Nation Food and Agriculture Organization (FAO), women are responsible for 50 percent of world food production, mainly for family consumption. The majority of rural women are "invisible" field workers in family plots, and as a result, they have no recognized independent status as farmers and their work is considered as secondary both in the family and in society. In sub-Saharan Africa, only 15 percent of landholders are women and they account for less than 10 percent of credit and 7 percent of extension services. According to estimates, policies that address gender inequalities could, conservatively, increase yields on women's farms by 2.5 to 4 percent as investing in rural women has been shown to increase productivity significantly and reduce hunger and malnutrition. The vulnerability of women to hunger and food insecurity has long been recognized in the human rights documents of the United Nations. One of the greatest concerns is the intergenerational transmission of malnutrition that is, pregnant women that are malnourished are more likely to have low-birth weight babies. As a result, their children are more susceptible to under nutrition and poor cognitive development, which in turn affects the children's ability to earn enough money to supportthemselves and their families when they become adults. Since women and children are especially vulnerable to food insecurity and to socioeconomic processes that cause it, ensuring women's rights is an important corollary to achieving the right to food.

In Nigeria the appalling food insecurity situation has degenerated to a level that it is listed among the 42 countries tagged "low-income food deficit countries." (Okunmadewa, 2003). Food insecurity disproportionately affects rural people particularly rural women, minorities and children, World Bank (2001) revealed that rural people face a high risk of food insecurity due to poverty, income inadequacies, limited access to resources, underemployment, and unemployment, and many barriers to self-sufficiency, which create family frailty and crisis.

Measurable Goals and Targets That Link Policies to Outcomes

A human rights framework applied to the right to food can interpret how food-related policies affect one's ability to purchase food and how such policies affect health and wellbeing. Although, many Nigerians have applauded the development in the nation's agricultural sector, many are also of the view that the claims to a drop in the nation's food import bill does not correspond with market development as the prices of food and agricultural commodities have remained on the rise despite claims to huge gains in the sector.

CONCLUSION AND RECOMMENDATIONS

The cost of inadequate diets to families and nations are considerably high. This includes increased vulnerability to diseases and parasites, reduced strength for tasks requiring physical effort, reduction of the benefit from schooling and training programmes and general lack of vigour, alertness and vitality. These results in a reduction in the productivity of people in the short and long terms, sacrifice in output and incomes, and increasing difficulty for families and nations to escape the cycle of food and nutrition insecurity.

A right-to-food approach requires that states fulfil their obligation to ensure that safe, nutritionally adequate and culturally acceptable food is available. They must also respect and protect consumers as well as promote good nutrition for the deprived. In order to ensure the progressive realization of the right to food at the domestic level, it is imperative that constitutional principles and framework laws are established as a means of providing an appropriate institutional structure. The refreshed strategy- APP under the present administration can use the recognition of the right to food to benefit all Nigerians and promote gender equity in their policy thrust and objectives which are to solve the challenges of food security, import substitution, job creation and economic diversification. Achieving these through rights based approach can thus reinforce the nation's fundamental goal of leveraging the right holders' capabilities to ensure food and nutrition security as well as income security.

Rights-based approach to food security should ensure programs that focus on vulnerable, marginalized, disadvantaged and excluded groups, and that aim to reduce disparities between these groups and others. It is important, however, to note that advocacy works on two levels highlighted below:

- With the right holders to help them claim their right in legitimate ways while working on protecting the rights of others. Working especially with the disadvantaged and marginalized right holders is crucially important as part of helping them regain confidence in their ability to equally and fully participate in the decision making process.
- With the Duty bearers – be them state, or non-state – to be aware of, and determined by, the rights of people in their determinations to respect, protect and fulfill these rights; and do their best to not only fulfill these rights, but also, avail themselves accountable and quick to respond to the people in this regard.

REFERENCES

Ecker & Breisinger. (2012). *The Food Security System*. Washington, DC: International Food Policy Research Institute. Retrieved from http://www.ifpri.org/sites/default/files/publications/ifpridp01166.pdf

FAO. (2006). *The right to Food: The Human Right to Adequate Food*. Retrieved from http://www.fao.org/righttofood/about-right-to-food/human-right-principles-panther/en/

FAO. (2009). *Declaration of the World Food Summit on Food Security*. Rome: Food and Agriculture Organization of the United Nations. Retrieved from http://www.fao.org/fileadmin/templates/wsfs/Summit/Docs/Final_Declaration/WSFS09_Declaration.pdf

FAO. (2010). *Food insecurity in the World: addressing food insecurity in protracted crises*. Rome: Food and Agriculture Organization of the United Nations.

FMARD. (2016). *The Agriculture Promotion Policy (2016-2020): Building on the successes of the ATA, Closing Key Gaps*. Policy and Strategy Document of the Federal Ministry of Agriculture and Rural Development.

Food and Agriculture Organization (FAO). (1997). The food system and factors affecting household food security and nutrition. In *Agriculture, food and nutrition for Africa: A resource book for teachers of agriculture*. Rome: Agriculture and Consumer Protection Department. Retrieved 15 October 2013 from http://www.fao.org/docrep/W0078E/W0078E00.htm

FRN. (1999). Constitution of the Federal Republic of Nigeria.

Garrett, J., & Ruel, M. (1999). *Are Determinants of Rural and Urban Food Security and Nutritional Status Different? Some Insights from Mozambique*. Washington, DC: International Food Policy Research Institute. Retrieved 15 October 2013 from http://www.ifpri.org/sites/default/files/pubs/divs/fcnd/dp/papers/dp65.pdf

Global Nutrition Report. (2015). *Basic Global Nutrition Report*. Retrieved from https://www.globalnutritionreport.org/files/2015/10/Basic-GNR-2015-presentation_102815.pdf

Godfray, H. C. J., Beddington, J. R., Crute, I. R., Haddad, L., Lawrence, D., Muir, J. F., & Toulmin, C. et al. (2010). Food Security: The Challenge of Feeding 9 Billion People. *Science*, *327*(5967), 812–818. doi:10.1126/science.1185383 PMID:20110467

Gregory, P. J., Ingram, J. S. I., & Brklacich, M. (2005). Climate change and food security. *Philosophical Transactions of the Royal Society of London. Series B, Biological Sciences*, *360*(1463), 2139–2148. doi:10.1098/rstb.2005.1745 PMID:16433099

Helen, H. J. (2002). Food Insecurity and the Food Stamp Programme. *American Journal of Agricultural Economics*, *84*(5), 1215–1218. doi:10.1111/1467-8276.00382

Hunger Task Force. (2003). *Halving Hunger by 2015: A Framework for Action*. Interim Report, Millennium Project, UNDP.

ICESCR. (1999). *The International Covenant on Economic, Social and Cultural Rights. General comments 12. Twentieth session, 12th May, 1999*. Retrieved from http://www.fao.org/fileadmin/templates/righttofood/documents/RTF_publications/EN/General_Comment_12_EN.pdf

King, F. S., & Burgess, A. (1993). Nutrition for Developing Countries (2nd ed.). Oxford, UK: Oxford University Press. doi:10.1093/acprof:oso/9780192622334.001.0001

Ljungman, C. M. (2004). Applying a Rights-Based Approach to Development: Concepts and Principles. *Conference Paper: The Winners and Losers from Rights-Based Approaches to Development*, 6.

Okunmadewa, F. (2003). Risk and Vulnerability in Agriculture: concept and Context. *Staff seminar paper in the Department of Agricultural Economics*. University of Ibadan.

Otaha I. J. (2013). Food Insecurity in Nigeria: Way Forward. *Journal of African Research Review*, *7*(4), 26-35.

Shetty P. (2003). Malnutrition and Undernutrition. Science. *Medicine*, *31*(4), 18-22. doi:10.1383/medc.31.4.18.27958

Tweeten, L. (1999). The Economics of Global Food Security. *Review of Agricultural Economics*, *21*(2), 473–488.

UN. (2002). *Rights-Based Approach to Development Programming: Training Manual*. Retrieved from http://www.un.org.ph/publications/RBAManual.pdf

UN. (2017). *What is a human rights-based approach? Human Rights-Based Approach Portal*. Retrieved from http://hrbaportal.org/faq/what-is-a-human-rights-based-approach

United Nations Committee on Economic, Social, and Cultural Rights. (1999). *The right to adequate food*. Geneva: United Nations.

WFP. (2002). Vulnerability Analysis and Mapping: Standard Analytical Framework. Rome: Author.

WFP. (2013). *Food and Nutrition Security in Democratic People's Republic of Korea. Vam Food security Analysis*. Retrieved from https://www.wfp.org/sites/.../files/DPRK_draft_FINAL%201%20October%202013.pdf

WFP. (2017). *Global Nutrition Report*. Retrieved from https://www.wfp.org/stories/10-fact-about-food-and-nutrition-nigeria

WHO. (2013). *Undernutrition: The Basics. Generation Nutrition 6*. Retrieved from http://www.who.int/nutgrowthdb/summary_jme_2013.pdf

World Bank. (2001). *Poverty and Hunger: Issues and Options for Food Security in Developing Countries*. Washington, DC: World Bank.

World Food Summit. (1996). *Rome Declaration on World Food Security*. Author.

World Population Prospects, (2016). *World Population Prospects: The 2015 Revision – Key Findings and Advance Tables*. United Nations Department of Economic and Social Affairs, Population Division.

Section 6

Farmland Conservation and Environmental Stewardship

This section deals with conservation of farmland and changes in the environment in the affecting food systems.

Chapter 11
Economics of Soil Fertility Management Practices in Nigeria

Ibukun Joyce Ogwu
University of Abuja, Nigeria

Olubunmi Abayomi Omotesho
University of Ilorin, Nigeria

Abdulazeez Muhammad-Lawal
University of Ilorin, Nigeria

ABSTRACT

The production of cereals, tubers, and vegetables largely depends on the application of organic and inorganic fertilizers to offset their nutrients requirement. In this chapter, the authors identify different soil fertility management practices the maize farmers are using and examine the economic benefits of such practices in maize production. To complete the study, 237 maize farmers across Kogi and Kwara States, Nigeria were investigated. Descriptive statistics, gross margin, and multinomial logit tools were used to analyze the data. The results show that majority of the maize farmers (41.40%) use only inorganic material. Labour employed in the application of fertility materials, distance to the source of fertility materials, the quantity of seed planted, educational attainment, and gender of the maize farmer were the determinants of the use of fertility management practices relative to integrated soil fertility management (ISFM). The analyses of the results show that the use of ISFM for maize production is the most profitable method with a profitability ratio of 2.29.

DOI: 10.4018/978-1-5225-3631-4.ch011

INTRODUCTION

Many factors influence the level of output every farmer obtains at the end of the production season. Some of these factors largely depend on soil management practices the farmer adopts. These soil management practices, either good or not, will influence the farmer's level of output. Some of the soil management practices include; use of agrochemicals, tillage system used, soil nutrient mining, removal or loss of vegetative cover, continuous cropping as well as use of soil fertility materials (Mtambanengwe and Kosina, 2007; NOAN, 2012; FAO, 2017). All these practices, when performed excessively may negatively affect the crop production and as a result, level of the crop output decreases. The use of these unfavourable soil management practices are referred as unsustainable.

The overall effect of these unsustainable practices may cause global warming or increasing global temperature as a result of green-house-gases (GHG) emission. Asides the militating impact of climate change on production, the existing and growing population needs their food sustenance to be met by adequate food production. The drive to increase food production in Nigeria for instance has resulted into nutrient mining due to continuous and perhaps mono-culture oriented cropping. A typical example in the Nigerian context is maize production. In the year 2014, a decline of 2.60 percent value in maize crop growth rate has been recorded. Showing that there was depletion in the level of maize output produced when compared to the previous year. Nigeria produces 7.7 percent of the world share of maize (FAOSTAT, 2016). The World Bank in 2013 however pointed that 16 kilograms of fertilizer was used per hectare of arable crop produced in Nigeria. There is need to emphasize the importance of soil fertility management as a means of improving agricultural productivity (USDA, 2014; FAOSTAT, 2016).

Soil Fertility Management in Crop Production

The application of organic and inorganic fertilizers solely or combined are found to have great influence on the vegetative growth and yield of the crop. A study carried out by Ullah, *et al,* (2008), showed the highest yield was obtained from the combined application of organic and inorganic sources of soil nutrients (ISFM). The organic matter content and availability of N, P, K and S in soil were increased by application of both organic materials and inorganic fertilizer application. It was also found that highest vegetative growth was recorded from the use of the ISFM and the lowest vegetative growth was found with the use of only organic fertilizer in form of cow dung used. Also, the application of only inorganic fertilizer was less effective than the combined application.

Integrated use of organic wastes and mineral NPK fertilizer showed promising potential for improving soil fertility, growth, and yield of maize in the Southern Guinea Savanna agro ecological zone of Nigeria. In the study carried out by Ogundare *et al.,* (2012) the results showed that integrated application of both agricultural wastes and NPK fertilizer enhanced maize yields over sole use of NPK and sole use of organic wastes. Rahman, *et al.* (1998) stated that vegetative growth and yield was the highest with the combined application of manures and fertilizers. Jose, *et al.* (1988) stated that an integrated use of urea and poultry manure also resulted in a higher nutrient uptake and yield.

Several studies have been conducted to assess the effects of organic and inorganic fertilizers on soil properties and crop yields, and different agronomic and environmental outcomes have been observed depending on the specific agro ecosystem (Son, *et al.* 2004; Havlin, *et al.* 2005; Saha, *et al.* 2008). Agricultural waste-turned manure has been found to increase the contents of nutrients in the soil and its uptake by maize crops are high allowing for a considerable yield of maize (Ndanu, *et al.* 2012; Adeniyan & Ojeniyi, 2005; Agele, *et al.* 2011).

Aguilera, *et al.* (2012) highlighted the importance of the use of integrated soil fertilization program using the diverse combination rates of both inorganic and organic soil amendments. It was concluded from the study that the practiced of the ISFM increased short and long term soil fertility for the usage of crops.

Saidu, *et al.* (2012) in a study on the use of ISFM practice indicated that 2kg/ha of poultry manure plus 260g N/ha could support the growth and yield of upland rice in the Southern Guinea Savanna Agro-ecological zone. A field experiment conducted continuously by Chand, *et al.* (2006) for seven years to evaluate the influence of combined applications of organic and chemical fertility buildup and nutrient uptake in a mint (*Mentha arvensis*) and mustard (*Brassica juncea*) cropping sequence resulted into the conclusion that integrated supply of plant nutrients through FYM (farmyard manure) and inorganic fertilizer-NPK, played a significant role in sustaining soil fertility and crop productivity.

Based on the evaluation of soil quality indicators Dutta, *et al.* (2003) reported that the use of organic fertilizers together with chemical fertilizers had a higher positive effect on microbial biomass and hence soil health when compared to the addition of organic fertilizers alone. Application of organic manure in combination with chemical fertilizer has also been reported to increase absorption of N, P and K in sugarcane leaf tissue in the sugar plant and ratoon crop, compared to the use of chemical fertilizer alone (Bokhtiar & Sakurai 2005).

Kaur, *et al.* (2005) compared the change of chemical and biological properties in soils receiving farm yard manure (FYM), poultry manure and sugarcane filter cake alone or in combination with chemical fertilizers for seven years under a cropping sequence of pearl millet and wheat. Results showed that all treatments except chemical fertilizer application alone improved the soil organic matter, total N, P, and K status. Increase in microbial biomass of carbon and nitrogen was observed in soils receiving organic manures only or with the combined application of organic manures and chemical fertilizers compared to soils receiving chemical fertilizers only. This study eventually showed that balanced fertilization using both organic and chemical fertilizers is important for maintenance of soil organic matter (OM) content and long-term soil productivity in the tropics where soil OM content is low.

The long term use of cow dung increased aggregate stability, pore space, bulk density and availability of water range of the soil (Vanlauwe, *et al.* 2001). Cow dung contains the undigested portion of the feed eaten by animals, and has higher Nitrogen and Potassium contents which are quickly available to plants (Fagbenro, 2001). Cow dung applied with inorganic Nitrogen (N) fertilizer was found to increase soil pH and ameliorated acidity (Olayinka & Ailenubhi, 2001). Mullens, *et al.* (2002) revealed that poultry litter contains a considerable amount of organic matter due to the combined manure and bedding materials. Poultry manure improves soil physical properties significantly by reducing soil bulk density, temperature, and increasing total soil porosity and moisture content in Nigeria (Agbede, *et al.* 2008; Saidu *et al.*, 2012).

Akande, *et al.* (2005) reported that complementary application of rock phosphate with poultry manure increased maize grain yield by 33% while cowpea yield was increased by 25%. Buri, *et al.* (2004) in an experiment with poultry manure, cattle manure and rice husk, applied singly or in combination with mineral fertilizer (using urea or sulphate of ammonia as nitrogen source), found that a combination of half rate of organic amendment and half rate of mineral fertilizer significantly contributed to the growth and yield of the cereal.

In contrast to mineral fertilizer, poultry manure adds organic matter to soil which improves soil structures, nutrient retention, soil moisture holding capacity, and water infiltration (Deksissa, *et al.* 2008). It was also indicated that poultry manure moves readily, supplies phosphorus to plants than other organic manure sources (Gary & Bahla, 2008). Most studies on the use of animal wastes dealt with cow dung and poultry droppings and their fertility improving value has been confirmed for many crops (Akanbi, 2002). The use of available and cheap cow dung by vegetable farmers in Nigeria ensure sustainability of production and balanced nutrition as reported by Moyin (2003). Soils treated with manure were found to contain enough soluble phosphoric acid, potash and lime (Pal, *et al.* 2001; Belay, *et al.* 2006).

Other such as Woomer and Swift (1994), Palm and Gachengo (2001), Vanlauwe *et al.* (2002), Vanlauwe and Giller (2006), Tittonell, *et al.* (2008), and Chivenge *et al.* (2011) have established that soil fertility is better improved by addition of both organic manure and inorganic fertilizers.

Soil Fertility Sustainability: A Conceptual Framework

Agricultural production in Africa is hampered by the predominance of fragile ecosystems such as is seen in the Derived Savanna zone. It is characterized by high rate of soil fertility loss, inherited soil fertility, and low use of modern inputs such as mineral fertilizers. Farmers in sub-Saharan Africa traditionally clear land, grow a few crops. They move on to clear more land, leaving the land fallow to regain fertility. Due to population pressure, farmers are forced to grow crop after crop, "mining" or depleting the soil of nutrients while giving nothing back into the soil for the plants to thrive on. With little access to fertilizers, the farmers are forced to bring less fertile soils into production (Henao & Baanante, (2006).

Escalating rates of soil nutrient mining make nutrient losses highly variable in agricultural areas in the sub-humid and humid savannas of West Africa, Nigeria inclusive. Depletion rates range from about 30 to 40 kilograms of nitrogen, phosphorus, and potassium (NPK) per hectare every year in the humid forests and wetlands of southern Central Africa and Sudan to more than 60 kilograms NPK per hectare yearly in the sub-humid savannas of West Africa. Nigeria has been reported to lose about 57 kilograms (NPK) per hectare yearly. Thus, the main factors contributing to nutrient depletion are loss of nitrogen, phosphorus and potassium through soil erosion by wind and water, and leaching. Nutrient loss due only to erosion in African soils range from of 10 to 45 kilograms of NPK per hectare per year. If erosion continues unabated, by 2020 the soil fetility loss could decrease production at about 10 million tons of cereals, 15 million tons of roots and tuber crops (Henao & Baanante, 2006; Bayu, *et al.* 2006).

The concept of sustaining soil fertility depends largely on the types and quantities of soil fertility materials for crop production. As such farmers must have some ways of evaluating different optional ways and making the decisions for appropriate choices. Whichever method of production chosen will influence the cost of production of the farmer. This makes most producers strive to determine the appropriate method of soil fertility practice that will result into least cost of production and also enhance maximum output. Thus, the choice of inputs for using during the crop production largely depends on the method of production used by the farmer. In rational behaviour, the producer makes decisions to attain certain objectives. From the economic viewpoint, four basic elements are involved in the making of economic decisions. These include;

1. The decision maker that is, the farmer,
2. The objective of the decision maker,
3. The conditions under which the decisions are made, and
4. The 'measuring stick' for determining how the objective is attained.

Vanlauwe *et al.* (2010) as well as Chivenge, *et al,* 2011, describes ISFM as the application of soil fertility management practices and the knowledge to adapt these applications to local conditions, which maximize organic and inorganic fertilizer resource use efficiency and crop productivity. This definition shows that there is an interrelationship between soil fertility material, its availability in sufficient quantities at cost efficient amount as well as the knowledge of the principle of using integrated application. It should also be noted that the application of the ISFM is a knowledge intensive process and requires the farmers to learn new knowledge and information to achieve appropriate application of the practice (Bekunda *et al*, 2010).

A framework for sustainability in the use of ISFM practice will therefore involve three aspects. These are:

1. Adequate farm output to be produced in order to ensure adequate and considerable return;
2. A profitable system for the producer in terms of minimum cost incurred during the production process; and
3. Responsible safeguard for the environment. The external environment is becoming increasingly important (Keeney, 1997).

The farmer's decision making of restoring soil fertility using the ISFM practices depends on the farmer's goals and the tendency of the farmer to apply ISFM system also depends largely on the other external controlling factors and likewise the socio-economic factors of the farmer.

The farmer's decision to use the ISFM system involves the choice of the quantity of resources, like land, to be allocated to the new technology if the technology is not divisible (machinery, irrigation). However, if the technology is divisible like seed and fertilizer, the farmer's decision process involves area allocation as well as level of usage or rate of application of the technology (Feder *et al* 1985).

Socioeconomic Characteristics of the Farmer

These are the characteristics that portray the farmers and give brief information about the farmers. These characteristics were measured as codes or numbers. Some of the characteristics that were looked into include:

1. Sex of the farmer.
2. Age of the farmer.
3. Level of education of the farmer.
4. The farmer's household size.
5. Membership in a cooperative organization etc.

Soil fertility usage describes the use of organic and/or inorganic soil fertility materials by the farmer. The use of any of these soil fertility materials will also depend on the benefit the farmer wants to achieve in the production process. The parameters considered in this aspect were;

1. Availability of the soil fertility materials.
2. The quantity of soil fertility material used by the farmer.
3. The cost incurred in the use of the soil fertility material.
4. Cost of delivering the soil fertility materials to the farm.
5. Labour used in the application of the soil fertility material.
6. Challenges faced in the use of the respective soil fertility materials etc.

Farm characteristics of the farmer described the farm attributes of the respective farmer. Some of the characteristics that were considered include:

1. Farm size.
2. Distance of the source of the soil fertility materials to farm.

Other external controlling factors that affect the usage of the fertility material are:

1. Availability of other soil fertility materials.
2. Sources of the soil fertility materials to the farmer.

Grilliches (1980) showed that aggregate usage is a function of economic variables such as profitability and the level of acceptance of the innovation. Studies by Rogers (2003) identified seven characteristics any basic innovation will have. These include, relative advantage, compatibility, complexity, divisibility, observability, variation in cost of usage and group action requirement of the innovation. However, while all the other six points might be relevant, group action requirement might not be binding on the farmers during the practice of the ISFM system on their various farms. Byerlee and Heisey (1986) examined the rate of usage of technology and various economic factors. The study revealed that the usage pattern of a particular technology is a function of five characteristics which are profitability, riskiness, divisibility or initial capital investment, complexity and availability. The study also showed that profitability and riskiness of a given innovation are a function of agro-climatic and socio-economic environment such as rainfall and prices. Interaction revealed during the usage of the innovation will also affect the rate of usage of the innovation (Hassan *et al.* 1998).

There is also arguement that the rate and the level of usage of an innovation depend on the availability of the materials needed by the farmers. Also farmers decision on the level of usage that farmers choose to use inputs sequentially and as a single component as a whole package (Leathers *et al.* 1995). It should be noted that innovations are classified into process and product innovations. The process innovation is an input to a production process while the product innovation is an end product for consumption. Technology on the other hand is any idea, object or practice that is perceived as new by the members of a social system. Thus, for this study, the process innovation will be meant by the term innovation.

The demand for fertilizer depends on its contribution to the value of the output. Two elements determine the returns to fertilizer use. These are the technical relationship between the different levels of fertilizer and the quantity of output produced holding all other factors constant. Also, the profit maximization assumption of the theory of firm, that an optimum level of fertilizer is achieved at the point where the value of additional output from the extra unit of fertilizer is equal to its cost (price of fertilizer).

Nowak (1992) summarized the inability to use a particular innovation to be influenced by these:

1. Lack of, or scarce information.
2. Too high costs of obtaining information.
3. Too great complexity of the system.
4. Too expensive.
5. Excessive labor requirements.
6. Too short planning horizon (benefits too far in the future).
7. Limited availability and accessibility of supporting resources.
8. Inadequate managerial skills.
9. Little or no control over the adoption decision.

Place (2003) argued that market opportunity act as a trigger that stimulated the usage of ISFM system. Also, Freeman and Coe (2002) obtained a positively correlated result in the amount of organic and inorganic fertilizer applied with the selling price of the farm product. Thus, this means that farmers might increase or reduce their demand for organic or inorganic fertilizers depending on the direction of demand for organically produced outputs.

Profitability of Using the Soil Fertility Management Practices

More profitability is ensured and sustainable in farms where organic soil amendments are adopted for crop production because there is a reduced costs on investments in chemical inputs, especially chemical fertilizers (Nwaiwu *et al.,* 2010). Arunah and Ibrahim (2004) found that the use of nitrogen fertilizer (urea) and poultry manure was more productive when compared to their individual effects. The gross margin analysis showed that the separate application of 80kg urea per hectare and 4ton of poultry manure gave lower gross margins per hectare as compared to the integrated application of 40kg urea and 4ton poultry manure per hectare on the same land. This implies that the combined application of organic and inorganic fertilizer will not only increase the output of the farmer but also increase the income of the farmers.

Merumba *et al.* (2012) studied the effect of fertilizers on crop productivity and profitability through application of ISFM on climbing beans, bush beans, soybeans, cassava and sweet potato. He found that there was significant difference in crops response to the application of different types of fertilizers within the treatments. Likewise, the result of a study to assess the effect of organic fertilizer inputs on soil nutrients, soil water availability and maize (Zea mays) productivity in semi-arid region of the Central Highlands of Kenya showed that sole use of manure, recorded less maize grain yield. Generally the maize grain yields were lower with the sole inorganic fertilizer treatments compared to the organics across the seasons and this was attributed to poorly distributed rainfall. The sole organics had higher yields during the short rainy seasons, while the combination of organic and inorganic fertilizer recorded higher yields during the long rainy seasons. Manure was superior in terms of improving soil chemical properties such as soil pH, magnesium, potassium, calcium and nitrogen across the sites (Mucheru-Muna *et al*, 2012). It was also asserted that it is uneconomical to cultivate crops without application of any form of fertilizers and that planting maize without the application of fertilizer will only result in an extremely low average in yield of maize of less than 1 ton per hectare instead of a reachable 6.0 tons per hectare.

In the study conducted by Kimaro *et al* (2009) on financial analysis of the farm budgets confirmed that the addition of organic and inorganic fertilizer increased net profit per hectare. Also Tittonell *et al.* (2008) in a study carried out in Kenya found that there was a significant profit realized in the integrated use of both types of fertilizer for maize in Kenyan field trials. The combined application of organic manure plus inorganic fertilizers gave the highest net cash returns and returns to labour also (Lunduka and Kelly, 2009).

Financial analysis of the farm budgets confirms that the combination of organic and inorganic fertilizer and intercropping with legumes increase net profit per hectare. Intercropping maize with pigeon peas also provides high net returns when combined with inorganic fertilizers (Lunduka and Kelly, 2009). In the study carried out by Segun-Olasanmi and Bamire (2010), to examine the profitability of the maize-legume intercrop system, it was indicated that maize-cowpea cultivation in the derived savanna zone of Oyo state, Nigeria is indeed a profitable farm enterprise.

Studies on the profitability of the maize enterprise by Omisore (2013) revealed that the use of only urea to improve soil fertility gave a gross margin of N87,683/ha while the combined use of organic manure and inorganic fertilizer gave a gross margin of N74,184/ha. The study also indicated that even though the use of only urea had high yield response, it was not as cost efficient as the combined use of organic manure and inorganic fertilizer. The use of only urea in the production of maize gave a variable cost of N114,893.00 and the combined use of organic manure and inorganic fertilizer incurred a lower variable cost of N101, 627.00 per hectare respectively. Hence, a rational maize farmer would like to produce optimally at the minimum cost of production and adopt the use of the integrated soil fertility management.

Where performance describes the measure of standards and completion, economic perforce describes the measure of certain economic parameters such as gross margin, profitability/benefit to centration and rate of return on capital investment. These measures give thorough analysis of the economics of the enterprise.

METHODOLOGY

Study Area and Sampling Technique

The sample survey was conducted during the 2014 maize cropping, season in both Kogi and Kwara State of Nigeria. These two states belong to the derived Savanna Zone also known as the grain belt of the Country. The derived Savanna Zone is also known to be most suitable compared to other vegetation belts due to the two distinct rainfall cycle that allow for the production of at least two maize crops in a year. The 237 farmers used for the study were selected using four staged sampling techniques. This helped to generate a list of maize farmers per state before a random sample using ballot system was conducted to choose the respondents. The farmers planted yellow and white varieties maize and they were the owners of the maize farm.

Sampling Tools

Gross Margin Analysis

Mathematically, gross margin is

$$GM=GI-TVC \quad (1)$$

where

GM= gross margin
GI= gross (farm) income

TVC= total variable cost

The gross (farm) income is obtained by multiplying total farm output with the prevailing price. This is also referred to as the value of output.

The benefit cost ratio or profitability ratio was obtained as the index of the value of output and the total variable cost incurred during production. The rate of return on capital investment was obtained by finding the index of the gross margin and the costs of production. This measure depicts how well the gross margin can cover the variable cost of production. For every rational farmer, the decision should be to keep producing as long as the variable cost of production is covered by the gross margins. The result of the profitability and other performances of the different soil fertility managment practices is shown in Table 3.

Multinomial Logit Regression Analysis

Multinomial logit regression analysis was the technique used to identify factors that deremine the use of soil fertility managment. The tool allows for the estimation of the probability that an event will occur or not. It predicts a dependent outcome from a set of independent variables. The Statistical Package for Social Sciences (SPSS) software version 16.0 was used to estimate the model.

The multinomial logit regression model expressed according to Gujarati (1988) and Babcock *et al* (1995) in Rahji *et al* (2008) as:

$$P_{ij} = \frac{\exp(\gamma jXiDi)}{1+\sum_{j-1}^{4}\exp(\gamma jXiDi)} \cdots \quad (2)$$

multinomial model for j =1,2,3,4 where:

P_{ij} = the probability associated with the four choices of soil fertility materials

We, therefore, have the following independent variables (X_i) as follows;

D_1 = sex (dummy)
X_1 = age (years)
X_2 = household size
X_3 = education (coded)
D_2 = membership of cooperative (dummy)
X_4 = frequency of extension visit
X_5 = farm size (hectare)

X_6 = total labour for application of fertility materials used (man-day)
X_7 = distance to fertility material sources (kilometer)
X_8 = income from other sources (off-farm income) (naira)
X_9 = quantity of seed (kilogram)
U = error term

These independent variables (X_i and D_i) provide the combined effects that prevent or promote the usage of any soil fertility system relative to the use of ISFM among maize farmers studied. These independent variables are hypothesized to influence farmer's decision to use any form of soil fertility management system in a positive or negative direction.

CIMMYT (1993) identified that farmer's objectives and their constraints influence their decision to use certain soil fertility management practices. Thus the signs attached to the variables are the presumed effect each factor may have on the usage of ISFM among maize farmers.

RESULTS AND DISCUSSION

Soil Fertility Management Practices: Classification and Yield Realized From Use

The methods of soil fertility management are mainly four types. Some of these methods have been transferred over generations and are in use due to experience gained over years. As such, the choice of fertility materials these farmers use based on the availability of these materials determine the classes of these soil fertility management practices. Hence, the classes of soil fertility management practices include,

1. Use of no soil fertility materials.
2. Use of only organic soil fertility materials.
3. Use of only inorganic soil fertility materials.
4. Use of ISFM (Paired use of organic and inorganic materials).

The use of no soil fertility materials by food crop formers involve the total dependence on the nutrients in the farmland alone. The farmer solely engages in other cultural practices, such as weeding, mulching among others, to support the crop production. In cases where the land is virgin, the farmer may gather more output in the early seasons of productions. However, continuous cultivation of such farmland without using any soil amendments will result into less and less output. Another perspective may be the inaccessibility of the farmers, to fertility materials that can be used to improve soil fertility. This system of use of no soil fertility material can be associated with the earliest manner of farming that is characterized with low output level just enough to sustain their families. However, this system is not sustainable with the geometric progression in population growth as well as the influence of urbanization. The on farm survey sample survery was conduct of a group of maize farmers presented in Table 1showing that 14percent of maize farmers do not use any farm of soil fertility materials. The average yield recorded of all four categories of soil fertility management practices used by the maize farmers is given in Table 1.

The use of only organic soil fertility materials in the field survey among maize farmers of Kogi and Kwara States falling into the grain belt of the country show that only 18.10 percent of the maize farmers practice the use of only organic materials to improve soil fertility. The organic materials considered for this study are animal wastes like poultry manure and cow dung. The use of only organic material produced maize yield of about 480kg/ha. This is significantly higher than the yield obtained by the farmers who did not use any soil fertility materials.

As expected, the sole use of inorganic soil materials was highly practiced among maize farmers at 41.40 percent. This implies that use of only inorganic soil fertility materials in maize production is predominant in Nigeria. This prominence however gave a higher level of yield of 537.78kg/ha of maize.

Integrated soil fertility management (ISFM) system can be broadly defined as the combination of adequate proportion of organic and inorganic fertility materials in the light of the farmer's socio economic factors, external controlling factors as well as high knowledge of ISFM system to promote soil fertility and hence increase crop production (Memudu, 2015). As seen from Table 1, 26.20 percent of the maize farmers studied practiced the use of ISFM system for their maize production. The yield produced however under the use of ISFM system is 738.77kg/ha. This practice obtained the highest level of yield. This portray, that the use of ISFM maize production can actually enhance yield in maize production.

Table 1. Use of fertility materials and the respective maize yield obtained

Soil Fertility ManagementPractices	Percentage (%)Use Among Maize Farmers	Maize Yield (kg/ha)Obtained
None use	14.30	287.9
Only organic	18.10	479.68
Only inorganic	41.40	537.78
Use of ISFM	26.20	783.77

Source: Field Report, 2015.

Difference in the Yields Obtained From Fertility Management Practices

Analysis of variance test was used to show the possibility of a difference in the yields obtained from each fertility management practices. The result indicate that, all the yields are significantly different from one another at P< 0.01. in other words, the maize yields obtained from the four categories of maize production are dissimilar at P<0.01 level of significance. This can be seen in Table 2.

Performance and Profitability of Soil Fertility Management Practices

The result of the gross margin analysis is presented in Table 3. This shows the performanace as well as profitability of maize-cassava production under the four different soil fertility management practices. Farmers that used ISFM practices had the best performance when compared to the other types of soil fertility management practice.

Gross margin analysis was used to determine the cost and benefits of maize enterprise under the different soil fertility management practices. This tool presents the difference between gross farm income and the total variable lost (Lunduka and Kelly 2009). It is used when the fixed cost is negligible portion of the farming enterprise as it is seen among small scale farmers where their fixed costs do not alter much with changes in production process.

The gross margin analysis result revealed that maize production enterprise was profitable under all the four types of soil fertility management practices employed in the maize production. These affirm the suitability of the Derived Savanna zone to maize production as it also called the grain belt of the nation. However, the use of ISFM had the highest level of profitability at NN75,784.75.

Table 2. Analysis of variance (ANOVA) in maize yield

		Sum of Square	DF	Mean Square	F-Value	Sig
Maize yield	Between Group	5944000.56	3	198133.521	1.16	0.00
	Within Group	4.136E7	233	177491.54		
	Total	4.730E7	236			

Data Analysis 2015

Table 3. Gross Margin Analysis of Maize Production Under Different Soil Fertility Management in Kogi and Kwara States

Items		Used ISFM	Used Only Inorganic	Used Only Organic	No Soil Fertility Materials used
		N/ha	N/ha	N/ha	N/ha
Outputs:	Maize	71,694.00	43,623.00	43,112.00	29,512.00
	Cassava	62,758.00	51,798.00	27,297.00	20,176.00
Variable costs:	Cost of seed	2,238.70	2,092.70	1,401.90	1,554.70
	Cost of stem cuttings	3,029.00	1,903.10	2,790.70	3,794.10
	Cost of NPK fertilizer	11,497.00	11,011.60	-	-
	Cost of urea	7.178.20	6,209.20	-	-
	Cost of poultry manure	342.86	-	382.35	-
	Cost of cow dung	100	-	0	-
	Cost of NPK fertilizer transport	194.52	232.35	-	-
	Cost of urea transport	170.32	208.88	-	-
	Cost of manure transport	143.71	-	592	-
	Cost of agrochemical	3,411.90	3,610.40	1,953.50	1,667.76
	Cost of hired labour	30,361.00	34,906.00	33,324.00	18,801.27
Total variable cost		58,667.21	60,265.63	40,444.45	25,817.83
Gross margin		75,784.74	35,155.77	29,964.55	23,870.17
Benefit-cost ratio/profitability ratio		2.29	1.58	1.74	1.92
Rate of return on capital investment		1.29	0.58	0.74	0.92

Source: Data Analysis, 2015

It is seen that labour had the highest cost incurred for all the different soil fertility management systems. This is in accordance with Olayide and Heady (1982) that labour accounted for the highest cost item in small scale agricultural production.

Maize farmers who used no form of soil fertility materials had the least gross margin. It is evident that maize farmers who used inorganic fertilizer incurred highest costs than the other types of soil fertility management practices. A similar assertion was made by Kayuki *et al.* (2012) that the use of inorganic fertilizer-NPK, was not as profitable in maize production when compared to other forms of soil fertility management practiced especially ISFM.

The gross margins realized from the use of soil fertility management practices was found to be significant at p=0.039 and with a t-value of 3.50. This implies that the gross margins of these four categories of soil fertility management practices were significantly different from one another at $p<0.05$ level of significance.

The rate of return on investment obtained as shown in Table 3 revealed that the use of ISFM had the highest return on investment of 1.29. This means that producing maize under the use of ISFM had the highest gain and it is the most lucrative maize enterprise to embark upon when compared to maize production under the other types of soil fertility management practices.

For the benefit-cost ratio of maize enterprise, the use of ISFM also had the highest benefit-cost ratio of 2.29. This means that the benefit obtained in maize enterprise under the use of ISFM doubles the cost incurred. With this, the maize enterprise under ISFM is the best of the four soil fertility management practices.

Determinants of Soil Fertility Management Practices

Certain factors have been identified by scientists to influence soil ferility and its management decisions and practices. Some of these factors are discussed in this section. The long term application of soil fertility materials, especially organic soil fertility materials, increased water soluble nutrients and also the average organic matter in the soil bringing about significant build up of N, P, and K nutrients. It was also noted that the increase in N, P, and K nutrients were brought about by increase in the rate of application and also the number of times of application of the soil fertility materials (Son *et al.* (2004) and Mohammadi et al (2009).

Age of the household head, rented land tenure and farmers' perception on soil erosion were identified to have negative influence on the usage intensity of ISFM practices. In an evaluation of the factors that determine the level of usage of the ISFM in Western Kenya, the percentage of usage of ISFM practices by various households was influenced by access to agricultural information, particularly from farmer groups who played the significant role in determining the intensity of use of ISFM by their various households (Nambiro *et al.*, 2012).

Marginal Effects and Quasi-Elasticities

The marginal effects or partial derivatives are obtained as the derivative from differentiating the dependent and the independent variables $\left(\frac{dP_j}{d\bar{X}_i}\right)$. The sign of the derivative does not have any influence on the magnitude of the marginal effects.

The quasi-elasticities, on the other hand, were obtained using the formula:

$$\eta = X_i \left(\frac{dP_j}{d\bar{X}_i} \right) \tag{2}$$

where $\bar{X}_i$ is the mean value of $\bar{X}_i$

The quasi-elasticity stands for the percentage point change in P_j upon a one percent increase in X_i (Rhaji *et al*, 2008).

Table 4. Determinants of Use of ISFM relative to the Use of Other Fertility Management Practices in Maize Production (Pooled)

	No Fertility Material Used	**Only Organic Manure Used**	**Only Inorganic Fertilizer Used**	**ISFM**
Intercept	-3.13	-0.38	-1.32	
Sex	-0.55	-0.1	-0.91**	1.56
	(-0.79)	(-0.49)	(-0.39)	
Age	0.06	0.02	0.01	-0.09
	(-0.05)	(-0.03)	(-0.02)	
Household size	0.60**	0.16	0.16	-0.92
	(-0.16)	(-0.1)	(-0.09)	
Education	-0.11	-0.36**	-0.02	0.49
	(-0.26)	(-0.17)	(-0.14)	
Membership in cooperative	-2.48	-1.52	-0.06	4.06
	(-1.78)	(-0.85)	(-0.76)	
Frequency of extension visit	-2.83	-0.51	-0.001	3.341
	(-1.8)	(-0.86)	(-0.77)	
Farm size	0.21	0.12	0.02	-0.35
	(-0.18)	(-0.14)	(-0.11)	
Total labour used (application)	-0.64**	-0.004	-0.03**	0.674
	(-0.14)	(-0.02)	(-0.01)	
Distance to fertility materials	-0.04**	-0.02**	-0.001	0.061
	(-0.01)	(-0.01)	(-0.001)	
Off farm income	0	0	0	0
	(0)	(0)	(0)	
Quantity of seed	-0.06**	-0.03	-0.06	0.15
	(-0.03)	(-0.02)	(-0.03)	
Log-Likelihood ratio (λ)	177.61**			
N	34	43	98	62

Source: Data Analysis (2015)
** Significant At P<0.05
Note: coefficient and (standard error) are the order of arrangement of each variable parameters

The multinomial logit model fitting gave a test statistics value lambda (λ) of 177.61. The critical values were significant at $p<0.05$ showing that the regression coefficients are significantly different from zero. This result implies that the soil fertility material groups are heterogeneous.

As shown in Table 4, four factors identified as significant affected the use of 'no soil fertility materials'. These are household sizes, total labour used for application, distance to fertility materials as well as the quantity of seed planted by the maize farmers. These four factors influenced the farmers' decision to use 'no soil fertility materials' relative to use of ISFM.

Small household size encouraged the use of 'no soil fertility materials'. This might be as a result of the family members not directing their labour efforts toward agriculture-maize production. This might have influenced the maize farmers' decision to produce without using any soil fertility materials since there is low labour for application of fertility materials during the maize production. In other words, household size increased the probability of using 'no soil fertility materials' by 0.60 relative to use of ISFM. Macharia *et al.* (2012) identified that household size influenced the use of integrated soil fertility management.

The labour employed for the application of soil fertility materials, distance to soil fertility materials as well as the quantity of seed input are seen to discourage the use of 'no soil fertility materials' at the rate of 0.64, 0.04 and 0.06 respectively. It is seen that availability of labour for the application of soil fertility materials also encouraged the probability to use ISFM by 0.64. Furthermore, the quantity of seed planted by the farmer motivated him to choose ISFM rather than not using any form of soil fertility materials at all. This means that the farmer planted more quantity of seed and put in all the necessary inputs, including the use of ISFM to ensure bounty harvest.

Table 4 also shows that the education and distance to soil fertility materials were the factors that influenced the use of organic soil fertility materials at $p<0.05$ level of significance. This means that for a level increase in the level of education of the maize farmers, the multinomial log odd of preferring only organic soil fertility materials relative to the use of ISFM is expected to reduce by 0.36 given that other variables are held constant. In other words, education reduces the probability of using only organic soil fertility materials relative to the use of ISFM by 0.36. This implies that increasing the level of education of the farmers exposed them to better technology, innovation as well as increased their ability to understand them. Macharia *et al.* (2012) identified that education and exposure to training through farmers' organizations influenced the use of ISFM.

In the case of distance to soil fertility materials, the log odd of using only organic soil fertility materials is expected to decrease by 0.02 relative to use of ISFM while holding other variables constant. This implies that distance to soil fertility materials only decreased the probability of using only organic manure compared to the use of ISFM.

The third column-section shown in Table 4 gave the result of the analysis - the use of only inorganic soil fertility materials relative to the use of ISFM. Sex of the farmers significantly influenced the use of only inorganic soil fertility materials at $p<0.05$. In other words, being a man reduced the log odds of preferring the use of only inorganic soil fertility material by 0.91. This means that men have the lesser probability of using only inorganic soil fertility materials in maize production. Men have the likelihood of using ISFM compared to using only inorganic soil fertility materials by 0.91. Macharia *et al.* (2012) identified that the sex of the farmer influenced the use of ISFM.

Labour used for the application of soil fertility materials is shown to reduce the probability of using only inorganic soil fertility materials by 0.03. This implies that availability of labour for application of the materials increased the tendency of the maize farmers to use ISFM instead of using only inorganic soil fertility materials. From previous studies, Nambiro *et al.* (2012) stated that the age of the farmer had a significant and negative influence on the use of ISFM and access to information farmers groups increased the use of ISFM. This is contrary to the findings of this study, however. This study, however, concurs with that of Makhoka *et al.* (2001) where it was indicated that age does not have a significant influence on the use of ISFM by maize farmers. Makhoka *et al.* (2001) reported that farm size had a negative but important effect on the use of ISFM, which in this study was negative too but not significant.

Table 5 shows the marginal effects as well as the quasi-elasticities of the relevant variables.

The quasi-elasticities of labour used as well as the quantity of seed planted were elastic at 14.680 and 4.444 respectively for those that did not use any soil fertility materials. For farmers that used only organic manure, distance to soil fertility materials was elastic at 1.167. For farmers who used only inorganic soil fertility materials, only total used for application was elastic at 15.986 while sex of the farmer was inelastic. Finally, for farmers who used ISFM, total used for application was elastic at 13.827, distance to soil fertility materials was elastic at 5.939 and quantity of seed was elastic at 5.612. The other quasi-elasticities were low and inelastic. From this result, it can be deduced that the elasticity of using any of the fertility materials is greatly affected by the marginal changes in the variables that are elastic. As such, these elastic variables are those that will lead to the widespread use of these fertility materials.

Table 5. Marginal Effects and Quasi-Elasticities Estimated

Variables	No Fertility Material Used	Only Organic Manure Used	Only Inorganic Fertilizer Used	ISFM
	P_1	P_2	P_3	P_0
Sex	-	-	0.049	0.0450
	-	-	(0.049)	(0.098)
Household size	(0.140)	-	-	-0.141
	(0.700)	-	-	(0.505)
Education	-	0.223	-	0.223
	-	(0.446)	-	(0.658)
Total labour used (application)	0.182	-	0.182	-0.182
	(14.680)	-	(15.986)	(13.827)
Distance to fertility materials	0.108	0.108	-	0.110
	(0.968)	(1.167)	-	(5.939)
Quantity of seed	0.239	-	-	0.240
	(4.444)	-	-	(5.612)

source: Data Analysis (2015) marginal effects are above while quasi-elasticities are in brackets

CONCLUSION

This chapter deals with the economics of soil fertility management practices in Nigeria. The results showed that only 26.2 percent of the maize farmers in the study area used ISFM practice. This implies that the use of the practice is not yet widely in use, among maize farmers in the Derived Savanna zone of Kogi and Kwara States. Majority of the maize farmers use only inorganic soil fertility materials. This is not a sustainable system of crop production because the consistent use of inorganic materials contribute to greenhouse gases emission.

The use of ISFM recorded the highest gross margin as well as profitability when compared to the other forms of soil fertility management practices. This established with affirmation that the use of ISFM performed more than the use of other soil fertility management practices. It is therefore pertinent to promote ISFM as the most appropriate soil fertility management practice that ensures highest return both in yield and profitability. This will encourage and ensure the attainment of food security in Nigeria.

REFERENCES

Adeniyan, O. N., & Ojeniyi, S. O. (2005). Effect of Poultry Manure NPK 15-15-15 and Combination of their Reduced Levels on Maize Growth and Soil Chemical Properties. *Niger Journal of Soil Science*, *15*, 34–41.

Agbede, T. M., Ojeniyi, S. O., & Adeyemo, A. J. (2008). Effect of Poultry Manure on Soil Physical and Chemical Properties, Growth and Grain Yield of Sorghum in South-West, Nigeria. *American-Eriession Journal of Sustainable Agriculture*, *2*(1), 72–77.

Agele, S. O., Adeyemo, A. J., & Famuwagun, I. B. (2011). Effects of Agricultural Wastes and Mineral Fertilizer on Soil and Plant Nutrient Status, Growth and Yield of Tomato. *Archives of Agronomy and Soil Science*, *57*(1), 91–104. doi:10.1080/03650340903225040

Aguilera, J., Motavalli, P. P., Gonzales, M. A., & Valdivia, C. (2012). Initial and ResidualEffectsof Organic and Inorganic Amendments on Soil Properties in a Potato-Based Cropping System in the Bolivian Andean Highlands. *American Journal of Experimental Agriculture*, *2*(4), 641–666. doi:10.9734/AJEA/2012/2006

Akanbi, W. B. (2002). *Growth, Nutrient Uptake and Yield of Maize and Okra as Influenced by Compost and Nitrogen Fertilizer under Different Cropping Systems* (Unpublished Ph.D. Thesis). University of Ibadan, Nigeria.

Akande, M. O., Adediran, J. A., & Oluwatoyinbo, F. I. (2005). Effects of Rock Phosphate Amended with Poultry Manure and Soil Available Phosphorus and Yield of Maize and Cowpea. *African Journal of Biotechnology*, *4*, 444–448.

Arunah, U. T., & Ibrahim, H. (2004). Effects of Inorganic Fertilizer and Poultry Manure on Sorghum Yield at Samaru in the Northern Guinea Savanna of Nigeria. *Agrosearch*, *6*(1), 49–54.

Babcock, B. A., Chaherli, N. M., & Lakshiminariayam, P. G. (1995). *Programme Participation and Farm-Level. Adoption of Conservation Tillage: Estimates from a Multinomial Logit Model.* Working paper 95-WP 136, Centre for Agricultural and Rural Development, Iowa State University.

Bayu, W., Rethman, N. F. G., Hammes, P. S., & Alamu, G. (2006). Effects of Farmyard Manure and Inorganic Fertilizers on Sorghum Growth, Yield and Nitrogen use in a Semi- Arid Area of Ethiopia. *Journal of Plant Nutrition*, *29*(2), 391–407. doi:10.1080/01904160500320962

Belay, A., Classens, A. S., Wehner, F. C., & De Beer, J. M. (2006). Influence of Residual Manure on Selected Nutrient Elements and Microbial Composition of Soil under Long-Term Crop Rotation. *South African Journal of Plant and Soil, 18*(1), 1–6. doi:10.1080/02571862.2001.10634392

Bekunda, M., Sanginga, N., & Woomer, P. L. (2010). Restoring Soil Fertility in Sub-Sahara Africa. Advances, 108(10), 183–236.

Bokhtiar, S. M., & Sakurai, K. (2005). Effects of Organic Manure and Chemical Fertilizer on Soil Fertility and Productivity of Plant and Ratoon Crops of Sugarcane. *Archives of Agronomy and Soil Science, 51*(3), 325–334. doi:10.1080/03650340500098006

Buri, M. M., Issaka, R. N., Wakatsuki, T. & Otoo, E. (2004). Soil Organic Amendments and Mineral Fertilizers: Options for Sustainable Lowland Rice Production in the Forest Agro – Ecology of Ghana. *Agricultural and Food Science Journal of Ghana, 3*, 237 – 248.

Byerlee, D., & Heisey, P. (1986). Past and Potential Impacts of Maize Research in Sub-Saharan Africa: A Critical Assessment. *Food Policy, 21*(3), 255–277. doi:10.1016/0306-9192(95)00076-3

Chand, S., Anwar, M., & Patra, D. D. (2006). Influence of Long-Term Application of Organic and Inorganic Fertilizer to Build up Soil Fertility and Nutrient Uptake in Mint-Mustard Cropping Sequence. *Communications in Soil Science and Plant Analysis, 37*(1-2), 63–76. doi:10.1080/00103620500408704

Chivenge, P., Vanlauwe, B., & Six, J. (2011). Does the Combined Application of Organic and Mineral Nutrient Sources Influence Maize Productivity? A Meta-Analysis Plant Soil. Springer.

CIMMYT. (1993). *The Adoption of Agricultural Technology: A Guide for Survey Design.* Author.

Mexico, D. F. Deksissa, T., Wyche-Moore, G. S. & Hare, W. W. (2008). Occurrence, Fate and Transport of 17 beta – estradiol and testosterone in the environment. *Proceeding of 2007 AWRA Summer Specialty Conference.*

Dutta, S., Pal, R., Chakeraborty, A., & Chakrabarti, K. (2003). Influence of Integrated Plant Nutrient Phosphorus and Sugarcane and Sugar Yields. *Field Crops Research, 77*, 43–49.

Fagbenro, J. A. (2001). Organic Fertilizer Raw Material Situation in Nigeria. *Proceedings of a Workshop on Organic Fertilizer Development*, 1-9.

FAO. (2017). *Soils. Key to unlocking the potential of mitigating and adopting to a changing climate. SDG goal 13*. FAO.

FAOSTAT. (2016). *Nigeria: Maize, green, production quality (tons)*. Retrieved from factfishwww.factfish.com

Feder, G., Richard, E. J., & David, Z. (1985). Adoption of Agricultural Innovations in Developing Countries: A Survey. *Economic Development and Cultural Change*, *33*(2), 255–298. doi:10.1086/451461

Freeman, H. A., & Coe, R. (2002). Smallholder Farmers' Use of Integrated Nutrient-management Strategies: Patterns and Possibilities in Machakos District of Eastern Kenya. In C. B. Barrett, F. Place, & A. Abdillahhi (Eds.), *Natural Resources Management in African Agriculture: Understanding and Improving Current Practices* (pp. 143–154). Wallingford, UK: CABI Publishing. doi:10.1079/9780851995847.0143

Gary, S., & Bahla, G. S. (2008). Phosphorus Availability to Maize as Influenced by Organic Manures and Fertilizers P Associated Phosphate Activity in Soils. *Bioresource Technology*, *99*(13), 5773–5777. doi:10.1016/j.biortech.2007.10.063 PMID:18325765

Grilliches, Z. vi. (1980). Hybrid Corn: An Exploration in the Economics of Technological Change. *Econometrica*, *24*, 501–522.

Gujarati, D. N. (1988). Basic Econometrics (2nd ed.). New York: McGraw Hill Publishers.

Hassan, R. M., Muriithi, F., & Kamau, G. (1998). Determinants of Fertilizer Use and the Gap between Farmers' Maize Yield and Potential Yields in Kenya. In R. M. Hassan (Ed.), *Maize Technology Development and Transfer: A GIS Application for Research Planning in Kenya*. Wallingford, UK: CAB International.

Havlin, J. L., Tisdale, S. L., Nelson, W. L., & Beaton, J. D. (2005). *Soil Fertility and Fertilizers. An Introduction to Nutrients Management*. Upper Saddle River, NJ: Pearson Educational Inc.

Henao, J., & Baanante, C. (2006). *Agricultural Production and Soil Nutrient Mining in Africa: Implications for Resource Conservation and Policy Development*. Muscle Shoals, AL: An International Center for Soil Fertility and Agricultural Development.

Jose, D., Shanmugavelu, K. G., & Thamburaj, S. (1988). Studies on the Efficiency of Organic versus Inorganic form of Nitrogen in Brinjal. *Indian Journal of Horticulture*, *45*, 100–103.

Kaur, K., Kapoor, K.K. & Gupta, A.P. (2005). Impact of Organic Manures with and without. *Mineral and Soil Science, 168*, 117-122.

Kayuki, C. K., John, B., Zimwanguyizza, W., Katwijukye, A. K., Mawanda, A., Kabuye, F. M., & Iceduna, C. et al. (2012). Maize response to application of inorganic fertilizers and manure in eastern and northern Uganda. In *12th Integrated Soil Fertility Management in Africa: From Microbes to Market. Conference Information, Program and Abstracts*. CIAT.

Keeney, D. R. (1997). Sustainable Agriculture and the Environment. In *Proceedings of the 1997 International Symposium on Sustainable Agricultural Development Compatible with Environmental Conservation in Asia*. JIRCAS.

Kimaro, A. A., Timmer, V. R., Chamshama, S. A. O., Ngaga, Y. N., & Kimaro, D. A. (2009). Competition between Maize and Pigeon-Pea in Semi-Arid Tanzania: Effect on Yields and Nutrition of Crops. *Agriculture, Ecosystems & Environment, 134*(1-2), 115–125. doi:10.1016/j.agee.2009.06.002

Leathers, H., Smale, M., & Heisey, P. (1995). Maize of the Ancestors and Modern Varieties: The Microeconomics of High-Yielding Variety Adoption in Malawi. *Economic Development and Cultural Change, 43*(2), 351–368. doi:10.1086/452154

Lunduka, R., & Kelly, V. (2009). *The Financial Benefits of Integrated Soil Fertility Management Using Inorganic Fertilizer in Combination with Other ISFM Practices. A Post-doctoral Fellow (Policy) Research, CABI Africa, ICRAF Complex, United Nations Avenue*. Nairobi, Kenya: Gigiri.

Macharia, J., Mugwe, J., Mucheru-Muna, M., Mang'uriu, D., & Mugendi, D. (2012). Factors Influencing ISFM Knowledge in the Central Highlands of Kenya. *12th Integrated Soil Fertility Management in Africa: From Microbes to Market. Conference Information, Program and Abstracts*. CIAT.

Makhokha, S., Kimani, S., Mwangi, W., Verkuijl, H., & Musembi, F. (2001). Determinants of Fertilizer and Manure Use for Maize Production in Kiambu District, Kenya. CIMMYT (International Maize and Wheat Improvement Centre) and Kenya Agricultural Research Institute (KARI)

Merumba, M. S., Ndyetabula, I. L., Nkuba, J. M., William, M. M., Mukandala, L., Sayi, B., & Shimwela, M. et al. (2012). Exploring the Effects of Different Fertilizers on Yields and Benefits of Important Crops Grown in Kagera, Tanzania Integrated Soil Fertility Management in Africa Conference Information. *Program with Abstracts (Geological Association of Canada)*, 44.

Memudu, I. J. (2015). *Economic Analysis of Integrated Soil Fertility Management for maize production in Kogi and Kwara States, Nigeria* (Unpublished Ph.D. Thesis). Department of Agricultural Economics, Faculty of Agriculture, University of Ilorin.

Mohammadi, S., Kalbasi, M., & Shariatmadari, H. (2009). Cumulative and Residual Effects of Organic Fertilizer on Selected soil Properties, Water Soluble P, Olsen-P and P-Sorbtion Index. *Journal of Agricultural Science and Technology*, *11*(4), 487–497.

Moyin, J. E. I. (2003). Incorporation of Agricultural Biomass and their Effects on Growth and Nutrient Content of Four Successive Crops of Amaranthus. *Conference Proceeding*, 102.

Mtambanengwe, F., & Kosina, P. (2007). *Cereal Knowledge Bank*. Retrieved from www.knowledgebank.irri.org/.../organic-and-inorganic-fertilizers.html

Mucheru-Muna, M., Ngetich, F., Mugendi, D., Mugwe, J., Pypers, P., Vanlauwe, B., & Merckx, R. et al. (2012). Organic Inputs and Mineral Fertilizer Effects on Soil Chemical Properties, Soil Water Content and Maize Productivity in Mbeere District, Kenya Integrated Soil Fertility Management in Africa Conference Information. *Program with Abstracts (Geological Association of Canada)*, 80.

Mullens, B. A., Szijj, C. E., & Hincle, N. C. (2002). Oviposition and Development of Fannia spp. (Diptera:Muscidae) on Poultry Manure of Low Moisture Levels. *Environmental Entomology*, *31*(4), 588–593. doi:10.1603/0046-225X-31.4.588

Nambiro, E., Okoth, P., Kinyanjui, S., Murua, E., Kibe, P., & Macharia, R. (2012). Which Households are Most Likely to Take Up ISFM Technology and Why? Case of Sidindi, Western Kenya Integrated Soil Fertility Management in Africa Conference Information. *Program with Abstracts (Geological Association of Canada)*, 86.

Ndanu, A., Esilaba, A., Kibunja, C., Kamoni, P., & Okoth, P. (2012). A Situational Analysis of Opportunities and Challenges of Using ISFM Information and Knowledge in Kenya Integrated Soil Fertility Management in Africa Conference Information. *Program with Abstracts (Geological Association of Canada)*, 86.

NOAN. (2012). *Nigerian Organic Agricultural Network*. Retrieved from www.noan.com

Nowak, P. J. (1992). Why Farmers Adopt Production Technology. *Journal of Soil and Water Conservation*, *47*(1), 14–16.

Nwaiwu, I. U., Ohajianju, D. O., Lemchi, J. I., Ibekwe, U. C., Nwosu, F. O., Ben-Chendo, N. G., & Kadiri, F. A. et al. (2010). Economics of Organic Manure Use by Food Crop Farmers in Ecologically Vulnerable Area of Imo State, Nigeria. *Researcher*, *2*(11), 56–61.

Ogundare, K., Agele, S., & Aiyelari, P. (2012). Organic Amendment of an Ultisol: Effects on Soil Properties, Growth, and Yield of Maize in Southern Guinea Savanna Zone of Nigeria. *International Journal of Recycling of Organic Waste in Agriculture*, *1*(1), 11. doi:10.1186/2251-7715-1-11

Olayide, S. O., & Heady, E. O. (1982). *Introduction to agricultural production economics*. Ibadan, Nigeria: Ibadan University Press, University of Ibadan.

Olayinka, A., & Ailenubhi, V. (2001). Influence of Combined Application of Cow Dung and Inorganic Nitrogen on Microbial Respiration and Nitrogen Transformation in an Alfisol. *Nigerian Journal of Soil Research*, *2*, 15–20.

Omisore, J. K. (2013). Response of Maize (Zea mays l.) to Organic and Inorganic Nitrogen Sources in the Southern Guinea Savanna of Nigeria (A Ph.D. Final Project Report). Department of Agronomy, University of Ilorin.

Pal, C. A., Gachengo, C. N., Delve, R. J., Cadisch, G., & Giller, K. E. (2001). Organic Inputs for Soil Fertility Management in Tropical Agro-Ecosystems. *Agriculture, Ecosystems & Environment*, *83*, 27–42. doi:10.1016/S0167-8809(00)00267-X

Palm, C. A., Gachengo, C. N., Delve, R. J., Cadisch, G., & Giller, K. E. (2001). Organic Inputs for Soil Fertility Management in Tropical Agroecosystems: Application of an Organic Resource Database. *Agriculture, Ecosystems & Environment*, *83*(1-2), 27–42. doi:10.1016/S0167-8809(00)00267-X

Pender, J., Nkonya, E., Jagger, P., Sserunkuuma, D., & Ssali, H. (2004). Strategies to Increase Agricultural Productivity and Reduce Land Degradation: Evidence from Uganda. *Agricultural Economics*, *31*(2-3), 181–195. doi:10.1111/j.1574-0862.2004.tb00256.x

Place, F., Barrett, C. B., Freeman, H. A., Ramisch, J. J., & Vanlauwe, B. (2003). Prospects for Integrated Soil Fertility Management Using Organic and Inorganic Inputs: Evidence from Smallholder African Agricultural Systems. *Food Policy*, *28*(4), 365–378. doi:10.1016/j.foodpol.2003.08.009

Rahji, M. A. Y., Fakayode, S. B., & Sanni, G. K. (2008). A Multinomial Logit Analysis of Agricultural Credit Rationing by Deposit Money Banks in Nigeria: A Case Study of Oyo State, South West, Nigeria. *Economic and Financial Review Central Bank of Nigeria*, *46*(2), 55–82.

Rahman, M. M., Sarker, M. A. M., Hasina, A., & Kashem, M. A. (1998). Role of some Indigenous Substances as Organic Fertilizers on the Growth and Yield of Brinjal Plants. *Bangladesh Journal of Scientific and Industrial Research*, *33*(2), 275–281.

Rogers, E. M. (2003). *Diffusion of innovations* (5th ed.). New York: Free Press.

Saha, S., Mina, B. L., Gopinath, K. A., Kundu, S., & Gupta, H. S. (2008). Organic Amendments Affect Biochemical Properties of a Sub-Temperate Soil of the Indian Himalayas. *Nutrient Cycling in Agroecosystems*, *80*(3), 233–242. doi:10.1007/s10705-007-9139-x

Saidu, A., Abayomi, Y. A. & Aduloju, M. O. (2012). Evaluation of Complementary Use of Organic and Inorganic Fertilizers on the Performance of Upland Rice (*Oryza sativa* l.). *IJABR*, *2*(3), 487-491.

Segun-Olasanmi, O. A., & Bamire, A. S. (2010). *Analysis of Costs and Returns to Maize-Cowpea Intercrop Production in Oyo state, Nigeria*. Poster presented at the Joint 3rd African Association of Agricultural Economists (AAAE) and 48th Agricultural Economists Association of South Africa (AEASA) Conference, Cape Town, South Africa.

Son, T. T. N., Thu, V., Man, L. H., Kobayashi, H., & Yamada, R. (2004). Effect of Long Term Application of Organic and Bio-Fertilizer on Soil Fertility Under Rice-Soyabeans-Rice Cropping Systems. *Omonrice*, *12*, 45–51.

Tittonell, P. M., Corbeels, M. T., van Wijk, M. T., Vanlauwe, B., & Giller, K. E. (2008). Combining Organic and Mineral Fertilizers for Intergrated Soil Fertility management in Smallholder Farming Systems of Kenya: Exploration Using the Crop-Soil Model FIELD. *Agronomy Journal*, *100*(5), 1511. doi:10.2134/agronj2007.0355

Ullah, M. S., Islam, M. S., Islam, M. A., & Haque, T. (2008). Effects of Organic Manures and Chemical Fertilizers on the Yield of Brinjal and Soil Properties. *Journal Bangladesh Agricultural University*, *6*(2), 271–276.

USDA (United States Department of Agriculture). (2014). *Maize production in Nigeria (2004-2014)*. USDA.

Vanlauwe, B., Wendt, J. W., & Diels, J. (2001). Combined Application of Organic Matter and Fertilizer. *Journal - American Society of Agronomy*, *50*, 247.

Vanlauwe, B., Diels, J., Aihou, K., Iwuafor, E. N. O., Lyasse, O., Sanginga, N., & Merckx, R. (2002). Direct Interactions between Nitrogen Fertilizer and Organic Matter: Evidence from Trials with 15 Nitrogen –Labelled Fertilizer. In B. Vanlauwe, J. Diels, N. Sanginga, & R. Merckx (Eds.), *Integrated Plant Nutrient Management in Sub-Saharan Africa* (pp. 173–184). Wallingford, UK: CAB International.

Vanlauwe, B., & Giller, K. E. (2006). Popular Myths around Soil Fertility Management in Sub-Saharan Africa. *Agriculture, Ecosystems & Environment, 116*(1-2), 34–46. doi:10.1016/j.agee.2006.03.016

Vanlauwe, B., Chianu, J., Giller, K. E., Merckx, R., Mokwunye, U., Pypers, P., . . . Sanginga, N. (2010). *Integrated Soil Fertility Management: Operational Definition and Consequences for Implementation and Dissemination.* 19th World Congress of Soil Science, Soil Solutions for a Changing World, Brisbane, Australia.

Woomer, P. L., & Swift, M. J. (1994). *The Biological Management of Tropical Soil Fertility*. Chichester, UK: John Wiley and Sons.

World Bank. (2013). Nigeria: Fertilizer consumption (%fertilizer production). *Factfish.* Retrieved from www.factfish.com

Chapter 12
Economic Impact Assessment of the National Fadama Development Project on Rural Farming Communities in Niger State of Nigeria

Godwin Anjeinu Abu
University of Agriculture – Makurdi, Nigeria.

Steve A. Okpachu
Federal College of Education (Technical), Nigeria

M. Mallam
University of Agriculture – Makurdi, Nigeria

ABSTRACT

The chapter tested the hypotheses that the National Fadama Development Project had no significant effect on the income of Fadama farmers and that there is no difference in the profit of sugar cane and rice, the two major crops cultivated in the project. Primary data were collected in 2007 through the use of questionnaire randomly administered to 150 farmers in Niger state who are Fadama beneficiaries. Data collected were analyzed using descriptive statistics, t-test, and gross margin analysis. The results of the study showed that sugar cane gave the highest gross margin of N93,460 and rice with a gross margin of N 51,051 gave the highest return per Naira invested of 1.77. The study showed that the difference between the pre-

DOI: 10.4018/978-1-5225-3631-4.ch012

project and post-project income was insignificant at 5%. Scarcity of production inputs and the attendant environmental degradation were the constraints experienced in the project. The chapter concludes that Fadama projects positively impacted on beneficiaries by sustainably increasing farm income. The chapter recommends that the National Fadama Development Project should make timely provision of subsidized production inputs.

INTRODUCTION

Fadama, a Hausa word adopted by World Bank, refers to the low lying swampy areas consisting of alluvial deposits and containing extensive exploitable aquifers. Fadama lands are among the world's most productive ecosystem, rich in bio-diversity of forest wildlife, fisheries, crops, livestock and water resources that are being competed for by fadama communities (Kutigi, 2005). Qureski (1989) defined it as alluvial lowlands formed by erosional and depositional actions of rivers and streams possessing fine texture and less acid which makes it a rich agricultural soil. In Nigeria, they are visible along the floodplains of Niger, Sokoto-Rima, Benue-Jemaari and Yobe rivers. They vary in width from a few hundred meters to as much as twenty hectares stretch and encompasses land and water resources that could be developed for irrigated agriculture (World Drop, 1993).

The development of the small-scale irrigation in the Fadamas using low-cost petrol driven pumps along with various types of driving technologies for tapping shallow ground water started on a pilot basis in the first state-wide Agricultural Development Projects (ADPs) of Bauchi, Kano and Sokoto states in the early 1980s. Following the success of the initiatives, a component of small-scale irrigation development was incorporated into the design of Kaduna. Katsina and Borno state-wide ADPs. Despite the success of these early attempts, rapid spread of small-scale irrigation in the Fadamas has been hampered by several constraints among which are poor infrastructure in the Fadamas, low investment in technology development and extension for irrigated agriculture, weak financial intermediation, poor organized Fadama farmers and limited access to foreign exchange for importation of irrigation equipment. The National Fadama Development Project (NFDP) was designed to tackle these constraints and accelerate the pace of small scale fadama farmers.

The glut of agricultural products at harvest during the cropping season with the attendant low prices that discourage crop production. The Ministry of Agriculture and National Resources introduced irrigation schemes across the state. The schemes at Badeggi and Edoz higi covered an area of 830 hectares and 746 hectares respectively. Most of these efforts were characterized by poor management, low yield and maintenance problem. A need to look at the maintenance practices of the structures put in place with the support of the National Fadama Development Project is important for sustainability.

The National Fadama Development Project emphasizes the exploitation of locations with shallow water table through tubewells and washbores, construction of access roads, formation of Fadama Users Association and agro processing. But, can these transform the poor farming techniques of the Fadama farmers? According to Onwueme (1979), farming in the Fadama is dominated by a large number of small scale farmers who rely mainly on their unimproved methods and poor farming techniques. So, has the National Fadama Development Project inculcated new farming techniques among Fadama farmers?

The broad objective of the study is to ascertain the impact of the National Fadama Development Project on the lives of the Niger state rural Fadama farmers. Specifically, the study seeks to: Assess the economic impact of the fadama project in the rural beneficiary communities of Niger state, Estimate the costs and returns of the major crops cultivated by the fadama farmers and identify the constraints in fadama farming.

METHODOLOGY

The Study Area

The study area is Niger state of Nigeria. It is situated in the middle belt zone of the country, between longitude $4^{0}30"$ and $8^{0}30"$ East and latitudes 8^{0} 50" and $11^{0}30"$ North (NSADP, 1999).

With a total land area of 58,500km and 8 Local Government Areas at inception (in 1976), the state has metamorphosed, resulting in increase in land area to about 75,000km and 25 L.G.As due to merger of Borgu L.G.A. formally of Kwara state to Niger in 1991. Niger state experiences distinct dry and wet seasons, the wet season decreasing in length and amount of rain from south to north. The mean annual rainfall varies from around 1100mm in the north to more than 1600mm in the east; and the duration of the wet season varies from 150 to 210 days north to south. Evaporation estimated by Panman's formula as potential evapo-transpiration, varies across the state from about 1300mm in the south to 1450mm in the north annually. The growing season for crops extends beyond the end of the rains because of residual soil moisture which takes sometimes to be consumed. Various climatic factors limit agricultural productivity in Niger State. These are the unreliability of rainfall in both timing and amount at the beginning of wet season; the risk of high water tables locally leading to water logging of soil later in the wet season; the sub-optimal range of temperatures in the wet season as far as so called "hot season crops" are concerned; and the risk of premature cessation of rains.

The state lies in the savannah vegetation zone of the country. The northern part falls within the Sudan savannah while the southern part falls into Guinea savannah zone. The northern part of the state has vast agricultural land characterized by sparse vegetation. It is popular for the production of sorghum, millet maize, groundnut and cotton as well as rearing of cattle, sheep and goat and poultry. Rice, yam, maize, sorghum, groundnut and melon are the predominant crops in the southern part. The flood-plains which form the southern boundary of the state and availability of extensive Fadama offer a good opportunity for production of pulses, rice, sugarcane and vegetables. The numerous streams and rivers also offer great opportunity for dry season irrigation farming and inland water fishing activities. (NSADP, 1999).

Sampling Technique

The study covered a sample size of 150 fadama farmers. Fifty (50) respondents were drawn from a local government in each of the three senatorial zones of the state. Both purposive and simple random samplings were used. Beneficiary local government that has the highest number of beneficiary Fadama User Association (FUAs) from each senatorial zone were purposively selected while simple random sampling was used to select FUAs. Simple random sampling was also used to select fadama farmers in a selected FUA.

Based on these criteria, Lavun L.G.A. was selected from zone A, Minna L.G.A. was selected from zone B and Kontagora L. G. A. was selected from zone C. Also Ndaloke Tifi and Ndaloke Tako FUAs were selected from Kotangora Local Government Area and Lokoto and Berikin sale FUAs were selected from Minna Local Government Area.

Data Collection

Both primary and secondary data were obtained for the study. The primary data was obtained from the questionnaire that seeks necessary information from the respondents on socio-economic characteristics, production details: farm size, crops grown, input used; cost and returns of the major crops: cost of input, total output, prize sold; changes in the commodity: new irrigation scheme used, increase in farm size, increase in income as well as a acquired knowledge of maintenance of irrigation structure.

The secondary data was obtained from the Niger State Agricultural Development Project since NSADP oversaw the first phase of the project. The data obtained include: the list of fadama user association, the list of tubewells and washbore constructed and the irrigation pump procured for the FUAs, the list of extension agents for NFDP as well as the number of TV and radio programmes on fadama.

RESULTS AND DISCUSSIONS

Socio-Economic Characteristics of Fadama Farmers in Niger State The respondents' socio-economic characteristics indicated that 91 percent of the fadama farmers were males and the highest number of respondents, 47 percent, came from the age bracket of 36 – 45. percent of the respondents were married, 83 percent of which came from a polygamous marriage. Household size of 11 – 15 gave the highest percentage of respondents; 50 percent. 82 percent of the respondents had no formal education, 55 percent of the respondents' major occupation was wet season farming while dry season farming was the major occupation of 28 percent of the respondents. Table 1 gave more details.

Crops Grown

The various crops grown by the respondents were rice, sugarcane, garden egg, banana, okra, amaranthus, tomatoes and pepper. Rice, sugarcane and garden egg were found to be the major crops grown by the farmers. The details of the total number of respondents and the total area cultivated for each were presented in Table 2.

Cost and Returns of the Major Crops

Analysis of costs and returns to these crops showed that sugarcane gave the highest gross margin of N93, 460 per hectare while garden egg gave the lowest gross margin of N22,960 per hectare. But rice gave the highest return per Naira invested of 1.77 followed by sugarcane and garden egg with Return per Naira invested of 1.40 and 1.06 respectively.

1. **Rice:** The cost of seed at N10,500/hectare representing 36.52 percent was the highest followed by that of fertilizer at N9,000/hectare representing 31.30 percent and that of labour at N3,500/hectare representing 12.17 percent. Chemical had a cost of N3,000/hectare representing 10.43 percent and transportation had a cost of N2,100/hectare representing 7.30 percent while the least cost of production, N649/hectare, representing 2.25 percent was recorded for irrigation. This may not only be due to the river diversion module used in areas of rice production but may also be due to the residual moisture which rakes time to be consumed after cessation of rains. The total variable cost per hectare was

Table 1. Distribution of Respondents According To Their Socio-Economic Characteristics

Characteristics	Number of Respondents	Percentage (%)
Age		
Up to 18	15	10
19 – 35	45	30
36 – 45	71	47
45 – 55	17	11
Above 55	3	2
Gender		
Male	136	91
Female	14	9
Marital Status		
Single	5	3
Married	127	85
Widow(rd)	18	12
Type of Marriage	**Number of Respondents**	**Percentage (%)**
Polygamy	105	83
Monogamy	22	17
House Hold Size		
Less than 5	9	6
5 – 10	31	21
11 – 15	75	50
16 and above	35	23
Educational Status		
No formal education	123	82
Primary education	15	10
Secondary education	12	8
Post secondary education	-	-
Major Occupation		
Wet season farming	82	55
Dry season farming	43	28
Fishing	12	8
Trading	9	6
Hunting	4	3

Table 2. Distribution of respondents According to Crops Grown

Crop/Crop	Frequency	Percentage	Total Area	Percentage
Mixture			(HA)	
Rice	82	54.7	89	52.5
Sugarcane	43	28.6	60	35.4
Garden egg	16	10.6	14	8.3
Banana	1	0.7	1	0.6
Okra/Amaranthus	4	2.7	2	1.2
Tomatoes/pepper	4	2.7	3	2.0

N28,749. The gross margin per hectare was N51,051 and the return per naira invested was 1.77.

2. **Sugarcane:** The cost of transport was the highest with a total cost of N38,000 per hectare, representing 57.10 percent, followed by Labour with a total cost of N10,000/hectare, representing 15.02 percent fertilizer had a total cost of N6,000 per hectare, representing 9.10 percent followed by seed with a total cost of N5,000/hectare, representing 7.51 percent had the least variable cost of production. The total variable cost per hectare was N66,540. The gross margin per hectare was N93,460, the highest among the three crops but with a return per Naira invested of N1.40, second to rice.
3. **Garden-Egg:** The cost of labour at n10,000 per hectare, representing 35.80 percent was the highest, followed by fertilizer at N6,000 per hectare, representing 21.447 percent and that of seed at N5,000 per hectare, representing 17.90 percent. Irrigation had a total cost of N3,040 per hectare, representing 10.88 percent, followed by transportation and chemical with a total cost of N2,400 per hectare, representing 8.58 percent and N1,500 per hectare, representing 5.36 percent respectively. The total variable cost is N27,940 per hectare. It gave a gross margin of 29,660 and Return per Naira invested of 1.06. Refer to Table 3.

Impact of Fadama Projects on Farming Communities

Irrigation Scheme Practiced Before and After NFDP

55 percent of the respondents did not practice irrigation scheme before the intervention of the National Fadama Development Project against the 100 percent that practiced various types of irrigation schemes after the intervention (Table 4).

Table 3. Cost and Returns of the Crops (Rice, Sugarcane and Garden egg) in Naira per Hectare.

	Rice	Garden Egg	Sugarcane
1. Revenue			
Yield (kg/ha)	2100	1920	Bundle/ha 400
Unit price (N/kg)	38	30	N/bundle 400
Revenue (Nha)	79,800	57600	Revenue 160,000
2. Variable cost			
Labour	3500	10,000	10,000
Seed	10,500	5000	5000
Fertilizer	9,000	6000	6000
Transport	2,100	2,400	3,800
Chemical	3,000	1,500	4,500
Irrigation	649	3040	3040
TVC	28,749	27,940	66,540
3. Gross Margin	51,051	29,660	93,460
4.Return to Naira Invested RNI	1.77	1.06	1.40

Table 4. Distribution of Respondents According to Irrigation Scheme Used Before and After National Fadama Development Project.

	Pre-Irrigation Scheme		Post-Project Irrigation Scheme	
Irrigation Scheme	Frequency	Percentage	Frequency	Percentage
None	82	55	-	-
Hand dug well	25	16	5	3
Tubewell	-	-	-	-
Washbore	-	-	-	-
Surface pumping	25	16	72	48
River diversion	18	13	73	49
Total	150	100	150	100

Farm Size Before and After NFDP

The size of fadama land cultivated by the respondent before the National Fadama Development Project was 38 hectares against the 169 hectares cultivated after the National Fadama Development Project. See Table 5.

Table 5. Distribution of Respondents According to Fadama Farm Size Before and After National Fadama Development Project.

	Pre-Project Farm Size			Post-Project Farm Size		
Farm Size (HA)	**No. of Respondents**	**Total Farm Size**	**Percentage**	**No. of Respondents**	**Total Farm Size**	**Percentage**
0	82	-	55	-	-	-
0.5	60	30	40	50	25	33
1.0	8	8	5	51	51	34
1.5	-	-	-	18	27	12
2.0	-	-	-	25	50	17
2.5	-	-	-	4	10	3
3.00	-	-	-	2	6	1
Total	150	38	100	150	169	100

Gross Margin of the Major Crops Before and After Intervention

Estimated Gross Margin of rice, sugarcane and garden egg before the National Fadama Development Project stood at N10,000/ha N22,000/ha and N19,000/ha respectively while the estimated gross margin for rice, sugarcane and garden egg after the intervention of the National Fadama Development Project was N15,000ha, N93,000/ha and N30,000/ha respectively. See Table 6.

Details of the cost and return is presented in Table 7

A considerable difference in the gross income of fadama farmers of rice, sugarcane and garden egg after the National Fadama Development Project was observed. This is an indication of increase in the farmers' income, which justifies the creation of the project. The gross margin of rice that was N10,000 per hectare moved to N50,000 per hectare, that of sugarcane that was N22,000 per hectare moved to N93,000 per hectare and that of garden egg that used to be N19,000 per hectare shifted to N30,000 per hectare. According to Ingawa (1988) improvement of irrigation performance critically depends on suitable macro-economic and political environment as well as development and utilization of technologies, however, in spite of the leap in the profit of these enterprises, data analysis showed that the difference between the pre-project and post-project income of the farmers from the major crops is not significant (t_{-cal}) < t_{-tab} (2.92). This could be linked to the fact that most farmers sold their produce at farm gate when no value had been added to their products.

Table 6. Distribution of Respondents According to Estimated Income, Gross Margin and RNI of the Major Crops Before and After NFDP.

Pre-Project Gross Margin & RNI		Post Project Gross Margin & RNI	
Rice	N/ha	Rice	N/ha
Production cost	18,000	Production cost	29,000
Income	28,000	Income	79,000
Gross margin	10,000	Gross margin	50,000
RNI	0.55	RNI	1.72
Sugarcane		Sugarcane	
Production cost	20,000	Production	69,000
Income	42,000	Income	160,000
Gross margin	22,000	Gross margin	93,000
RNI	1.10	RNI	1.38
Garden egg		Garden egg	
Production cost	15,000	Production cost	28,000
Income	34,000	Income	58,000
Gross margin	19,000	Gross margin	30,000
RNI	1.26	RNI	1.07

Table 7. Cost and Returns of the Crops (Rice, Sugarcane and Garden egg) in Naira per Hectare.

	Rice	Garden Egg	Sugarcane
1. Revenue			
Yield (kg/ha)	2100	1920	Bundle/ha 400
Unit price (N/kg)	38	30	N/bundle 400
Revenue (Nha)	79,800	57600	Revenue 160,000
2. Variable cost			
Labour	3500	10,000	10,000
Seed	10,500	5000	5000
Fertilizer	9,000	6000	6000
Transport	2,100	2,400	3,800
Chemical	3,000	1,500	4,500
Irrigation	649	3040	3040
TVC	28,749	27,940	66,540
3. Gross Margin	51,051	29,660	93,460
4. RNI	1.77	1.06	1.40

Table 8. Distribution ofRespondents According to Problems encountered In Fadama Farming.

Problems	Number of Respondents	Percentage
Scarcity of fertilizer	142	95
High cost of farm input	53	35
Poor rural roads	42	28
Cost of fuel	38	25
Lack of rice milling machine	34	23
In adequate pumping machine	23	15

Constraints in Fadama Farming

Scarcity of fertilizer had the highest respondents (95), followed by high cost of inputs with 53 percent. Refer to Table 8.

CONCLUSION

The study concludes that Fadama projects positively impacted on beneficiaries by sustainably increasing farm income. Giving Gross Margin of rice, sugarcane and garden egg before the National Fadama Development Project at N10,000/ha N22,000/ha and N19,000/ha respectively while the gross margin for rice, sugarcane and garden egg after the intervention of the National Fadama Development Project was N15,000ha, N93,000/ha and N30,000/ha respectively.

RECOMMENDATIONS

The study recommends that:

1. Production inputs be provided to the farmers on time.
2. Farmer's capacities be built on technical management of water pumps and other irrigation facilities.

REFERENCES

Ingawa, D. N. (1988). *Water and land use rights under the National Fadama Development Project: A matter of Conservation*. Paper presented at the annual planning, Monitoring and evaluation Workshop in soil and water conservation Institute.

Kutigi, A. U. (2005). *State of Fadama development in Niger state*. Paper presented at the agricultural week, Minna.

Niger State Agricultural Development Programme. (1999). *Annual Evaluation Report*. Author.

Onwueme, I. C. (1979). *Cassell's Tropical Agriculture Series*. Southampton, UK: The Camelot Press ltd.

Qureski, A. A. (1989). *Water management in Fadama areas*. Paper presented at the National workshop on Fadama and Irrigation Development in Nigeria, Bauchi, Nigeria.

World Drop Engineering, Inc. (1993). *Final Report on the Study of Irrigation Potential of Shallow Aquifer in Fadama Areas*. Benue Agricultural and Rural Development Authority BNARDA.

Related References

To continue our tradition of advancing academic research, we have compiled a list of recommended IGI Global readings. These references will provide additional information and guidance to further enrich your knowledge and assist you with your own research and future publications.

Adeyemo, O. (2013). The nationwide health information network: A biometric approach to prevent medical identity theft. In *User-driven healthcare: Concepts, methodologies, tools, and applications* (pp. 1636–1649). Hershey, PA: IGI Global. doi:10.4018/978-1-4666-2770-3.ch081

Adler, M., & Henman, P. (2009). Justice beyond the courts: The implications of computerisation for procedural justice in social security. In A. Martínez & P. Abat (Eds.), *E-justice: Using information communication technologies in the court system* (pp. 65–86). Hershey, PA: IGI Global. doi:10.4018/978-1-59904-998-4.ch005

Aflalo, E., & Gabay, E. (2013). An information system for coping with student dropout. In L. Tomei (Ed.), *Learning tools and teaching approaches through ICT advancements* (pp. 176–187). Hershey, PA: IGI Global. doi:10.4018/978-1-4666-2017-9.ch016

Ahmed, M. A., Janssen, M., & van den Hoven, J. (2012). Value sensitive transfer (VST) of systems among countries: Towards a framework. *International Journal of Electronic Government Research*, *8*(1), 26–42. doi:10.4018/jegr.2012010102

Aikins, S. K. (2008). Issues and trends in internet-based citizen participation. In G. Garson & M. Khosrow-Pour (Eds.), *Handbook of research on public information technology* (pp. 31–40). Hershey, PA: IGI Global. doi:10.4018/978-1-59904-857-4.ch004

Aikins, S. K. (2009). A comparative study of municipal adoption of internet-based citizen participation. In C. Reddick (Ed.), *Handbook of research on strategies for local e-government adoption and implementation: Comparative studies* (pp. 206–230). Hershey, PA: IGI Global. doi:10.4018/978-1-60566-282-4.ch011

Aikins, S. K. (2012). Improving e-government project management: Best practices and critical success factors. In *Digital democracy: Concepts, methodologies, tools, and applications* (pp. 1314–1332). Hershey, PA: IGI Global. doi:10.4018/978-1-4666-1740-7.ch065

Akabawi, M. S. (2011). Ghabbour group ERP deployment: Learning from past technology failures. In E. Business Research and Case Center (Ed.), Cases on business and management in the MENA region: New trends and opportunities (pp. 177-203). Hershey, PA: IGI Global. doi:10.4018/978-1-60960-583-4.ch012

Akabawi, M. S. (2013). Ghabbour group ERP deployment: Learning from past technology failures. In *Industrial engineering: Concepts, methodologies, tools, and applications* (pp. 933–958). Hershey, PA: IGI Global. doi:10.4018/978-1-4666-1945-6.ch051

Akbulut, A. Y., & Motwani, J. (2008). Integration and information sharing in e-government. In G. Putnik & M. Cruz-Cunha (Eds.), *Encyclopedia of networked and virtual organizations* (pp. 729–734). Hershey, PA: IGI Global. doi:10.4018/978-1-59904-885-7.ch096

Akers, E. J. (2008). Technology diffusion in public administration. In G. Garson & M. Khosrow-Pour (Eds.), *Handbook of research on public information technology* (pp. 339–348). Hershey, PA: IGI Global. doi:10.4018/978-1-59904-857-4.ch033

Al-Shafi, S. (2008). Free wireless internet park services: An investigation of technology adoption in Qatar from a citizens' perspective. *Journal of Cases on Information Technology, 10*(3), 21–34. doi:10.4018/jcit.2008070103

Al-Shafi, S., & Weerakkody, V. (2009). Implementing free wi-fi in public parks: An empirical study in Qatar. *International Journal of Electronic Government Research, 5*(3), 21–35. doi:10.4018/jegr.2009070102

Aladwani, A. M. (2002). Organizational actions, computer attitudes and end-user satisfaction in public organizations: An empirical study. In C. Snodgrass & E. Szewczak (Eds.), *Human factors in information systems* (pp. 153–168). Hershey, PA: IGI Global. doi:10.4018/978-1-931777-10-0.ch012

Aladwani, A. M. (2002). Organizational actions, computer attitudes, and end-user satisfaction in public organizations: An empirical study. *Journal of Organizational and End User Computing*, *14*(1), 42–49. doi:10.4018/joeuc.2002010104

Allen, B., Juillet, L., Paquet, G., & Roy, J. (2005). E-government and private-public partnerships: Relational challenges and strategic directions. In M. Khosrow-Pour (Ed.), *Practicing e-government: A global perspective* (pp. 364–382). Hershey, PA: IGI Global. doi:10.4018/978-1-59140-637-2.ch016

Alshawaf, A., & Knalil, O. E. (2008). IS success factors and IS organizational impact: Does ownership type matter in Kuwait? *International Journal of Enterprise Information Systems*, *4*(2), 13–33. doi:10.4018/jeis.2008040102

Ambali, A. R. (2009). Digital divide and its implication on Malaysian e-government: Policy initiatives. In H. Rahman (Ed.), *Social and political implications of data mining: Knowledge management in e-government* (pp. 267–287). Hershey, PA: IGI Global. doi:10.4018/978-1-60566-230-5.ch016

Amoretti, F. (2007). Digital international governance. In A. Anttiroiko & M. Malkia (Eds.), *Encyclopedia of digital government* (pp. 365–370). Hershey, PA: IGI Global. doi:10.4018/978-1-59140-789-8.ch056

Amoretti, F. (2008). Digital international governance. In A. Anttiroiko (Ed.), *Electronic government: Concepts, methodologies, tools, and applications* (pp. 688–696). Hershey, PA: IGI Global. doi:10.4018/978-1-59904-947-2.ch058

Amoretti, F. (2008). E-government at supranational level in the European Union. In A. Anttiroiko (Ed.), *Electronic government: Concepts, methodologies, tools, and applications* (pp. 1047–1055). Hershey, PA: IGI Global. doi:10.4018/978-1-59904-947-2.ch079

Amoretti, F. (2008). E-government regimes. In A. Anttiroiko (Ed.), *Electronic government: Concepts, methodologies, tools, and applications* (pp. 3846–3856). Hershey, PA: IGI Global. doi:10.4018/978-1-59904-947-2.ch280

Amoretti, F. (2009). Electronic constitution: A Braudelian perspective. In F. Amoretti (Ed.), *Electronic constitution: Social, cultural, and political implications* (pp. 1–19). Hershey, PA: IGI Global. doi:10.4018/978-1-60566-254-1.ch001

Amoretti, F., & Musella, F. (2009). Institutional isomorphism and new technologies. In M. Khosrow-Pour (Ed.), *Encyclopedia of information science and technology* (2nd ed.; pp. 2066–2071). Hershey, PA: IGI Global. doi:10.4018/978-1-60566-026-4.ch325

Andersen, K. V., & Henriksen, H. Z. (2007). E-government research: Capabilities, interaction, orientation, and values. In D. Norris (Ed.), *Current issues and trends in e-government research* (pp. 269–288). Hershey, PA: IGI Global. doi:10.4018/978-1-59904-283-1.ch013

Anderson, K. V., & Henriksen, H. Z. (2005). The first leg of e-government research: Domains and application areas 1998-2003. *International Journal of Electronic Government Research, 1*(4), 26–44. doi:10.4018/jegr.2005100102

Anttiroiko, A. (2009). Democratic e-governance. In M. Khosrow-Pour (Ed.), *Encyclopedia of information science and technology* (2nd ed.; pp. 990–995). Hershey, PA: IGI Global. doi:10.4018/978-1-60566-026-4.ch158

Association, I. R. (2010). *Networking and telecommunications: Concepts, methodologies, tools and applications* (Vols. 1–3). Hershey, PA: IGI Global. doi:10.4018/978-1-60566-986-1

Association, I. R. (2010). *Web-based education: Concepts, methodologies, tools and applications* (Vols. 1–3). Hershey, PA: IGI Global. doi:10.4018/978-1-61520-963-7

Baker, P. M., Bell, A., & Moon, N. W. (2009). Accessibility issues in municipal wireless networks. In C. Reddick (Ed.), *Handbook of research on strategies for local e-government adoption and implementation: Comparative studies* (pp. 569–588). Hershey, PA: IGI Global. doi:10.4018/978-1-60566-282-4.ch030

Becker, S. A., Keimer, R., & Muth, T. (2010). A case on university and community collaboration: The sci-tech entrepreneurial training services (ETS) program. In S. Becker & R. Niebuhr (Eds.), *Cases on technology innovation: Entrepreneurial successes and pitfalls* (pp. 68–90). Hershey, PA: IGI Global. doi:10.4018/978-1-61520-609-4.ch003

Becker, S. A., Keimer, R., & Muth, T. (2012). A case on university and community collaboration: The sci-tech entrepreneurial training services (ETS) program. In Regional development: Concepts, methodologies, tools, and applications (pp. 947-969). Hershey, PA: IGI Global. doi:10.4018/978-1-4666-0882-5.ch507

Bernardi, R. (2012). Information technology and resistance to public sector reforms: A case study in Kenya. In T. Papadopoulos & P. Kanellis (Eds.), *Public sector reform using information technologies: Transforming policy into practice* (pp. 59–78). Hershey, PA: IGI Global. doi:10.4018/978-1-60960-839-2.ch004

Bernardi, R. (2013). Information technology and resistance to public sector reforms: A case study in Kenya. In *User-driven healthcare: Concepts, methodologies, tools, and applications* (pp. 14–33). Hershey, PA: IGI Global. doi:10.4018/978-1-4666-2770-3.ch002

Bolívar, M. P., Pérez, M. D., & Hernández, A. M. (2012). Municipal e-government services in emerging economies: The Latin-American and Caribbean experiences. In Y. Chen & P. Chu (Eds.), *Electronic governance and cross-boundary collaboration: Innovations and advancing tools* (pp. 198–226). Hershey, PA: IGI Global. doi:10.4018/978-1-60960-753-1.ch011

Borycki, E. M., & Kushniruk, A. W. (2010). Use of clinical simulations to evaluate the impact of health information systems and ubiquitous computing devices upon health professional work. In S. Mohammed & J. Fiaidhi (Eds.), *Ubiquitous health and medical informatics: The ubiquity 2.0 trend and beyond* (pp. 552–573). Hershey, PA: IGI Global. doi:10.4018/978-1-61520-777-0.ch026

Borycki, E. M., & Kushniruk, A. W. (2011). Use of clinical simulations to evaluate the impact of health information systems and ubiquitous computing devices upon health professional work. In *Clinical technologies: Concepts, methodologies, tools and applications* (pp. 532–553). Hershey, PA: IGI Global. doi:10.4018/978-1-60960-561-2.ch220

Buchan, J. (2011). Developing a dynamic and responsive online learning environment: A case study of a large Australian university. In B. Czerkawski (Ed.), *Free and open source software for e-learning: Issues, successes and challenges* (pp. 92–109). Hershey, PA: IGI Global. doi:10.4018/978-1-61520-917-0.ch006

Buenger, A. W. (2008). Digital convergence and cybersecurity policy. In G. Garson & M. Khosrow-Pour (Eds.), *Handbook of research on public information technology* (pp. 395–405). Hershey, PA: IGI Global. doi:10.4018/978-1-59904-857-4.ch038

Burn, J. M., & Loch, K. D. (2002). The societal impact of world wide web - Key challenges for the 21st century. In A. Salehnia (Ed.), *Ethical issues of information systems* (pp. 88–106). Hershey, PA: IGI Global. doi:10.4018/978-1-931777-15-5.ch007

Burn, J. M., & Loch, K. D. (2003). The societal impact of the world wide web-Key challenges for the 21st century. In M. Khosrow-Pour (Ed.), *Advanced topics in information resources management* (Vol. 2, pp. 32–51). Hershey, PA: IGI Global. doi:10.4018/978-1-59140-062-2.ch002

Bwalya, K. J., Du Plessis, T., & Rensleigh, C. (2012). The "quicksilver initiatives" as a framework for e-government strategy design in developing economies. In K. Bwalya & S. Zulu (Eds.), *Handbook of research on e-government in emerging economies: Adoption, e-participation, and legal frameworks* (pp. 605–623). Hershey, PA: IGI Global. doi:10.4018/978-1-4666-0324-0.ch031

Cabotaje, C. E., & Alampay, E. A. (2013). Social media and citizen engagement: Two cases from the Philippines. In S. Saeed & C. Reddick (Eds.), *Human-centered system design for electronic governance* (pp. 225–238). Hershey, PA: IGI Global. doi:10.4018/978-1-4666-3640-8.ch013

Camillo, A., Di Pietro, L., Di Virgilio, F., & Franco, M. (2013). Work-groups conflict at PetroTech-Italy, S.R.L.: The influence of culture on conflict dynamics. In B. Christiansen, E. Turkina, & N. Williams (Eds.), *Cultural and technological influences on global business* (pp. 272–289). Hershey, PA: IGI Global. doi:10.4018/978-1-4666-3966-9.ch015

Capra, E., Francalanci, C., & Marinoni, C. (2008). Soft success factors for m-government. In A. Anttiroiko (Ed.), *Electronic government: Concepts, methodologies, tools, and applications* (pp. 1213–1233). Hershey, PA: IGI Global. doi:10.4018/978-1-59904-947-2.ch089

Cartelli, A. (2009). The implementation of practices with ICT as a new teaching-learning paradigm. In A. Cartelli & M. Palma (Eds.), *Encyclopedia of information communication technology* (pp. 413–417). Hershey, PA: IGI Global. doi:10.4018/978-1-59904-845-1.ch055

Charalabidis, Y., Lampathaki, F., & Askounis, D. (2010). Investigating the landscape in national interoperability frameworks. *International Journal of E-Services and Mobile Applications, 2*(4), 28–41. doi:10.4018/jesma.2010100103

Charalabidis, Y., Lampathaki, F., & Askounis, D. (2012). Investigating the landscape in national interoperability frameworks. In A. Scupola (Ed.), *Innovative mobile platform developments for electronic services design and delivery* (pp. 218–231). Hershey, PA: IGI Global. doi:10.4018/978-1-4666-1568-7.ch013

Chen, I. (2005). Distance education associations. In C. Howard, J. Boettcher, L. Justice, K. Schenk, P. Rogers, & G. Berg (Eds.), *Encyclopedia of distance learning* (pp. 599–612). Hershey, PA: IGI Global. doi:10.4018/978-1-59140-555-9.ch087

Chen, I. (2008). Distance education associations. In L. Tomei (Ed.), *Online and distance learning: Concepts, methodologies, tools, and applications* (pp. 562–579). Hershey, PA: IGI Global. doi:10.4018/978-1-59904-935-9.ch048

Chen, Y. (2008). Managing IT outsourcing for digital government. In A. Anttiroiko (Ed.), *Electronic government: Concepts, methodologies, tools, and applications* (pp. 3107–3114). Hershey, PA: IGI Global. doi:10.4018/978-1-59904-947-2.ch229

Chen, Y., & Dimitrova, D. V. (2006). Electronic government and online engagement: Citizen interaction with government via web portals. *International Journal of Electronic Government Research, 2*(1), 54–76. doi:10.4018/jegr.2006010104

Chen, Y., & Knepper, R. (2005). Digital government development strategies: Lessons for policy makers from a comparative perspective. In W. Huang, K. Siau, & K. Wei (Eds.), *Electronic government strategies and implementation* (pp. 394–420). Hershey, PA: IGI Global. doi:10.4018/978-1-59140-348-7.ch017

Chen, Y., & Knepper, R. (2008). Digital government development strategies: Lessons for policy makers from a comparative perspective. In H. Rahman (Ed.), *Developing successful ICT strategies: Competitive advantages in a global knowledge-driven society* (pp. 334–356). Hershey, PA: IGI Global. doi:10.4018/978-1-59904-654-9.ch017

Cherian, E. J., & Ryan, T. W. (2014). Incongruent needs: Why differences in the iron-triangle of priorities make health information technology adoption and use difficult. In C. El Morr (Ed.), *Research perspectives on the role of informatics in health policy and management* (pp. 209–221). Hershey, PA: IGI Global. doi:10.4018/978-1-4666-4321-5.ch012

Cho, H. J., & Hwang, S. (2010). Government 2.0 in Korea: Focusing on e-participation services. In C. Reddick (Ed.), *Politics, democracy and e-government: Participation and service delivery* (pp. 94–114). Hershey, PA: IGI Global. doi:10.4018/978-1-61520-933-0.ch006

Chorus, C., & Timmermans, H. (2010). Ubiquitous travel environments and travel control strategies: Prospects and challenges. In M. Wachowicz (Ed.), *Movement-aware applications for sustainable mobility: Technologies and approaches* (pp. 30–51). Hershey, PA: IGI Global. doi:10.4018/978-1-61520-769-5.ch003

Chuanshen, R. (2007). E-government construction and China's administrative litigation act. In A. Anttiroiko & M. Malkia (Eds.), *Encyclopedia of digital government* (pp. 507–510). Hershey, PA: IGI Global. doi:10.4018/978-1-59140-789-8.ch077

Ciaghi, A., & Villafiorita, A. (2012). Law modeling and BPR for public administration improvement. In K. Bwalya & S. Zulu (Eds.), *Handbook of research on e-government in emerging economies: Adoption, e-participation, and legal frameworks* (pp. 391–410). Hershey, PA: IGI Global. doi:10.4018/978-1-4666-0324-0.ch019

Ciaramitaro, B. L., & Skrocki, M. (2012). mHealth: Mobile healthcare. In B. Ciaramitaro (Ed.), Mobile technology consumption: Opportunities and challenges (pp. 99-109). Hershey, PA: IGI Global. doi:10.4018/978-1-61350-150-4.ch007

Comite, U. (2012). Innovative processes and managerial effectiveness of e-procurement in healthcare. In A. Manoharan & M. Holzer (Eds.), *Active citizen participation in e-government: A global perspective* (pp. 206–229). Hershey, PA: IGI Global. doi:10.4018/978-1-4666-0116-1.ch011

Cordella, A. (2013). E-government success: How to account for ICT, administrative rationalization, and institutional change. In J. Gil-Garcia (Ed.), *E-government success factors and measures: Theories, concepts, and methodologies* (pp. 40–51). Hershey, PA: IGI Global. doi:10.4018/978-1-4666-4058-0.ch003

Cropf, R. A. (2009). ICT and e-democracy. In M. Khosrow-Pour (Ed.), *Encyclopedia of information science and technology* (2nd ed.; pp. 1789–1793). Hershey, PA: IGI Global. doi:10.4018/978-1-60566-026-4.ch281

Cropf, R. A. (2009). The virtual public sphere. In M. Pagani (Ed.), *Encyclopedia of multimedia technology and networking* (2nd ed.; pp. 1525–1530). Hershey, PA: IGI Global. doi:10.4018/978-1-60566-014-1.ch206

D'Abundo, M. L. (2013). Electronic health record implementation in the United States healthcare industry: Making the process of change manageable. In V. Wang (Ed.), *Handbook of research on technologies for improving the 21st century workforce: Tools for lifelong learning* (pp. 272–286). Hershey, PA: IGI Global. doi:10.4018/978-1-4666-2181-7.ch018

Damurski, L. (2012). E-participation in urban planning: Online tools for citizen engagement in Poland and in Germany. *International Journal of E-Planning Research*, *1*(3), 40–67. doi:10.4018/ijepr.2012070103

de Almeida, M. O. (2007). E-government strategy in Brazil: Increasing transparency and efficiency through e-government procurement. In M. Gascó-Hernandez (Ed.), *Latin America online: Cases, successes and pitfalls* (pp. 34–82). Hershey, PA: IGI Global. doi:10.4018/978-1-59140-974-8.ch002

de Juana Espinosa, S. (2008). Empirical study of the municipalitites' motivations for adopting online presence. In A. Anttiroiko (Ed.), *Electronic government: Concepts, methodologies, tools, and applications* (pp. 3593–3608). Hershey, PA: IGI Global. doi:10.4018/978-1-59904-947-2.ch262

de Souza Dias, D. (2002). Motivation for using information technology. In C. Snodgrass & E. Szewczak (Eds.), *Human factors in information systems* (pp. 55–60). Hershey, PA: IGI Global. doi:10.4018/978-1-931777-10-0.ch005

Demediuk, P. (2006). Government procurement ICT's impact on the sustainability of SMEs and regional communities. In S. Marshall, W. Taylor, & X. Yu (Eds.), *Encyclopedia of developing regional communities with information and communication technology* (pp. 321–324). Hershey, PA: IGI Global. doi:10.4018/978-1-59140-575-7.ch056

Devonshire, E., Forsyth, H., Reid, S., & Simpson, J. M. (2013). The challenges and opportunities of online postgraduate coursework programs in a traditional university context. In B. Tynan, J. Willems, & R. James (Eds.), *Outlooks and opportunities in blended and distance learning* (pp. 353–368). Hershey, PA: IGI Global. doi:10.4018/978-1-4666-4205-8.ch026

Di Cerbo, F., Scotto, M., Sillitti, A., Succi, G., & Vernazza, T. (2007). Toward a GNU/Linux distribution for corporate environments. In S. Sowe, I. Stamelos, & I. Samoladas (Eds.), *Emerging free and open source software practices* (pp. 215–236). Hershey, PA: IGI Global. doi:10.4018/978-1-59904-210-7.ch010

Diesner, J., & Carley, K. M. (2005). Revealing social structure from texts: Meta-matrix text analysis as a novel method for network text analysis. In V. Narayanan & D. Armstrong (Eds.), *Causal mapping for research in information technology* (pp. 81–108). Hershey, PA: IGI Global. doi:10.4018/978-1-59140-396-8.ch004

Dologite, D. G., Mockler, R. J., Bai, Q., & Viszhanyo, P. F. (2006). IS change agents in practice in a US-Chinese joint venture. In M. Hunter & F. Tan (Eds.), *Advanced topics in global information management* (Vol. 5, pp. 331–352). Hershey, PA: IGI Global. doi:10.4018/978-1-59140-923-6.ch015

Drnevich, P., Brush, T. H., & Luckock, G. T. (2011). Process and structural implications for IT-enabled outsourcing. *International Journal of Strategic Information Technology and Applications*, *2*(4), 30–43. doi:10.4018/jsita.2011100103

Dwivedi, A. N. (2009). *Handbook of research on information technology management and clinical data administration in healthcare* (Vols. 1–2). Hershey, PA: IGI Global. doi:10.4018/978-1-60566-356-2

Elbeltagi, I., McBride, N., & Hardaker, G. (2006). Evaluating the factors affecting DSS usage by senior managers in local authorities in Egypt. In M. Hunter & F. Tan (Eds.), *Advanced topics in global information management* (Vol. 5, pp. 283–307). Hershey, PA: IGI Global. doi:10.4018/978-1-59140-923-6.ch013

Eom, S., & Fountain, J. E. (2013). Enhancing information services through public-private partnerships: Information technology knowledge transfer underlying structures to develop shared services in the U.S. and Korea. In J. Gil-Garcia (Ed.), *E-government success around the world: Cases, empirical studies, and practical recommendations* (pp. 15–40). Hershey, PA: IGI Global. doi:10.4018/978-1-4666-4173-0.ch002

Esteves, T., Leuenberger, D., & Van Leuven, N. (2012). Reaching citizen 2.0: How government uses social media to send public messages during times of calm and times of crisis. In K. Kloby & M. D'Agostino (Eds.), *Citizen 2.0: Public and governmental interaction through web 2.0 technologies* (pp. 250–268). Hershey, PA: IGI Global. doi:10.4018/978-1-4666-0318-9.ch013

Estevez, E., Fillottrani, P., Janowski, T., & Ojo, A. (2012). Government information sharing: A framework for policy formulation. In Y. Chen & P. Chu (Eds.), *Electronic governance and cross-boundary collaboration: Innovations and advancing tools* (pp. 23–55). Hershey, PA: IGI Global. doi:10.4018/978-1-60960-753-1.ch002

Ezz, I. E. (2008). E-governement emerging trends: Organizational challenges. In A. Anttiroiko (Ed.), *Electronic government: Concepts, methodologies, tools, and applications* (pp. 3721–3737). Hershey, PA: IGI Global. doi:10.4018/978-1-59904-947-2.ch269

Fabri, M. (2009). The Italian style of e-justice in a comparative perspective. In A. Martínez & P. Abat (Eds.), *E-justice: Using information communication technologies in the court system* (pp. 1–19). Hershey, PA: IGI Global. doi:10.4018/978-1-59904-998-4.ch001

Fagbe, T., & Adekola, O. D. (2010). Workplace safety and personnel well-being: The impact of information technology. *International Journal of Green Computing*, *1*(1), 28–33. doi:10.4018/jgc.2010010103

Fagbe, T., & Adekola, O. D. (2011). Workplace safety and personnel well-being: The impact of information technology. In *Global business: Concepts, methodologies, tools and applications* (pp. 1438–1444). Hershey, PA: IGI Global. doi:10.4018/978-1-60960-587-2.ch509

Farmer, L. (2008). Affective collaborative instruction with librarians. In S. Kelsey & K. St.Amant (Eds.), *Handbook of research on computer mediated communication* (pp. 15–24). Hershey, PA: IGI Global. doi:10.4018/978-1-59904-863-5.ch002

Favier, L., & Mekhantar, J. (2007). Use of OSS by local e-administration: The French situation. In K. St.Amant & B. Still (Eds.), *Handbook of research on open source software: Technological, economic, and social perspectives* (pp. 428–444). Hershey, PA: IGI Global. doi:10.4018/978-1-59140-999-1.ch033

Fernando, S. (2009). Issues of e-learning in third world countries. In M. Khosrow-Pour (Ed.), *Encyclopedia of information science and technology* (2nd ed.; pp. 2273–2277). Hershey, PA: IGI Global. doi:10.4018/978-1-60566-026-4.ch360

Filho, J. R., & dos Santos, J. R. Jr. (2009). Local e-government in Brazil: Poor interaction and local politics as usual. In C. Reddick (Ed.), *Handbook of research on strategies for local e-government adoption and implementation: Comparative studies* (pp. 863–878). Hershey, PA: IGI Global. doi:10.4018/978-1-60566-282-4.ch045

Fletcher, P. D. (2004). Portals and policy: Implications of electronic access to U.S. federal government information services. In A. Pavlichev & G. Garson (Eds.), *Digital government: Principles and best practices* (pp. 52–62). Hershey, PA: IGI Global. doi:10.4018/978-1-59140-122-3.ch004

Fletcher, P. D. (2008). Portals and policy: Implications of electronic access to U.S. federal government information services. In A. Anttiroiko (Ed.), *Electronic government: Concepts, methodologies, tools, and applications* (pp. 3970–3979). Hershey, PA: IGI Global. doi:10.4018/978-1-59904-947-2.ch289

Forlano, L. (2004). The emergence of digital government: International perspectives. In A. Pavlichev & G. Garson (Eds.), *Digital government: Principles and best practices* (pp. 34–51). Hershey, PA: IGI Global. doi:10.4018/978-1-59140-122-3.ch003

Franzel, J. M., & Coursey, D. H. (2004). Government web portals: Management issues and the approaches of five states. In A. Pavlichev & G. Garson (Eds.), *Digital government: Principles and best practices* (pp. 63–77). Hershey, PA: IGI Global. doi:10.4018/978-1-59140-122-3.ch005

Gaivéo, J. M. (2013). Security of ICTs supporting healthcare activities. In M. Cruz-Cunha, I. Miranda, & P. Gonçalves (Eds.), *Handbook of research on ICTs for human-centered healthcare and social care services* (pp. 208–228). Hershey, PA: IGI Global. doi:10.4018/978-1-4666-3986-7.ch011

Garson, G. D. (1999). *Information technology and computer applications in public administration: Issues and trends*. Hershey, PA: IGI Global. doi:10.4018/978-1-87828-952-0

Garson, G. D. (2003). Toward an information technology research agenda for public administration. In G. Garson (Ed.), *Public information technology: Policy and management issues* (pp. 331–357). Hershey, PA: IGI Global. doi:10.4018/978-1-59140-060-8.ch014

Garson, G. D. (2004). The promise of digital government. In A. Pavlichev & G. Garson (Eds.), *Digital government: Principles and best practices* (pp. 2–15). Hershey, PA: IGI Global. doi:10.4018/978-1-59140-122-3.ch001

Garson, G. D. (2007). An information technology research agenda for public administration. In G. Garson (Ed.), *Modern public information technology systems: Issues and challenges* (pp. 365–392). Hershey, PA: IGI Global. doi:10.4018/978-1-59904-051-6.ch018

Gasco, M. (2007). Civil servants' resistance towards e-government development. In A. Anttiroiko & M. Malkia (Eds.), *Encyclopedia of digital government* (pp. 190–195). Hershey, PA: IGI Global. doi:10.4018/978-1-59140-789-8.ch028

Gasco, M. (2008). Civil servants' resistance towards e-government development. In A. Anttiroiko (Ed.), *Electronic government: Concepts, methodologies, tools, and applications* (pp. 2580–2588). Hershey, PA: IGI Global. doi:10.4018/978-1-59904-947-2.ch190

Ghere, R. K. (2010). Accountability and information technology enactment: Implications for social empowerment. In E. Ferro, Y. Dwivedi, J. Gil-Garcia, & M. Williams (Eds.), *Handbook of research on overcoming digital divides: Constructing an equitable and competitive information society* (pp. 515–532). Hershey, PA: IGI Global. doi:10.4018/978-1-60566-699-0.ch028

Gibson, I. W. (2012). Simulation modeling of healthcare delivery. In A. Kolker & P. Story (Eds.), *Management engineering for effective healthcare delivery: Principles and applications* (pp. 69–89). Hershey, PA: IGI Global. doi:10.4018/978-1-60960-872-9.ch003

Gil-Garcia, J. R. (2007). Exploring e-government benefits and success factors. In A. Anttiroiko & M. Malkia (Eds.), *Encyclopedia of digital government* (pp. 803–811). Hershey, PA: IGI Global. doi:10.4018/978-1-59140-789-8.ch122

Gil-Garcia, J. R., & González Miranda, F. (2010). E-government and opportunities for participation: The case of the Mexican state web portals. In C. Reddick (Ed.), *Politics, democracy and e-government: Participation and service delivery* (pp. 56–74). Hershey, PA: IGI Global. doi:10.4018/978-1-61520-933-0.ch004

Goldfinch, S. (2012). Public trust in government, trust in e-government, and use of e-government. In Z. Yan (Ed.), *Encyclopedia of cyber behavior* (pp. 987–995). Hershey, PA: IGI Global. doi:10.4018/978-1-4666-0315-8.ch081

Goodyear, M. (2012). Organizational change contributions to e-government project transitions. In S. Aikins (Ed.), *Managing e-government projects: Concepts, issues, and best practices* (pp. 1–21). Hershey, PA: IGI Global. doi:10.4018/978-1-4666-0086-7.ch001

Gordon, S., & Mulligan, P. (2003). Strategic models for the delivery of personal financial services: The role of infocracy. In S. Gordon (Ed.), *Computing information technology: The human side* (pp. 220–232). Hershey, PA: IGI Global. doi:10.4018/978-1-93177-752-0.ch014

Gordon, T. F. (2007). Legal knowledge systems. In A. Anttiroiko & M. Malkia (Eds.), *Encyclopedia of digital government* (pp. 1161–1166). Hershey, PA: IGI Global. doi:10.4018/978-1-59140-789-8.ch175

Graham, J. E., & Semich, G. W. (2008). Integrating technology to transform pedagogy: Revisiting the progress of the three phase TUI model for faculty development. In L. Tomei (Ed.), *Adapting information and communication technologies for effective education* (pp. 1–12). Hershey, PA: IGI Global. doi:10.4018/978-1-59904-922-9.ch001

Grandinetti, L., & Pisacane, O. (2012). Web services for healthcare management. In D. Prakash Vidyarthi (Ed.), *Technologies and protocols for the future of internet design: Reinventing the web* (pp. 60–94). Hershey, PA: IGI Global. doi:10.4018/978-1-4666-0203-8.ch004

Groenewegen, P., & Wagenaar, F. P. (2008). VO as an alternative to hierarchy in the Dutch police sector. In G. Putnik & M. Cruz-Cunha (Eds.), *Encyclopedia of networked and virtual organizations* (pp. 1851–1857). Hershey, PA: IGI Global. doi:10.4018/978-1-59904-885-7.ch245

Gronlund, A. (2001). Building an infrastructure to manage electronic services. In S. Dasgupta (Ed.), *Managing internet and intranet technologies in organizations: Challenges and opportunities* (pp. 71–103). Hershey, PA: IGI Global. doi:10.4018/978-1-878289-95-7.ch006

Gronlund, A. (2002). Introduction to electronic government: Design, applications and management. In Å. Grönlund (Ed.), *Electronic government: Design, applications and management* (pp. 1–21). Hershey, PA: IGI Global. doi:10.4018/978-1-930708-19-8.ch001

Gupta, A., Woosley, R., Crk, I., & Sarnikar, S. (2009). An information technology architecture for drug effectiveness reporting and post-marketing surveillance. In J. Tan (Ed.), *Medical informatics: Concepts, methodologies, tools, and applications* (pp. 631–646). Hershey, PA: IGI Global. doi:10.4018/978-1-60566-050-9.ch047

Hallin, A., & Lundevall, K. (2007). mCity: User focused development of mobile services within the city of Stockholm. In I. Kushchu (Ed.), Mobile government: An emerging direction in e-government (pp. 12-29). Hershey, PA: IGI Global. doi:10.4018/978-1-59140-884-0.ch002

Hallin, A., & Lundevall, K. (2009). mCity: User focused development of mobile services within the city of Stockholm. In S. Clarke (Ed.), Evolutionary concepts in end user productivity and performance: Applications for organizational progress (pp. 268-280). Hershey, PA: IGI Global. doi:10.4018/978-1-60566-136-0.ch017

Hallin, A., & Lundevall, K. (2009). mCity: User focused development of mobile services within the city of Stockholm. In D. Taniar (Ed.), Mobile computing: Concepts, methodologies, tools, and applications (pp. 3455-3467). Hershey, PA: IGI Global. doi:10.4018/978-1-60566-054-7.ch253

Hanson, A. (2005). Overcoming barriers in the planning of a virtual library. In M. Khosrow-Pour (Ed.), *Encyclopedia of information science and technology* (pp. 2255–2259). Hershey, PA: IGI Global. doi:10.4018/978-1-59140-553-5.ch397

Haque, A. (2008). Information technology and surveillance: Implications for public administration in a new word order. In T. Loendorf & G. Garson (Eds.), *Patriotic information systems* (pp. 177–185). Hershey, PA: IGI Global. doi:10.4018/978-1-59904-594-8.ch008

Hauck, R. V., Thatcher, S. M., & Weisband, S. P. (2012). Temporal aspects of information technology use: Increasing shift work effectiveness. In J. Wang (Ed.), *Advancing the service sector with evolving technologies: Techniques and principles* (pp. 87–104). Hershey, PA: IGI Global. doi:10.4018/978-1-4666-0044-7.ch006

Hawk, S., & Witt, T. (2006). Telecommunications courses in information systems programs. *International Journal of Information and Communication Technology Education, 2*(1), 79–92. doi:10.4018/jicte.2006010107

Helms, M. M., Moore, R., & Ahmadi, M. (2009). Information technology (IT) and the healthcare industry: A SWOT analysis. In J. Tan (Ed.), *Medical informatics: Concepts, methodologies, tools, and applications* (pp. 134–152). Hershey, PA: IGI Global. doi:10.4018/978-1-60566-050-9.ch012

Hendrickson, S. M., & Young, M. E. (2014). Electronic records management at a federally funded research and development center. In J. Krueger (Ed.), *Cases on electronic records and resource management implementation in diverse environments* (pp. 334–350). Hershey, PA: IGI Global. doi:10.4018/978-1-4666-4466-3.ch020

Henman, P. (2010). Social policy and information communication technologies. In J. Martin & L. Hawkins (Eds.), *Information communication technologies for human services education and delivery: Concepts and cases* (pp. 215–229). Hershey, PA: IGI Global. doi:10.4018/978-1-60566-735-5.ch014

Hismanoglu, M. (2011). Important issues in online education: E-pedagogy and marketing. In U. Demiray & S. Sever (Eds.), *Marketing online education programs: Frameworks for promotion and communication* (pp. 184–209). Hershey, PA: IGI Global. doi:10.4018/978-1-60960-074-7.ch012

Ho, K. K. (2008). The e-government development, IT strategies, and portals of the Hong Kong SAR government. In A. Anttiroiko (Ed.), *Electronic government: Concepts, methodologies, tools, and applications* (pp. 715–733). Hershey, PA: IGI Global. doi:10.4018/978-1-59904-947-2.ch060

Holden, S. H. (2003). The evolution of information technology management at the federal level: Implications for public administration. In G. Garson (Ed.), *Public information technology: Policy and management issues* (pp. 53–73). Hershey, PA: IGI Global. doi:10.4018/978-1-59140-060-8.ch003

Holden, S. H. (2007). The evolution of federal information technology management literature: Does IT finally matter? In G. Garson (Ed.), *Modern public information technology systems: Issues and challenges* (pp. 17–34). Hershey, PA: IGI Global. doi:10.4018/978-1-59904-051-6.ch002

Holland, J. W. (2009). Automation of American criminal justice. In M. Khosrow-Pour (Ed.), *Encyclopedia of information science and technology* (2nd ed.; pp. 300–302). Hershey, PA: IGI Global. doi:10.4018/978-1-60566-026-4.ch051

Holloway, K. (2013). Fair use, copyright, and academic integrity in an online academic environment. In *Digital rights management: Concepts, methodologies, tools, and applications* (pp. 917–928). Hershey, PA: IGI Global. doi:10.4018/978-1-4666-2136-7.ch044

Horiuchi, C. (2005). E-government databases. In L. Rivero, J. Doorn, & V. Ferraggine (Eds.), *Encyclopedia of database technologies and applications* (pp. 206–210). Hershey, PA: IGI Global. doi:10.4018/978-1-59140-560-3.ch035

Horiuchi, C. (2006). Creating IS quality in government settings. In E. Duggan & J. Reichgelt (Eds.), *Measuring information systems delivery quality* (pp. 311–327). Hershey, PA: IGI Global. doi:10.4018/978-1-59140-857-4.ch014

Hsiao, N., Chu, P., & Lee, C. (2012). Impact of e-governance on businesses: Model development and case study. In *Digital democracy: Concepts, methodologies, tools, and applications* (pp. 1407–1425). Hershey, PA: IGI Global. doi:10.4018/978-1-4666-1740-7.ch070

Huang, T., & Lee, C. (2010). Evaluating the impact of e-government on citizens: Cost-benefit analysis. In C. Reddick (Ed.), *Citizens and e-government: Evaluating policy and management* (pp. 37–52). Hershey, PA: IGI Global. doi:10.4018/978-1-61520-931-6.ch003

Hunter, M. G., Diochon, M., Pugsley, D., & Wright, B. (2002). Unique challenges for small business adoption of information technology: The case of the Nova Scotia ten. In S. Burgess (Ed.), *Managing information technology in small business: Challenges and solutions* (pp. 98–117). Hershey, PA: IGI Global. doi:10.4018/978-1-930708-35-8.ch006

Hurskainen, J. (2003). Integration of business systems and applications in merger and alliance: Case metso automation. In T. Reponen (Ed.), *Information technology enabled global customer service* (pp. 207–225). Hershey, PA: IGI Global. doi:10.4018/978-1-59140-048-6.ch012

Iazzolino, G., & Pietrantonio, R. (2011). The soveria.it project: A best practice of e-government in southern Italy. In D. Piaggesi, K. Sund, & W. Castelnovo (Eds.), *Global strategy and practice of e-governance: Examples from around the world* (pp. 34–56). Hershey, PA: IGI Global. doi:10.4018/978-1-60960-489-9.ch003

Imran, A., & Gregor, S. (2012). A process model for successful e-government adoption in the least developed countries: A case of Bangladesh. In F. Tan (Ed.), *International comparisons of information communication technologies: Advancing applications* (pp. 321–350). Hershey, PA: IGI Global. doi:10.4018/978-1-61350-480-2.ch014

Inoue, Y., & Bell, S. T. (2005). Electronic/digital government innovation, and publishing trends with IT. In M. Khosrow-Pour (Ed.), *Encyclopedia of information science and technology* (pp. 1018–1023). Hershey, PA: IGI Global. doi:10.4018/978-1-59140-553-5.ch180

Islam, M. M., & Ehsan, M. (2013). Understanding e-governance: A theoretical approach. In M. Islam & M. Ehsan (Eds.), *From government to e-governance: Public administration in the digital age* (pp. 38–49). Hershey, PA: IGI Global. doi:10.4018/978-1-4666-1909-8.ch003

Jaeger, B. (2009). E-government and e-democracy in the making. In M. Khosrow-Pour (Ed.), *Encyclopedia of information science and technology* (2nd ed.; pp. 1318–1322). Hershey, PA: IGI Global. doi:10.4018/978-1-60566-026-4.ch208

Jain, R. B. (2007). Revamping the administrative structure and processes in India for online diplomacy. In A. Anttiroiko & M. Malkia (Eds.), *Encyclopedia of digital government* (pp. 1418–1423). Hershey, PA: IGI Global. doi:10.4018/978-1-59140-789-8.ch217

Jain, R. B. (2008). Revamping the administrative structure and processes in India for online diplomacy. In A. Anttiroiko (Ed.), *Electronic government: Concepts, methodologies, tools, and applications* (pp. 3142–3149). Hershey, PA: IGI Global. doi:10.4018/978-1-59904-947-2.ch233

Jauhiainen, J. S., & Inkinen, T. (2009). E-governance and the information society in periphery. In C. Reddick (Ed.), *Handbook of research on strategies for local e-government adoption and implementation: Comparative studies* (pp. 497–514). Hershey, PA: IGI Global. doi:10.4018/978-1-60566-282-4.ch026

Jensen, M. J. (2009). Electronic democracy and citizen influence in government. In C. Reddick (Ed.), *Handbook of research on strategies for local e-government adoption and implementation: Comparative studies* (pp. 288–305). Hershey, PA: IGI Global. doi:10.4018/978-1-60566-282-4.ch015

Jiao, Y., Hurson, A. R., Potok, T. E., & Beckerman, B. G. (2009). Integrating mobile-based systems with healthcare databases. In J. Erickson (Ed.), *Database technologies: Concepts, methodologies, tools, and applications* (pp. 484–504). Hershey, PA: IGI Global. doi:10.4018/978-1-60566-058-5.ch031

Joia, L. A. (2002). A systematic model to integrate information technology into metabusinesses: A case study in the engineering realms. In F. Tan (Ed.), *Advanced topics in global information management* (Vol. 1, pp. 250–267). Hershey, PA: IGI Global. doi:10.4018/978-1-930708-43-3.ch016

Jones, T. H., & Song, I. (2000). Binary equivalents of ternary relationships in entity-relationship modeling: A logical decomposition approach. *Journal of Database Management, 11*(2), 12–19. doi:10.4018/jdm.2000040102

Juana-Espinosa, S. D. (2007). Empirical study of the municipalitites' motivations for adopting online presence. In L. Al-Hakim (Ed.), *Global e-government: Theory, applications and benchmarking* (pp. 261–279). Hershey, PA: IGI Global. doi:10.4018/978-1-59904-027-1.ch015

Jun, K., & Weare, C. (2012). Bridging from e-government practice to e-government research: Past trends and future directions. In K. Bwalya & S. Zulu (Eds.), *Handbook of research on e-government in emerging economies: Adoption, e-participation, and legal frameworks* (pp. 263–289). Hershey, PA: IGI Global. doi:10.4018/978-1-4666-0324-0.ch013

Junqueira, A., Diniz, E. H., & Fernandez, M. (2010). Electronic government implementation projects with multiple agencies: Analysis of the electronic invoice project under PMBOK framework. In J. Cordoba-Pachon & A. Ochoa-Arias (Eds.), *Systems thinking and e-participation: ICT in the governance of society* (pp. 135–153). Hershey, PA: IGI Global. doi:10.4018/978-1-60566-860-4.ch009

Juntunen, A. (2009). Joint service development with the local authorities. In C. Reddick (Ed.), *Handbook of research on strategies for local e-government adoption and implementation: Comparative studies* (pp. 902–920). Hershey, PA: IGI Global. doi:10.4018/978-1-60566-282-4.ch047

Kamel, S. (2001). *Using DSS for crisis management.* Hershey, PA: IGI Global. doi:10.4018/978-1-87828-961-2.ch020

Kamel, S. (2006). DSS for strategic decision making. In M. Khosrow-Pour (Ed.), *Cases on information technology and organizational politics & culture* (pp. 230–246). Hershey, PA: IGI Global. doi:10.4018/978-1-59904-411-8.ch015

Kamel, S. (2009). The software industry in Egypt as a potential contributor to economic growth. In M. Khosrow-Pour (Ed.), *Encyclopedia of information science and technology* (2nd ed.; pp. 3531–3537). Hershey, PA: IGI Global. doi:10.4018/978-1-60566-026-4.ch562

Kamel, S., & Husscin, M. (2008). Xceed: Pioneering the contact center industry in Egypt. *Journal of Cases on Information Technology, 10*(1), 67–91. doi:10.4018/jcit.2008010105

Kamel, S., & Wahba, K. (2003). The use of a hybrid model in web-based education: "The Global campus project. In A. Aggarwal (Ed.), *Web-based education: Learning from experience* (pp. 331–346). Hershey, PA: IGI Global. doi:10.4018/978-1-59140-102-5.ch020

Kardaras, D. K., & Papathanassiou, E. A. (2008). An exploratory study of the e-government services in Greece. In G. Garson & M. Khosrow-Pour (Eds.), *Handbook of research on public information technology* (pp. 162–174). Hershey, PA: IGI Global. doi:10.4018/978-1-59904-857-4.ch016

Kassahun, A. E., Molla, A., & Sarkar, P. (2012). Government process reengineering: What we know and what we need to know. In *Digital democracy: Concepts, methodologies, tools, and applications* (pp. 1730–1752). Hershey, PA: IGI Global. doi:10.4018/978-1-4666-1740-7.ch086

Khan, B. (2005). Technological issues. In B. Khan (Ed.), *Managing e-learning strategies: Design, delivery, implementation and evaluation* (pp. 154–180). Hershey, PA: IGI Global. doi:10.4018/978-1-59140-634-1.ch004

Khasawneh, A., Bsoul, M., Obeidat, I., & Al Azzam, I. (2012). Technology fears: A study of e-commerce loyalty perception by Jordanian customers. In J. Wang (Ed.), *Advancing the service sector with evolving technologies: Techniques and principles* (pp. 158–165). Hershey, PA: IGI Global. doi:10.4018/978-1-4666-0044-7.ch010

Khatibi, V., & Montazer, G. A. (2012). E-research methodology. In A. Juan, T. Daradoumis, M. Roca, S. Grasman, & J. Faulin (Eds.), *Collaborative and distributed e-research: Innovations in technologies, strategies and applications* (pp. 62–81). Hershey, PA: IGI Global. doi:10.4018/978-1-4666-0125-3.ch003

Kidd, T. (2011). The dragon in the school's backyard: A review of literature on the uses of technology in urban schools. In L. Tomei (Ed.), *Online courses and ICT in education: Emerging practices and applications* (pp. 242–257). Hershey, PA: IGI Global. doi:10.4018/978-1-60960-150-8.ch019

Kidd, T. T. (2010). My experience tells the story: Exploring technology adoption from a qualitative perspective - A pilot study. In H. Song & T. Kidd (Eds.), *Handbook of research on human performance and instructional technology* (pp. 247–262). Hershey, PA: IGI Global. doi:10.4018/978-1-60566-782-9.ch015

Kieley, B., Lane, G., Paquet, G., & Roy, J. (2002). e-Government in Canada: Services online or public service renewal? In Å. Grönlund (Ed.), Electronic government: Design, applications and management (pp. 340-355). Hershey, PA: IGI Global. doi:10.4018/978-1-930708-19-8.ch016

Kim, P. (2012). "Stay out of the way! My kid is video blogging through a phone!": A lesson learned from math tutoring social media for children in underserved communities. In *Wireless technologies: Concepts, methodologies, tools and applications* (pp. 1415–1428). Hershey, PA: IGI Global. doi:10.4018/978-1-61350-101-6.ch517

Kirlidog, M. (2010). Financial aspects of national ICT strategies. In S. Kamel (Ed.), *E-strategies for technological diffusion and adoption: National ICT approaches for socioeconomic development* (pp. 277–292). Hershey, PA: IGI Global. doi:10.4018/978-1-60566-388-3.ch016

Kisielnicki, J. (2006). Transfer of information and knowledge in the project management. In E. Coakes & S. Clarke (Eds.), *Encyclopedia of communities of practice in information and knowledge management* (pp. 544–551). Hershey, PA: IGI Global. doi:10.4018/978-1-59140-556-6.ch091

Kittner, M., & Van Slyke, C. (2006). Reorganizing information technology services in an academic environment. In M. Khosrow-Pour (Ed.), *Cases on the human side of information technology* (pp. 49–66). Hershey, PA: IGI Global. doi:10.4018/978-1-59904-405-7.ch004

Knoell, H. D. (2008). Semi virtual workplaces in German financial service enterprises. In P. Zemliansky & K. St.Amant (Eds.), *Handbook of research on virtual workplaces and the new nature of business practices* (pp. 570–581). Hershey, PA: IGI Global. doi:10.4018/978-1-59904-893-2.ch041

Koh, S. L., & Maguire, S. (2009). Competing in the age of information technology in a developing economy: Experiences of an Indian bank. In S. Koh & S. Maguire (Eds.), *Information and communication technologies management in turbulent business environments* (pp. 326–350). Hershey, PA: IGI Global. doi:10.4018/978-1-60566-424-8.ch018

Kollmann, T., & Häsel, M. (2009). Competence of information technology professionals in internet-based ventures. In I. Lee (Ed.), *Electronic business: Concepts, methodologies, tools, and applications* (pp. 1905–1919). Hershey, PA: IGI Global. doi:10.4018/978-1-60566-056-1.ch118

Kollmann, T., & Häsel, M. (2009). Competence of information technology professionals in internet-based ventures. In A. Cater-Steel (Ed.), *Information technology governance and service management: Frameworks and adaptations* (pp. 239–253). Hershey, PA: IGI Global. doi:10.4018/978-1-60566-008-0.ch013

Kollmann, T., & Häsel, M. (2010). Competence of information technology professionals in internet-based ventures. In *Electronic services: Concepts, methodologies, tools and applications* (pp. 1551–1565). Hershey, PA: IGI Global. doi:10.4018/978-1-61520-967-5.ch094

Kraemer, K., & King, J. L. (2006). Information technology and administrative reform: Will e-government be different? *International Journal of Electronic Government Research*, *2*(1), 1–20. doi:10.4018/jegr.2006010101

Kraemer, K., & King, J. L. (2008). Information technology and administrative reform: Will e-government be different? In D. Norris (Ed.), *E-government research: Policy and management* (pp. 1–20). Hershey, PA: IGI Global. doi:10.4018/978-1-59904-913-7.ch001

Lampathaki, F., Tsiakaliaris, C., Stasis, A., & Charalabidis, Y. (2011). National interoperability frameworks: The way forward. In Y. Charalabidis (Ed.), *Interoperability in digital public services and administration: Bridging e-government and e-business* (pp. 1–24). Hershey, PA: IGI Global. doi:10.4018/978-1-61520-887-6.ch001

Lan, Z., & Scott, C. R. (1996). The relative importance of computer-mediated information versus conventional non-computer-mediated information in public managerial decision making. *Information Resources Management Journal, 9*(1), 27–0. doi:10.4018/irmj.1996010103

Law, W. (2004). *Public sector data management in a developing economy*. Hershey, PA: IGI Global. doi:10.4018/978-1-59140-259-6.ch034

Law, W. K. (2005). Information resources development challenges in a cross-cultural environment. In M. Khosrow-Pour (Ed.), *Encyclopedia of information science and technology* (pp. 1476–1481). Hershey, PA: IGI Global. doi:10.4018/978-1-59140-553-5.ch259

Law, W. K. (2009). Cross-cultural challenges for information resources management. In M. Khosrow-Pour (Ed.), *Encyclopedia of information science and technology* (2nd ed.; pp. 840–846). Hershey, PA: IGI Global. doi:10.4018/978-1-60566-026-4.ch136

Law, W. K. (2011). Cross-cultural challenges for information resources management. In *Global business: Concepts, methodologies, tools and applications* (pp. 1924–1932). Hershey, PA: IGI Global. doi:10.4018/978-1-60960-587-2.ch704

Malkia, M., & Savolainen, R. (2004). eTransformation in government, politics and society: Conceptual framework and introduction. In M. Malkia, A. Anttiroiko, & R. Savolainen (Eds.), eTransformation in governance: New directions in government and politics (pp. 1-21). Hershey, PA: IGI Global. doi:10.4018/978-1-59140-130-8.ch001

Mandujano, S. (2011). Network manageability security. In D. Kar & M. Syed (Eds.), *Network security, administration and management: Advancing technology and practice* (pp. 158–181). Hershey, PA: IGI Global. doi:10.4018/978-1-60960-777-7.ch009

Marich, M. J., Schooley, B. L., & Horan, T. A. (2012). A normative enterprise architecture for guiding end-to-end emergency response decision support. In M. Jennex (Ed.), *Managing crises and disasters with emerging technologies: Advancements* (pp. 71–87). Hershey, PA: IGI Global. doi:10.4018/978-1-4666-0167-3.ch006

Markov, R., & Okujava, S. (2008). Costs, benefits, and risks of e-government portals. In G. Putnik & M. Cruz-Cunha (Eds.), *Encyclopedia of networked and virtual organizations* (pp. 354–363). Hershey, PA: IGI Global. doi:10.4018/978-1-59904-885-7.ch047

Martin, N., & Rice, J. (2013). Evaluating and designing electronic government for the future: Observations and insights from Australia. In V. Weerakkody (Ed.), *E-government services design, adoption, and evaluation* (pp. 238–258). Hershey, PA: IGI Global. doi:10.4018/978-1-4666-2458-0.ch014

i. Martinez, A. C. (2008). Accessing administration's information via internet in Spain. In F. Tan (Ed.), *Global information technologies: Concepts, methodologies, tools, and applications* (pp. 2558–2573). Hershey, PA: IGI Global. doi:10.4018/978-1-59904-939-7.ch186

Mbarika, V. W., Meso, P. N., & Musa, P. F. (2006). A disconnect in stakeholders' perceptions from emerging realities of teledensity growth in Africa's least developed countries. In M. Hunter & F. Tan (Eds.), *Advanced topics in global information management* (Vol. 5, pp. 263–282). Hershey, PA: IGI Global. doi:10.4018/978-1-59140-923-6.ch012

Mbarika, V. W., Meso, P. N., & Musa, P. F. (2008). A disconnect in stakeholders' perceptions from emerging realities of teledensity growth in Africa's least developed countries. In F. Tan (Ed.), *Global information technologies: Concepts, methodologies, tools, and applications* (pp. 2948–2962). Hershey, PA: IGI Global. doi:10.4018/978-1-59904-939-7.ch209

Means, T., Olson, E., & Spooner, J. (2013). Discovering ways that don't work on the road to success: Strengths and weaknesses revealed by an active learning studio classroom project. In A. Benson, J. Moore, & S. Williams van Rooij (Eds.), *Cases on educational technology planning, design, and implementation: A project management perspective* (pp. 94–113). Hershey, PA: IGI Global. doi:10.4018/978-1-4666-4237-9.ch006

Melitski, J., Holzer, M., Kim, S., Kim, C., & Rho, S. (2008). Digital government worldwide: An e-government assessment of municipal web sites. In G. Garson & M. Khosrow-Pour (Eds.), *Handbook of research on public information technology* (pp. 790–804). Hershey, PA: IGI Global. doi:10.4018/978-1-59904-857-4.ch069

Memmola, M., Palumbo, G., & Rossini, M. (2009). Web & RFID technology: New frontiers in costing and process management for rehabilitation medicine. In L. Al-Hakim & M. Memmola (Eds.), *Business web strategy: Design, alignment, and application* (pp. 145–169). Hershey, PA: IGI Global. doi:10.4018/978-1-60566-024-0.ch008

Meng, Z., Fahong, Z., & Lei, L. (2008). Information technology and environment. In Y. Kurihara, S. Takaya, H. Harui, & H. Kamae (Eds.), *Information technology and economic development* (pp. 201–212). Hershey, PA: IGI Global. doi:10.4018/978-1-59904-579-5.ch014

Mentzingen de Moraes, A. J., Ferneda, E., Costa, I., & Spinola, M. D. (2011). Practical approach for implementation of governance process in IT: Information technology areas. In N. Shi & G. Silvius (Eds.), *Enterprise IT governance, business value and performance measurement* (pp. 19–40). Hershey, PA: IGI Global. doi:10.4018/978-1-60566-346-3.ch002

Merwin, G. A. Jr, McDonald, J. S., & Odera, L. C. (2008). Economic development: Government's cutting edge in IT. In M. Raisinghani (Ed.), *Handbook of research on global information technology management in the digital economy* (pp. 1–37). Hershey, PA: IGI Global. doi:10.4018/978-1-59904-875-8.ch001

Meso, P., & Duncan, N. (2002). Can national information infrastructures enhance social development in the least developed countries? An empirical investigation. In M. Dadashzadeh (Ed.), *Information technology management in developing countries* (pp. 23–51). Hershey, PA: IGI Global. doi:10.4018/978-1-931777-03-2.ch002

Meso, P. N., & Duncan, N. B. (2002). Can national information infrastructures enhance social development in the least developed countries? In F. Tan (Ed.), *Advanced topics in global information management* (Vol. 1, pp. 207–226). Hershey, PA: IGI Global. doi:10.4018/978-1-930708-43-3.ch014

Middleton, M. (2008). Evaluation of e-government web sites. In G. Garson & M. Khosrow-Pour (Eds.), *Handbook of research on public information technology* (pp. 699–710). Hershey, PA: IGI Global. doi:10.4018/978-1-59904-857-4.ch063

Mingers, J. (2010). Pluralism, realism, and truth: The keys to knowledge in information systems research. In D. Paradice (Ed.), *Emerging systems approaches in information technologies: Concepts, theories, and applications* (pp. 86–98). Hershey, PA: IGI Global. doi:10.4018/978-1-60566-976-2.ch006

Mital, K. M. (2012). ICT, unique identity and inclusive growth: An Indian perspective. In A. Manoharan & M. Holzer (Eds.), *E-governance and civic engagement: Factors and determinants of e-democracy* (pp. 584–612). Hershey, PA: IGI Global. doi:10.4018/978-1-61350-083-5.ch029

Mizell, A. P. (2008). Helping close the digital divide for financially disadvantaged seniors. In F. Tan (Ed.), *Global information technologies: Concepts, methodologies, tools, and applications* (pp. 2396–2402). Hershey, PA: IGI Global. doi:10.4018/978-1-59904-939-7.ch173

Molinari, F., Wills, C., Koumpis, A., & Moumtzi, V. (2011). A citizen-centric platform to support networking in the area of e-democracy. In H. Rahman (Ed.), *Cases on adoption, diffusion and evaluation of global e-governance systems: Impact at the grass roots* (pp. 282–302). Hershey, PA: IGI Global. doi:10.4018/978-1-61692-814-8.ch014

Molinari, F., Wills, C., Koumpis, A., & Moumtzi, V. (2013). A citizen-centric platform to support networking in the area of e-democracy. In H. Rahman (Ed.), *Cases on progressions and challenges in ICT utilization for citizen-centric governance* (pp. 265–297). Hershey, PA: IGI Global. doi:10.4018/978-1-4666-2071-1.ch013

Monteverde, F. (2010). The process of e-government public policy inclusion in the governmental agenda: A framework for assessment and case study. In J. Cordoba-Pachon & A. Ochoa-Arias (Eds.), *Systems thinking and e-participation: ICT in the governance of society* (pp. 233–245). Hershey, PA: IGI Global. doi:10.4018/978-1-60566-860-4.ch015

Moodley, S. (2008). Deconstructing the South African government's ICT for development discourse. In A. Anttiroiko (Ed.), *Electronic government: Concepts, methodologies, tools, and applications* (pp. 622–631). Hershey, PA: IGI Global. doi:10.4018/978-1-59904-947-2.ch053

Moodley, S. (2008). Deconstructing the South African government's ICT for development discourse. In C. Van Slyke (Ed.), *Information communication technologies: Concepts, methodologies, tools, and applications* (pp. 816–825). Hershey, PA: IGI Global. doi:10.4018/978-1-59904-949-6.ch052

Mora, M., Cervantes-Perez, F., Gelman-Muravchik, O., Forgionne, G. A., & Mejia-Olvera, M. (2003). DMSS implementation research: A conceptual analysis of the contributions and limitations of the factor-based and stage-based streams. In G. Forgionne, J. Gupta, & M. Mora (Eds.), *Decision-making support systems: Achievements and challenges for the new decade* (pp. 331–356). Hershey, PA: IGI Global. doi:10.4018/978-1-59140-045-5.ch020

Mörtberg, C., & Elovaara, P. (2010). Attaching people and technology: Between e and government. In S. Booth, S. Goodman, & G. Kirkup (Eds.), *Gender issues in learning and working with information technology: Social constructs and cultural contexts* (pp. 83–98). Hershey, PA: IGI Global. doi:10.4018/978-1-61520-813-5.ch005

Murphy, J., Harper, E., Devine, E. C., Burke, L. J., & Hook, M. L. (2011). Case study: Lessons learned when embedding evidence-based knowledge in a nurse care planning and documentation system. In A. Cashin & R. Cook (Eds.), *Evidence-based practice in nursing informatics: Concepts and applications* (pp. 174–190). Hershey, PA: IGI Global. doi:10.4018/978-1-60960-034-1.ch014

Mutula, S. M. (2013). E-government's role in poverty alleviation: Case study of South Africa. In H. Rahman (Ed.), *Cases on progressions and challenges in ICT utilization for citizen-centric governance* (pp. 44–68). Hershey, PA: IGI Global. doi:10.4018/978-1-4666-2071-1.ch003

Nath, R., & Angeles, R. (2005). Relationships between supply characteristics and buyer-supplier coupling in e-procurement: An empirical analysis. *International Journal of E-Business Research, 1*(2), 40–55. doi:10.4018/jebr.2005040103

Nissen, M. E. (2006). Application cases in government. In M. Nissen (Ed.), *Harnessing knowledge dynamics: Principled organizational knowing & learning* (pp. 152–181). Hershey, PA: IGI Global. doi:10.4018/978-1-59140-773-7.ch008

Norris, D. F. (2003). Leading-edge information technologies and American local governments. In G. Garson (Ed.), *Public information technology: Policy and management issues* (pp. 139–169). Hershey, PA: IGI Global. doi:10.4018/978-1-59140-060-8.ch007

Norris, D. F. (2008). Information technology among U.S. local governments. In G. Garson & M. Khosrow-Pour (Eds.), *Handbook of research on public information technology* (pp. 132–144). Hershey, PA: IGI Global. doi:10.4018/978-1-59904-857-4.ch013

Northrop, A. (1999). The challenge of teaching information technology in public administration graduate programs. In G. Garson (Ed.), *Information technology and computer applications in public administration: Issues and trends* (pp. 1–22). Hershey, PA: IGI Global. doi:10.4018/978-1-87828-952-0.ch001

Northrop, A. (2003). Information technology and public administration: The view from the profession. In G. Garson (Ed.), *Public information technology: Policy and management issues* (pp. 1–19). Hershey, PA: IGI Global. doi:10.4018/978-1-59140-060-8.ch001

Northrop, A. (2007). Lip service? How PA journals and textbooks view information technology. In G. Garson (Ed.), *Modern public information technology systems: Issues and challenges* (pp. 1–16). Hershey, PA: IGI Global. doi:10.4018/978-1-59904-051-6.ch001

Null, E. (2013). Legal and political barriers to municipal networks in the United States. In A. Abdelaal (Ed.), *Social and economic effects of community wireless networks and infrastructures* (pp. 27–56). Hershey, PA: IGI Global. doi:10.4018/978-1-4666-2997-4.ch003

Okunoye, A., Frolick, M., & Crable, E. (2006). ERP implementation in higher education: An account of pre-implementation and implementation phases. *Journal of Cases on Information Technology*, *8*(2), 110–132. doi:10.4018/jcit.2006040106

Olasina, G. (2012). A review of egovernment services in Nigeria. In A. Tella & A. Issa (Eds.), *Library and information science in developing countries: Contemporary issues* (pp. 205–221). Hershey, PA: IGI Global. doi:10.4018/978-1-61350-335-5.ch015

Orgeron, C. P. (2008). A model for reengineering IT job classes in state government. In G. Garson & M. Khosrow-Pour (Eds.), *Handbook of research on public information technology* (pp. 735–746). Hershey, PA: IGI Global. doi:10.4018/978-1-59904-857-4.ch066

Owsinski, J. W., & Pielak, A. M. (2011). Local authority websites in rural areas: Measuring quality and functionality, and assessing the role. In Z. Andreopoulou, B. Manos, N. Polman, & D. Viaggi (Eds.), *Agricultural and environmental informatics, governance and management: Emerging research applications* (pp. 39–60). Hershey, PA: IGI Global. doi:10.4018/978-1-60960-621-3.ch003

Owsiński, J. W., Pielak, A. M., Sęp, K., & Stańczak, J. (2014). Local web-based networks in rural municipalities: Extension, density, and meaning. In Z. Andreopoulou, V. Samathrakis, S. Louca, & M. Vlachopoulou (Eds.), *E-innovation for sustainable development of rural resources during global economic crisis* (pp. 126–151). Hershey, PA: IGI Global. doi:10.4018/978-1-4666-4550-9.ch011

Pagani, M., & Pasinetti, C. (2008). Technical and functional quality in the development of t-government services. In A. Anttiroiko (Ed.), *Electronic government: Concepts, methodologies, tools, and applications* (pp. 2943–2965). Hershey, PA: IGI Global. doi:10.4018/978-1-59904-947-2.ch220

Pani, A. K., & Agrahari, A. (2005). On e-markets in emerging economy: An Indian experience. In M. Khosrow-Pour (Ed.), *Advanced topics in electronic commerce* (Vol. 1, pp. 287–299). Hershey, PA: IGI Global. doi:10.4018/978-1-59140-819-2.ch015

Papadopoulos, T., Angelopoulos, S., & Kitsios, F. (2011). A strategic approach to e-health interoperability using e-government frameworks. In A. Lazakidou, K. Siassiakos, & K. Ioannou (Eds.), *Wireless technologies for ambient assisted living and healthcare: Systems and applications* (pp. 213–229). Hershey, PA: IGI Global. doi:10.4018/978-1-61520-805-0.ch012

Papadopoulos, T., Angelopoulos, S., & Kitsios, F. (2013). A strategic approach to e-health interoperability using e-government frameworks. In *User-driven healthcare: Concepts, methodologies, tools, and applications* (pp. 791–807). Hershey, PA: IGI Global. doi:10.4018/978-1-4666-2770-3.ch039

Papaleo, G., Chiarella, D., Aiello, M., & Caviglione, L. (2012). Analysis, development and deployment of statistical anomaly detection techniques for real e-mail traffic. In T. Chou (Ed.), *Information assurance and security technologies for risk assessment and threat management: Advances* (pp. 47–71). Hershey, PA: IGI Global. doi:10.4018/978-1-61350-507-6.ch003

Papp, R. (2003). Information technology & FDA compliance in the pharmaceutical industry. In M. Khosrow-Pour (Ed.), *Annals of cases on information technology* (Vol. 5, pp. 262–273). Hershey, PA: IGI Global. doi:10.4018/978-1-59140-061-5.ch017

Parsons, T. W. (2007). Developing a knowledge management portal. In A. Tatnall (Ed.), *Encyclopedia of portal technologies and applications* (pp. 223–227). Hershey, PA: IGI Global. doi:10.4018/978-1-59140-989-2.ch039

Passaris, C. E. (2007). Immigration and digital government. In A. Anttiroiko & M. Malkia (Eds.), *Encyclopedia of digital government* (pp. 988–994). Hershey, PA: IGI Global. doi:10.4018/978-1-59140-789-8.ch148

Pavlichev, A. (2004). The e-government challenge for public administration. In A. Pavlichev & G. Garson (Eds.), *Digital government: Principles and best practices* (pp. 276–290). Hershey, PA: IGI Global. doi:10.4018/978-1-59140-122-3.ch018

Penrod, J. I., & Harbor, A. F. (2000). Designing and implementing a learning organization-oriented information technology planning and management process. In L. Petrides (Ed.), *Case studies on information technology in higher education: Implications for policy and practice* (pp. 7–19). Hershey, PA: IGI Global. doi:10.4018/978-1-878289-74-2.ch001

Planas-Silva, M. D., & Joseph, R. C. (2011). Perspectives on the adoption of electronic resources for use in clinical trials. In M. Guah (Ed.), *Healthcare delivery reform and new technologies: Organizational initiatives* (pp. 19–28). Hershey, PA: IGI Global. doi:10.4018/978-1-60960-183-6.ch002

Pomazalová, N., & Rejman, S. (2013). The rationale behind implementation of new electronic tools for electronic public procurement. In N. Pomazalová (Ed.), *Public sector transformation processes and internet public procurement: Decision support systems* (pp. 85–117). Hershey, PA: IGI Global. doi:10.4018/978-1-4666-2665-2.ch006

Postorino, M. N. (2012). City competitiveness and airport: Information science perspective. In M. Bulu (Ed.), *City competitiveness and improving urban subsystems: Technologies and applications* (pp. 61–83). Hershey, PA: IGI Global. doi:10.4018/978-1-61350-174-0.ch004

Poupa, C. (2002). Electronic government in Switzerland: Priorities for 2001-2005 - Electronic voting and federal portal. In Å. Grönlund (Ed.), *Electronic government: Design, applications and management* (pp. 356–369). Hershey, PA: IGI Global. doi:10.4018/978-1-930708-19-8.ch017

Powell, S. R. (2010). Interdisciplinarity in telecommunications and networking. In *Networking and telecommunications: Concepts, methodologies, tools and applications* (pp. 33–40). Hershey, PA: IGI Global. doi:10.4018/978-1-60566-986-1.ch004

Priya, P. S., & Mathiyalagan, N. (2011). A study of the implementation status of two e-governance projects in land revenue administration in India. In M. Shareef, V. Kumar, U. Kumar, & Y. Dwivedi (Eds.), *Stakeholder adoption of e-government services: Driving and resisting factors* (pp. 214–230). Hershey, PA: IGI Global. doi:10.4018/978-1-60960-601-5.ch011

Prysby, C., & Prysby, N. (2000). Electronic mail, employee privacy and the workplace. In L. Janczewski (Ed.), *Internet and intranet security management: Risks and solutions* (pp. 251–270). Hershey, PA: IGI Global. doi:10.4018/978-1-878289-71-1.ch009

Prysby, C. L., & Prysby, N. D. (2003). Electronic mail in the public workplace: Issues of privacy and public disclosure. In G. Garson (Ed.), *Public information technology: Policy and management issues* (pp. 271–298). Hershey, PA: IGI Global. doi:10.4018/978-1-59140-060-8.ch012

Prysby, C. L., & Prysby, N. D. (2007). You have mail, but who is reading it? Issues of e-mail in the public workplace. In G. Garson (Ed.), *Modern public information technology systems: Issues and challenges* (pp. 312–336). Hershey, PA: IGI Global. doi:10.4018/978-1-59904-051-6.ch016

Radl, A., & Chen, Y. (2005). Computer security in electronic government: A state-local education information system. *International Journal of Electronic Government Research*, *1*(1), 79–99. doi:10.4018/jegr.2005010105

Rahman, H. (2008). Information dynamics in developing countries. In C. Van Slyke (Ed.), *Information communication technologies: Concepts, methodologies, tools, and applications* (pp. 104–114). Hershey, PA: IGI Global. doi:10.4018/978-1-59904-949-6.ch008

Ramanathan, J. (2009). Adaptive IT architecture as a catalyst for network capability in government. In P. Saha (Ed.), *Advances in government enterprise architecture* (pp. 149–172). Hershey, PA: IGI Global. doi:10.4018/978-1-60566-068-4.ch007

Ramos, I., & Berry, D. M. (2006). Social construction of information technology supporting work. In M. Khosrow-Pour (Ed.), *Cases on information technology: Lessons learned* (Vol. 7, pp. 36–52). Hershey, PA: IGI Global. doi:10.4018/978-1-59140-673-0.ch003

Ray, D., Gulla, U., Gupta, M. P., & Dash, S. S. (2009). Interoperability and constituents of interoperable systems in public sector. In V. Weerakkody, M. Janssen, & Y. Dwivedi (Eds.), *Handbook of research on ICT-enabled transformational government: A global perspective* (pp. 175–195). Hershey, PA: IGI Global. doi:10.4018/978-1-60566-390-6.ch010

Reddick, C. G. (2007). E-government and creating a citizen-centric government: A study of federal government CIOs. In G. Garson (Ed.), *Modern public information technology systems: Issues and challenges* (pp. 143–165). Hershey, PA: IGI Global. doi:10.4018/978-1-59904-051-6.ch008

Reddick, C. G. (2010). Citizen-centric e-government. In C. Reddick (Ed.), *Homeland security preparedness and information systems: Strategies for managing public policy* (pp. 45–75). Hershey, PA: IGI Global. doi:10.4018/978-1-60566-834-5.ch002

Reddick, C. G. (2010). E-government and creating a citizen-centric government: A study of federal government CIOs. In C. Reddick (Ed.), *Homeland security preparedness and information systems: Strategies for managing public policy* (pp. 230–250). Hershey, PA: IGI Global. doi:10.4018/978-1-60566-834-5.ch012

Reddick, C. G. (2010). Perceived effectiveness of e-government and its usage in city governments: Survey evidence from information technology directors. In C. Reddick (Ed.), *Homeland security preparedness and information systems: Strategies for managing public policy* (pp. 213–229). Hershey, PA: IGI Global. doi:10.4018/978-1-60566-834-5.ch011

Reddick, C. G. (2012). Customer relationship management adoption in local governments in the United States. In S. Chhabra & M. Kumar (Eds.), *Strategic enterprise resource planning models for e-government: Applications and methodologies* (pp. 111–124). Hershey, PA: IGI Global. doi:10.4018/978-1-60960-863-7.ch008

Reeder, F. S., & Pandy, S. M. (2008). Identifying effective funding models for e-government. In A. Anttiroiko (Ed.), *Electronic government: Concepts, methodologies, tools, and applications* (pp. 1108–1138). Hershey, PA: IGI Global. doi:10.4018/978-1-59904-947-2.ch083

Riesco, D., Acosta, E., & Montejano, G. (2003). An extension to a UML activity graph from workflow. In L. Favre (Ed.), *UML and the unified process* (pp. 294–314). Hershey, PA: IGI Global. doi:10.4018/978-1-93177-744-5.ch015

Ritzhaupt, A. D., & Gill, T. G. (2008). A hybrid and novel approach to teaching computer programming in MIS curriculum. In S. Negash, M. Whitman, A. Woszczynski, K. Hoganson, & H. Mattord (Eds.), *Handbook of distance learning for real-time and asynchronous information technology education* (pp. 259–281). Hershey, PA: IGI Global. doi:10.4018/978-1-59904-964-9.ch014

Roche, E. M. (1993). International computing and the international regime. *Journal of Global Information Management, 1*(2), 33–44. doi:10.4018/jgim.1993040103

Rocheleau, B. (2007). Politics, accountability, and information management. In G. Garson (Ed.), *Modern public information technology systems: Issues and challenges* (pp. 35–71). Hershey, PA: IGI Global. doi:10.4018/978-1-59904-051-6.ch003

Rodrigues Filho, J. (2010). E-government in Brazil: Reinforcing dominant institutions or reducing citizenship? In C. Reddick (Ed.), *Politics, democracy and e-government: Participation and service delivery* (pp. 347–362). Hershey, PA: IGI Global. doi:10.4018/978-1-61520-933-0.ch021

Rodriguez, S. R., & Thorp, D. A. (2013). eLearning for industry: A case study of the project management process. In A. Benson, J. Moore, & S. Williams van Rooij (Eds.), Cases on educational technology planning, design, and implementation: A project management perspective (pp. 319-342). Hershey, PA: IGI Global. doi:10.4018/978-1-4666-4237-9.ch017

Roman, A. V. (2013). Delineating three dimensions of e-government success: Security, functionality, and transformation. In J. Gil-Garcia (Ed.), *E-government success factors and measures: Theories, concepts, and methodologies* (pp. 171–192). Hershey, PA: IGI Global. doi:10.4018/978-1-4666-4058-0.ch010

Ross, S. C., Tyran, C. K., & Auer, D. J. (2008). Up in smoke: Rebuilding after an IT disaster. In H. Nemati (Ed.), *Information security and ethics: Concepts, methodologies, tools, and applications* (pp. 3659–3675). Hershey, PA: IGI Global. doi:10.4018/978-1-59904-937-3.ch248

Ross, S. C., Tyran, C. K., Auer, D. J., Junell, J. M., & Williams, T. G. (2005). Up in smoke: Rebuilding after an IT disaster. *Journal of Cases on Information Technology*, *7*(2), 31–49. doi:10.4018/jcit.2005040103

Roy, J. (2008). Security, sovereignty, and continental interoperability: Canada's elusive balance. In T. Loendorf & G. Garson (Eds.), *Patriotic information systems* (pp. 153–176). Hershey, PA: IGI Global. doi:10.4018/978-1-59904-594-8.ch007

Rubeck, R. F., & Miller, G. A. (2009). vGOV: Remote video access to government services. In A. Scupola (Ed.), Cases on managing e-services (pp. 253-268). Hershey, PA: IGI Global. doi:10.4018/978-1-60566-064-6.ch017

Saekow, A., & Boonmee, C. (2011). The challenges of implementing e-government interoperability in Thailand: Case of official electronic correspondence letters exchange across government departments. In Y. Charalabidis (Ed.), *Interoperability in digital public services and administration: Bridging e-government and e-business* (pp. 40–61). Hershey, PA: IGI Global. doi:10.4018/978-1-61520-887-6.ch003

Saekow, A., & Boonmee, C. (2012). The challenges of implementing e-government interoperability in Thailand: Case of official electronic correspondence letters exchange across government departments. In *Digital democracy: Concepts, methodologies, tools, and applications* (pp. 1883–1905). Hershey, PA: IGI Global. doi:10.4018/978-1-4666-1740-7.ch094

Sagsan, M., & Medeni, T. (2012). Understanding "knowledge management (KM) paradigms" from social media perspective: An empirical study on discussion group for KM at professional networking site. In M. Cruz-Cunha, P. Gonçalves, N. Lopes, E. Miranda, & G. Putnik (Eds.), *Handbook of research on business social networking: Organizational, managerial, and technological dimensions* (pp. 738–755). Hershey, PA: IGI Global. doi:10.4018/978-1-61350-168-9.ch039

Sahi, G., & Madan, S. (2013). Information security threats in ERP enabled e-governance: Challenges and solutions. In *Enterprise resource planning: Concepts, methodologies, tools, and applications* (pp. 825–837). Hershey, PA: IGI Global. doi:10.4018/978-1-4666-4153-2.ch048

Sanford, C., & Bhattacherjee, A. (2008). IT implementation in a developing country municipality: A sociocognitive analysis. *International Journal of Technology and Human Interaction*, *4*(3), 68–93. doi:10.4018/jthi.2008070104

Schelin, S. H. (2003). E-government: An overview. In G. Garson (Ed.), *Public information technology: Policy and management issues* (pp. 120–138). Hershey, PA: IGI Global. doi:10.4018/978-1-59140-060-8.ch006

Schelin, S. H. (2004). Training for digital government. In A. Pavlichev & G. Garson (Eds.), *Digital government: Principles and best practices* (pp. 263–275). Hershey, PA: IGI Global. doi:10.4018/978-1-59140-122-3.ch017

Schelin, S. H. (2007). E-government: An overview. In G. Garson (Ed.), *Modern public information technology systems: Issues and challenges* (pp. 110–126). Hershey, PA: IGI Global. doi:10.4018/978-1-59904-051-6.ch006

Schelin, S. H., & Garson, G. (2004). Theoretical justification of critical success factors. In G. Garson & S. Schelin (Eds.), *IT solutions series: Humanizing information technology: Advice from experts* (pp. 4–15). Hershey, PA: IGI Global. doi:10.4018/978-1-59140-245-9.ch002

Scime, A. (2002). Information systems and computer science model curricula: A comparative look. In M. Dadashzadeh, A. Saber, & S. Saber (Eds.), *Information technology education in the new millennium* (pp. 146–158). Hershey, PA: IGI Global. doi:10.4018/978-1-931777-05-6.ch018

Scime, A. (2009). Computing curriculum analysis and development. In M. Khosrow-Pour (Ed.), *Encyclopedia of information science and technology* (2nd ed.; pp. 667–671). Hershey, PA: IGI Global. doi:10.4018/978-1-60566-026-4.ch108

Scime, A., & Wania, C. (2008). Computing curricula: A comparison of models. In C. Van Slyke (Ed.), *Information communication technologies: Concepts, methodologies, tools, and applications* (pp. 1270–1283). Hershey, PA: IGI Global. doi:10.4018/978-1-59904-949-6.ch088

Seidman, S. B. (2009). An international perspective on professional software engineering credentials. In H. Ellis, S. Demurjian, & J. Naveda (Eds.), *Software engineering: Effective teaching and learning approaches and practices* (pp. 351–361). Hershey, PA: IGI Global. doi:10.4018/978-1-60566-102-5.ch018

Seifert, J. W. (2007). E-government act of 2002 in the United States. In A. Anttiroiko & M. Malkia (Eds.), *Encyclopedia of digital government* (pp. 476–481). Hershey, PA: IGI Global. doi:10.4018/978-1-59140-789-8.ch072

Seifert, J. W., & Relyea, H. C. (2008). E-government act of 2002 in the United States. In A. Anttiroiko (Ed.), *Electronic government: Concepts, methodologies, tools, and applications* (pp. 154–161). Hershey, PA: IGI Global. doi:10.4018/978-1-59904-947-2.ch013

Seufert, S. (2002). E-learning business models: Framework and best practice examples. In M. Raisinghani (Ed.), *Cases on worldwide e-commerce: Theory in action* (pp. 70–94). Hershey, PA: IGI Global. doi:10.4018/978-1-930708-27-3.ch004

Shareef, M. A., & Archer, N. (2012). E-government service development. In M. Shareef, N. Archer, & S. Dutta (Eds.), *E-government service maturity and development: Cultural, organizational and technological perspectives* (pp. 1–14). Hershey, PA: IGI Global. doi:10.4018/978-1-60960-848-4.ch001

Shareef, M. A., & Archer, N. (2012). E-government initiatives: Review studies on different countries. In M. Shareef, N. Archer, & S. Dutta (Eds.), *E-government service maturity and development: Cultural, organizational and technological perspectives* (pp. 40–76). Hershey, PA: IGI Global. doi:10.4018/978-1-60960-848-4.ch003

Shareef, M. A., Kumar, U., & Kumar, V. (2011). E-government development: Performance evaluation parameters. In M. Shareef, V. Kumar, U. Kumar, & Y. Dwivedi (Eds.), *Stakeholder adoption of e-government services: Driving and resisting factors* (pp. 197–213). Hershey, PA: IGI Global. doi:10.4018/978-1-60960-601-5.ch010

Shareef, M. A., Kumar, U., Kumar, V., & Niktash, M. (2012). Electronic-government vision: Case studies for objectives, strategies, and initiatives. In M. Shareef, N. Archer, & S. Dutta (Eds.), *E-government service maturity and development: Cultural, organizational and technological perspectives* (pp. 15–39). Hershey, PA: IGI Global. doi:10.4018/978-1-60960-848-4.ch002

Shukla, P., Kumar, A., & Anu Kumar, P. B. (2013). Impact of national culture on business continuity management system implementation. *International Journal of Risk and Contingency Management*, *2*(3), 23–36. doi:10.4018/ijrcm.2013070102

Shulman, S. W. (2007). The federal docket management system and the prospect for digital democracy in U S rulemaking. In G. Garson (Ed.), *Modern public information technology systems: Issues and challenges* (pp. 166–184). Hershey, PA: IGI Global. doi:10.4018/978-1-59904-051-6.ch009

Simonovic, S. (2007). Problems of offline government in e-Serbia. In A. Anttiroiko & M. Malkia (Eds.), *Encyclopedia of digital government* (pp. 1342–1351). Hershey, PA: IGI Global. doi:10.4018/978-1-59140-789-8.ch205

Simonovic, S. (2008). Problems of offline government in e-Serbia. In A. Anttiroiko (Ed.), *Electronic government: Concepts, methodologies, tools, and applications* (pp. 2929–2942). Hershey, PA: IGI Global. doi:10.4018/978-1-59904-947-2.ch219

Singh, A. M. (2005). Information systems and technology in South Africa. In M. Khosrow-Pour (Ed.), *Encyclopedia of information science and technology* (pp. 1497–1502). Hershey, PA: IGI Global. doi:10.4018/978-1-59140-553-5.ch263

Singh, S., & Naidoo, G. (2005). Towards an e-government solution: A South African perspective. In W. Huang, K. Siau, & K. Wei (Eds.), *Electronic government strategies and implementation* (pp. 325–353). Hershey, PA: IGI Global. doi:10.4018/978-1-59140-348-7.ch014

Snoke, R., & Underwood, A. (2002). Generic attributes of IS graduates: An analysis of Australian views. In F. Tan (Ed.), *Advanced topics in global information management* (Vol. 1, pp. 370–384). Hershey, PA: IGI Global. doi:10.4018/978-1-930708-43-3.ch023

Sommer, L. (2006). Revealing unseen organizations in higher education: A study framework and application example. In A. Metcalfe (Ed.), *Knowledge management and higher education: A critical analysis* (pp. 115–146). Hershey, PA: IGI Global. doi:10.4018/978-1-59140-509-2.ch007

Song, H., Kidd, T., & Owens, E. (2011). Examining technological disparities and instructional practices in English language arts classroom: Implications for school leadership and teacher training. In L. Tomei (Ed.), *Online courses and ICT in education: Emerging practices and applications* (pp. 258–274). Hershey, PA: IGI Global. doi:10.4018/978-1-60960-150-8.ch020

Speaker, P. J., & Kleist, V. F. (2003). Using information technology to meet electronic commerce and MIS education demands. In A. Aggarwal (Ed.), *Web-based education: Learning from experience* (pp. 280–291). Hershey, PA: IGI Global. doi:10.4018/978-1-59140-102-5.ch017

Spitler, V. K. (2007). Learning to use IT in the workplace: Mechanisms and masters. In M. Mahmood (Ed.), *Contemporary issues in end user computing* (pp. 292–323). Hershey, PA: IGI Global. doi:10.4018/978-1-59140-926-7.ch013

Stellefson, M. (2011). Considerations for marketing distance education courses in health education: Five important questions to examine before development. In U. Demiray & S. Sever (Eds.), *Marketing online education programs: Frameworks for promotion and communication* (pp. 222–234). Hershey, PA: IGI Global. doi:10.4018/978-1-60960-074-7.ch014

Straub, D. W., & Loch, K. D. (2006). Creating and developing a program of global research. *Journal of Global Information Management, 14*(2), 1–28. doi:10.4018/jgim.2006040101

Straub, D. W., Loch, K. D., & Hill, C. E. (2002). Transfer of information technology to the Arab world: A test of cultural influence modeling. In M. Dadashzadeh (Ed.), *Information technology management in developing countries* (pp. 92–134). Hershey, PA: IGI Global. doi:10.4018/978-1-931777-03-2.ch005

Straub, D. W., Loch, K. D., & Hill, C. E. (2003). Transfer of information technology to the Arab world: A test of cultural influence modeling. In F. Tan (Ed.), *Advanced topics in global information management* (Vol. 2, pp. 141–172). Hershey, PA: IGI Global. doi:10.4018/978-1-59140-064-6.ch009

Suki, N. M., Ramayah, T., Ming, M. K., & Suki, N. M. (2013). Factors enhancing employed job seekers intentions to use social networking sites as a job search tool. In A. Mesquita (Ed.), *User perception and influencing factors of technology in everyday life* (pp. 265–281). Hershey, PA: IGI Global. doi:10.4018/978-1-4666-1954-8.ch018

Suomi, R. (2006). Introducing electronic patient records to hospitals: Innovation adoption paths. In T. Spil & R. Schuring (Eds.), *E-health systems diffusion and use: The innovation, the user and the use IT model* (pp. 128–146). Hershey, PA: IGI Global. doi:10.4018/978-1-59140-423-1.ch008

Swim, J., & Barker, L. (2012). Pathways into a gendered occupation: Brazilian women in IT. *International Journal of Social and Organizational Dynamics in IT, 2*(4), 34–51. doi:10.4018/ijsodit.2012100103

Tarafdar, M., & Vaidya, S. D. (2006). Adoption and implementation of IT in developing nations: Experiences from two public sector enterprises in India. In M. Khosrow-Pour (Ed.), *Cases on information technology planning, design and implementation* (pp. 208–233). Hershey, PA: IGI Global. doi:10.4018/978-1-59904-408-8.ch013

Tarafdar, M., & Vaidya, S. D. (2008). Adoption and implementation of IT in developing nations: Experiences from two public sector enterprises in India. In G. Garson & M. Khosrow-Pour (Eds.), *Handbook of research on public information technology* (pp. 905–924). Hershey, PA: IGI Global. doi:10.4018/978-1-59904-857-4.ch076

Thesing, Z. (2007). Zarina thesing, pumpkin patch. In M. Hunter (Ed.), *Contemporary chief information officers: Management experiences* (pp. 83–94). Hershey, PA: IGI Global. doi:10.4018/978-1-59904-078-3.ch007

Thomas, J. C. (2004). Public involvement in public administration in the information age: Speculations on the effects of technology. In M. Malkia, A. Anttiroiko, & R. Savolainen (Eds.), *eTransformation in governance: New directions in government and politics* (pp. 67–84). Hershey, PA: IGI Global. doi:10.4018/978-1-59140-130-8.ch004

Treiblmaier, H., & Chong, S. (2013). Trust and perceived risk of personal information as antecedents of online information disclosure: Results from three countries. In F. Tan (Ed.), *Global diffusion and adoption of technologies for knowledge and information sharing* (pp. 341–361). Hershey, PA: IGI Global. doi:10.4018/978-1-4666-2142-8.ch015

van Grembergen, W., & de Haes, S. (2008). IT governance in practice: Six case studies. In W. van Grembergen & S. De Haes (Eds.), *Implementing information technology governance: Models, practices and cases* (pp. 125–237). Hershey, PA: IGI Global. doi:10.4018/978-1-59904-924-3.ch004

van Os, G., Homburg, V., & Bekkers, V. (2013). Contingencies and convergence in European social security: ICT coordination in the back office of the welfare state. In M. Cruz-Cunha, I. Miranda, & P. Gonçalves (Eds.), *Handbook of research on ICTs and management systems for improving efficiency in healthcare and social care* (pp. 268–287). Hershey, PA: IGI Global. doi:10.4018/978-1-4666-3990-4.ch013

Velloso, A. B., Gassenferth, W., & Machado, M. A. (2012). Evaluating IBMEC-RJ's intranet usability using fuzzy logic. In M. Cruz-Cunha, P. Gonçalves, N. Lopes, E. Miranda, & G. Putnik (Eds.), *Handbook of research on business social networking: Organizational, managerial, and technological dimensions* (pp. 185–205). Hershey, PA: IGI Global. doi:10.4018/978-1-61350-168-9.ch010

Villablanca, A. C., Baxi, H., & Anderson, K. (2009). Novel data interface for evaluating cardiovascular outcomes in women. In A. Dwivedi (Ed.), *Handbook of research on information technology management and clinical data administration in healthcare* (pp. 34–53). Hershey, PA: IGI Global. doi:10.4018/978-1-60566-356-2.ch003

Villablanca, A. C., Baxi, H., & Anderson, K. (2011). Novel data interface for evaluating cardiovascular outcomes in women. In *Clinical technologies: Concepts, methodologies, tools and applications* (pp. 2094–2113). Hershey, PA: IGI Global. doi:10.4018/978-1-60960-561-2.ch806

Virkar, S. (2011). Information and communication technologies in administrative reform for development: Exploring the case of property tax systems in Karnataka, India. In J. Steyn, J. Van Belle, & E. Mansilla (Eds.), *ICTs for global development and sustainability: Practice and applications* (pp. 127–149). Hershey, PA: IGI Global. doi:10.4018/978-1-61520-997-2.ch006

Virkar, S. (2013). Designing and implementing e-government projects: Actors, influences, and fields of play. In S. Saeed & C. Reddick (Eds.), *Human-centered system design for electronic governance* (pp. 88–110). Hershey, PA: IGI Global. doi:10.4018/978-1-4666-3640-8.ch007

Wallace, A. (2009). E-justice: An Australian perspective. In A. Martínez & P. Abat (Eds.), *E-justice: Using information communication technologies in the court system* (pp. 204–228). Hershey, PA: IGI Global. doi:10.4018/978-1-59904-998-4.ch014

Wang, G. (2012). E-democratic administration and bureaucratic responsiveness: A primary study of bureaucrats' perceptions of the civil service e-mail box in Taiwan. In K. Kloby & M. D'Agostino (Eds.), *Citizen 2.0: Public and governmental interaction through web 2.0 technologies* (pp. 146–173). Hershey, PA: IGI Global. doi:10.4018/978-1-4666-0318-9.ch009

Wangpipatwong, S., Chutimaskul, W., & Papasratorn, B. (2011). Quality enhancing the continued use of e-government web sites: Evidence from e-citizens of Thailand. In V. Weerakkody (Ed.), *Applied technology integration in governmental organizations: New e-government research* (pp. 20–36). Hershey, PA: IGI Global. doi:10.4018/978-1-60960-162-1.ch002

Wedemeijer, L. (2006). Long-term evolution of a conceptual schema at a life insurance company. In M. Khosrow-Pour (Ed.), *Cases on database technologies and applications* (pp. 202–226). Hershey, PA: IGI Global. doi:10.4018/978-1-59904-399-9.ch012

Whybrow, E. (2008). Digital access, ICT fluency, and the economically disadvantages: Approaches to minimize the digital divide. In F. Tan (Ed.), *Global information technologies: Concepts, methodologies, tools, and applications* (pp. 1409–1422). Hershey, PA: IGI Global. doi:10.4018/978-1-59904-939-7.ch102

Whybrow, E. (2008). Digital access, ICT fluency, and the economically disadvantages: Approaches to minimize the digital divide. In C. Van Slyke (Ed.), *Information communication technologies: Concepts, methodologies, tools, and applications* (pp. 764–777). Hershey, PA: IGI Global. doi:10.4018/978-1-59904-949-6.ch049

Wickramasinghe, N., & Geisler, E. (2010). Key considerations for the adoption and implementation of knowledge management in healthcare operations. In M. Saito, N. Wickramasinghe, M. Fuji, & E. Geisler (Eds.), *Redesigning innovative healthcare operation and the role of knowledge management* (pp. 125–142). Hershey, PA: IGI Global. doi:10.4018/978-1-60566-284-8.ch009

Wickramasinghe, N., & Geisler, E. (2012). Key considerations for the adoption and implementation of knowledge management in healthcare operations. In *Organizational learning and knowledge: Concepts, methodologies, tools and applications* (pp. 1316–1328). Hershey, PA: IGI Global. doi:10.4018/978-1-60960-783-8.ch405

Wickramasinghe, N., & Goldberg, S. (2007). A framework for delivering m-health excellence. In L. Al-Hakim (Ed.), *Web mobile-based applications for healthcare management* (pp. 36–61). Hershey, PA: IGI Global. doi:10.4018/978-1-59140-658-7.ch002

Wickramasinghe, N., & Goldberg, S. (2008). Critical success factors for delivering m-health excellence. In N. Wickramasinghe & E. Geisler (Eds.), *Encyclopedia of healthcare information systems* (pp. 339–351). Hershey, PA: IGI Global. doi:10.4018/978-1-59904-889-5.ch045

Wyld, D. (2009). Radio frequency identification (RFID) technology. In J. Symonds, J. Ayoade, & D. Parry (Eds.), *Auto-identification and ubiquitous computing applications* (pp. 279–293). Hershey, PA: IGI Global. doi:10.4018/978-1-60566-298-5.ch017

Yaghmaei, F. (2010). Understanding computerised information systems usage in community health. In J. Rodrigues (Ed.), *Health information systems: Concepts, methodologies, tools, and applications* (pp. 1388–1399). Hershey, PA: IGI Global. doi:10.4018/978-1-60566-988-5.ch088

Yee, G., El-Khatib, K., Korba, L., Patrick, A. S., Song, R., & Xu, Y. (2005). Privacy and trust in e-government. In W. Huang, K. Siau, & K. Wei (Eds.), *Electronic government strategies and implementation* (pp. 145–190). Hershey, PA: IGI Global. doi:10.4018/978-1-59140-348-7.ch007

Yeh, S., & Chu, P. (2010). Evaluation of e-government services: A citizen-centric approach to citizen e-complaint services. In C. Reddick (Ed.), *Citizens and e-government: Evaluating policy and management* (pp. 400–417). Hershey, PA: IGI Global. doi:10.4018/978-1-61520-931-6.ch022

Young-Jin, S., & Seang-tae, K. (2008). E-government concepts, measures, and best practices. In A. Anttiroiko (Ed.), *Electronic government: Concepts, methodologies, tools, and applications* (pp. 32–57). Hershey, PA: IGI Global. doi:10.4018/978-1-59904-947-2.ch004

Yun, H. J., & Opheim, C. (2012). New technology communication in American state governments: The impact on citizen participation. In K. Bwalya & S. Zulu (Eds.), *Handbook of research on e-government in emerging economies: Adoption, e-participation, and legal frameworks* (pp. 573–590). Hershey, PA: IGI Global. doi:10.4018/978-1-4666-0324-0.ch029

Zhang, N., Guo, X., Chen, G., & Chau, P. Y. (2011). User evaluation of e-government systems: A Chinese cultural perspective. In F. Tan (Ed.), *International enterprises and global information technologies: Advancing management practices* (pp. 63–84). Hershey, PA: IGI Global. doi:10.4018/978-1-60960-605-3.ch004

Zuo, Y., & Hu, W. (2011). Trust-based information risk management in a supply chain network. In J. Wang (Ed.), *Supply chain optimization, management and integration: Emerging applications* (pp. 181–196). Hershey, PA: IGI Global. doi:10.4018/978-1-60960-135-5.ch013

Compilation of References

Abea, S. S., Yamasakib, T., & Wakatsukib, C. (2016). Assessing Silicon Availability in Soils of Rice-Growing Lowlands and Neighbouring Uplands in Benin and Nigeria. *Rice Science*, *23*(4), 196–202. doi:10.1016/j.rsci.2016.06.002

Abebe, M. B., & Geta, E. (2014). *Farmers Willingness To Pay For Irrigation Water Use: The Case Of Agarfa District, Bale Zone, Oromiya National Regional State*. Haramaya University.

Aberra, Y. (2004). Problems Of The Solution: Intervention Into Small-Scale Irrigation For Drought Proofing In The Mekele Plateau Of Northern Ethiopia. *The Geographical Journal*, *170*(3), 226–237. doi:10.1111/j.0016-7398.2004.00122.x

Adam, J. N., Al-Hassan, S., & Akolgo, D. A. (2016). Small-Scale Irrigation And Rural Poverty Reduction In The Upper-East Region Of Ghana. *African Journal of Science and Research*, *2*(5), 38–42.

Adeniyan, O. N., & Ojeniyi, S. O. (2005). Effect of Poultry Manure NPK 15-15-15 and Combination of their Reduced Levels on Maize Growth and Soil Chemical Properties. *Niger Journal of Soil Science*, *15*, 34–41.

Adesina, A. (2012). *Agricultural Transformation Agenda: Repositioning Agriculture to Drive Nigeria's Economy*. Federal Ministry of Agriculture and Rural Development.

Adewumi, J. K., Olayanju, M. A., & Adewuyi, S. A. (2007). Support for Small Rice Threshers in Nigeria. London: DFID.

Adugna, E., Ermias, A., Mekonnen, A., & Mihret, D. (2014). The Role Of Small Scale Irrigation In Poverty Reduction. *Journal of Development and Agricultural Economics*, *6*(1), 12–21. doi:10.5897/JDAE2013.0499

Agbede, T. M., Ojeniyi, S. O., & Adeyemo, A. J. (2008). Effect of Poultry Manure on Soil Physical and Chemical Properties, Growth and Grain Yield of Sorghum in South-West, Nigeria. *American-Eriession Journal of Sustainable Agriculture*, *2*(1), 72–77.

Agele, S. O., Adeyemo, A. J., & Famuwagun, I. B. (2011). Effects of Agricultural Wastes and Mineral Fertilizer on Soil and Plant Nutrient Status, Growth and Yield of Tomato. *Archives of Agronomy and Soil Science*, *57*(1), 91–104. doi:10.1080/03650340903225040

Agnew, J. (2005). *Hegemony The New Shape of Global Power*. Philadelphia: Temple University Press.

Agrarian Research and Development Agency (ARDA). (2012). Sugar Beet Report. ARDA.

Aguilera, J., Motavalli, P. P., Gonzales, M. A., & Valdivia, C. (2012). Initial and ResidualEffectsof Organic and Inorganic Amendments on Soil Properties in a Potato-Based Cropping System in the Bolivian Andean Highlands. *American Journal of Experimental Agriculture*, *2*(4), 641–666. doi:10.9734/AJEA/2012/2006

Ahmad, M. (2003). Agricultural Productivity, Efficiency, And Rural Poverty In Irrigated Pakistan: A Stochastic Production Frontier Analysis. *Pakistan Development Review*, *42*(2), 219–248.

Ahmadu, J., & Alufohai, G. O. (2012). Estimation Of Technical Efficiency Of Irrigated Rice Farmers In Niger State, Nigeria. *American-Eurasian J. Agric. & Environ. Sci.*, *12*(12), 1610–1616.

Ajala, A. S., & Gana, A. (2015). Analysis of Challenges Facing Rice Processing in Nigeria. *Journal of Food Processing*, *6*. doi:10.1155/2015/893673

Akanbi, W. B. (2002). *Growth, Nutrient Uptake and Yield of Maize and Okra as Influenced by Compost and Nitrogen Fertilizer under Different Cropping Systems* (Unpublished Ph.D. Thesis). University of Ibadan, Nigeria.

Akande, M. O., Adediran, J. A., & Oluwatoyinbo, F. I. (2005). Effects of Rock Phosphate Amended with Poultry Manure and Soil Available Phosphorus and Yield of Maize and Cowpea. *African Journal of Biotechnology*, *4*, 444–448.

Ake, C. (1981). *A Political Economy of Africa*. London: Longman.

Akehurst, G., Afonso, C., & Gonçalves, H. (2012). Re-examining green purchase behavior and the green consumer profile: New evidences. *Management Decision*, *50*(5), 972–988. doi:10.1108/00251741211227726

Akenji, L., & Begsson, M. (2014). *Making Sustainable Consumption and Production the Core of the Sustainable Development Goals*. Discussion Paper Institute for Environmental Global Strategies.

Akinnagbe, O.M., & Olaolu, M.O. (2016). Policy issues for improving monitoring and evaluation of agricultural extension programmes in Nigeria. *African Evaluation Journal*, *3*(2), a122. doi: .v3i2.12210.4102/aej

Akter, S. (2007). Farmers' Willingness To Pay For Irrigation Water Under Government Managed Small Scale Irrigation Projects In Bangladesh. *J. Bangladesh Stud*, *9*, 21–31.

Akudugu, M. A., Nyamadi, B. V., & Dittoh, S., & Others. (2016). Transforming Smallholder Agriculture In Africa Through Irrigation: An Assessment Of Irrigation Impact Pathways In Ghana. In *2016 Aaae Fifth International Conference*. African Association Of Agricultural Economists (Aaae).

Alemayehu, T. (2014). Smallholder Farmer's Willingness To Pay For Improved Irrigation Water: A Contingent Valuation Study In Koga Irrigation Project, Ethiopia. *Journal Of Economics And Sustainable Development*, *5*(19), 5–15.

Alemu, A. E., Kassa, B., & Yihdego, G. (2011). Determinants Of Land Allocation To Irrigation And Its Wealth Effect: Evidence From Northern Ethiopia. *Journal Of The Drylands*, *4*(2), 310–319.

Alemu, Z. G. (2010). *Livelihood strategies in rural South Africa: Implications for poverty reduction. International Association of Agricultural Economists (IAAE).* Foz do Iguacu, Brazil: Triennial Conference.

Alexandros, Lee, von Maltitz, Mathai, de Oliveira, & Willis. (2013). Biofuels in Africa Impacts on Ecosystem Services, Biodiversity and Human Well-being. UNU-IAS Policy Report.

Alex, G., Byerlee, D., Helene-Collion, M., & Rivera, W. (2004). *Extension and rural development: Converging views on institutional approaches?* Agriculture & Rural Development Department of the World Bank.

Alhassan, M., Loomis, J., Frasier, M., Davies, S., & Andales, A. (2013). Estimating Farmers' Willingness To Pay For Improved Irrigation: An Economic Study Of The Bontanga Irrigation Scheme In Northern Ghana. *The Journal of Agricultural Science*, *5*(4).

Alila, P. O., & Atieno, R. (2006). *Agricultural Policy in Kenya: Issues and Processes.* Paper presented for the Future Agricultures Consortium workshop, Nairobi, Kenya.

Alinov, L. (2010). Livelihoods strategies and household resilience to food insecurity: An empirical analysis to Kenya. In Promoting Resilience through Social Protection in Sub-Saharan Africa. Dakar, Senegal: Academic Press.

Al-Karablieh, E. K., Salman, A. Z., Al-Omari, A. S., Wolff, H.-P., Al-Assa'd, T. A., Hunaiti, D. A., & Subah, A. M. (2012). Estimation Of The Economic Value Of Irrigation Water In Jordan. *Journal of Agricultural Science and Technology B*, *2*(5b), 487.

Allen C.R., Fontaine J.J., Pope K.L., Garmestani A.S. (2011) Adaptive management for a turbulent future. *Journal of Environmental Management*, *92*, 1339-1345.

Altieri, M. (2008). *Small farms as a planetary ecological asset: Five key reasons why we should support the revitalization of small farms in the Global South.* Third World Network.

Amadi, L. (2012). Africa: Beyond the "new" dependency: A political economy. *African Journal of Political Science and International Relations*, *6*(8), 191–203. doi:10.5897/AJPSIR12.022

Amadi, L., & Igwe, P. (2016). Maximizing the Eco Tourism Potentials of the Wetland Regions through Sustainable Environmental Consumption: A Case of the Niger Delta, Nigeria. *The Journal of Social Sciences Research*, *2*(1), 13–22.

Amadi, L., Igwe, P., & Wordu, S. (2014). Sustainable Development, Greening and Eco-efficiency. A Political Ecology. *Journal of Sustainable Development Studies*, *7*(2), 161–196.

Amadi, L., & Imoh-Ita, I. (2017). Intellectual capital and environmental sustainability measurement nexus: A review of the literature *Int. J. Learning and Intellectual Capital*, *14*(2), 154–172. doi:10.1504/IJLIC.2017.084071

Amb, M. K., & Ahluwalia, A. S. (2016). Allelopathy: Potential Role to Achieve New Milestones in Rice Cultivation. *Rice Science*, *23*(4), 165–183. doi:10.1016/j.rsci.2016.06.001

Amendah, D. D., Buigut, S., & Mohamed, S. (2014). Coping strategies among urban poor: Evidence from Nairobi, Kenya. *PLoS One*, *9*(1), 1–8. doi:10.1371/journal.pone.0083428 PMID:24427272

Amin, S. (1972). Underdevelopment and Dependency in Black Africa Origins and Contemporary Form. *The Journal of Modern African Studies*, *10*(4), 503–524. doi:10.1017/S0022278X00022801

Anang, B. T., Bäckman, S., & Sipiläinen, T. (2016). Technical Efficiency And Its Determinants In Smallholder Rice Production In Northern Ghana. *Journal of Developing Areas*, *50*(2), 311–328. doi:10.1353/jda.2016.0072

Apam, L. S. (2012). An Assessment Of The Impact Of Small-Scale Irrigation Schemes On Livelihoods And Poverty Reduction. In *The Upper East Region*. Ghana: Kwame Nkrumah University.

Argonne National Laboratory. (2013). *Greenhouse Gases, Regulated Emissions, and Energy Use in Transportation Model (GREET). GREET version 2.7.* Author.

Arndt, C., Benfica, R., Tarp, F., Thurlow, J., & Uaiene, R. (2010). Biofuels, poverty and growth: A computable general equilibrium analysis of Mozambique. *Environment and Development Economics*, *15*(1), 81–105. doi:10.1017/S1355770X09990027

Arunah, U. T., & Ibrahim, H. (2004). Effects of Inorganic Fertilizer and Poultry Manure on Sorghum Yield at Samaru in the Northern Guinea Savanna of Nigeria. *Agrosearch*, *6*(1), 49–54.

Aseyehegu, K., Yirga, C., & Rajan, S. (2012). Effect Of Small-Scale Irrigation On The Income Of Rural Farm Households: The Case Of Laelay Maichew District, Central Tigray, Ethiopia. *Journal of Agricultural Sciences*, *7*(1). doi:10.4038/jas.v7i1.4066

Asfaw, S., Shiferaw, B., Simtowe, F., & Lipper, L. (2012). Impact of Modern Agricultural Technologies on Smallholder Welfare: Evidence from Tanzania and Ethiopia. *Food Policy*, *37*(3), 283–295. doi:10.1016/j.foodpol.2012.02.013

Ashfaq, M., Jabeen, S., & Baig, I. A. (2005). Estimation Of The Economic Value Of Irrigation Water. *Journal Of Agriculture And Soil Science*, *1*(3), 270–272.

Asongu, S. A. (2016). Law and Investment in Africa. *Institutions, Etc.*, *8*(2), 91–120.

AspenTech. (2011). *AspenEngineeringSuiteV7.3.* Author.

Assefa, N. (2012). Valuing The Economic Benefit Of Irrigation Water: Application Of Choice Experiment And Contingent Valuation Methods To Ribb Irrigation And Drainage Project. In *South Gonder, Ethiopia*. Ethiopia: Addis Ababa University.

Åström, S., Roth, S., Wranne, J., Jelse, K., & Lindblad, M. (2013). *Food consumption choices and climate change*. Swedish Environmental Research Institute.

Auestad, N., & Fulgoni, V. L. (2015). What current literature tells us about sustainable diets: Emerging research linking dietary patterns, environmental sustainability, and economics. *Advances in Nutrition*, *6*(1), 19–36. doi:10.3945/an.114.005694 PMID:25593141

Awlachew, S. B., Loulseged, M., & Yilma, A. (2007). Impact Of Irrigation On Poverty And Environment In Ethiopia. In *Draft Proceedings Of The Symposium And Exhibition*. Ghion Hotel, Addis Ababa, Ethiopia.

Awulachew, S. B., & Ayana, M. (2011). Performance Of Irrigation: An Assessment At Different Scales In Ethiopia. *Experimental Agriculture*, *47*(S1), 57–69. doi:10.1017/S0014479710000955

Aydogdu, M. H. (2016). Evaluation Of Willingness To Pay For Irrigation Water: Harran Plain Sampling In Gap Region - Turkey. *Applied Ecology and Environmental Research*, *14*(1), 349–365. doi:10.15666/aeer/1401_349365

Aydogdu, M. H., & Bilgic, A. (2016). An Evaluation Of Farmers' Willingness To Pay For Efficient Irrigation For Sustainable Usage Of Resources: The Gap-Harran Plain Case, Turkey. *Journal of Integrative Environmental Sciences*, 1–12. doi:10.1080/1943815X.2016.1241808

Ayele, G. K. (2011). The Impact Of Selected Small-Scale Irrigation Schemes On Household Income And The Likelihood Of Poverty. In *The Lake Tana Basin Of Ethiopia (Master Of Professional Studies)*. Cornell University.

Babatunde, R. O., Salami, M. F., & Mohammed, B. A. (2016). Determinants Of Yield Gap In Rain Fed And Irrigated Rice Production: Evidence From Household Survey In Kwara State. In *2016 Aaae Fifth International Conference*. African Association Of Agricultural Economists (Aaae).

Babcock, B. A., Chaherli, N. M., & Lakshiminariayam, P. G. (1995). *Programme Participation and Farm-Level. Adoption of Conservation Tillage: Estimates from a Multinomial Logit Model*. Working paper 95-WP 136, Centre for Agricultural and Rural Development, Iowa State University.

Bacha, D., Namara, R., Bogale, A., & Tesfaye, A. (2011). Impact Of Small-Scale Irrigation On Household Poverty: Empirical Evidence From The Ambo District In Ethiopia. *Irrigation and Drainage*, *60*(1), 1–10. doi:10.1002/ird.550

Bagson, E., & Kuuder, C.-J. W. (2013). Assessment Of A Small-Scale Irrigation Scheme On Household Food Security And Leisure In Kokoligu. Academic Press.

Bakan, J. (2004). *The Corporation: The Pathological Pursuit of Profit & Power*. Toronto: Penguin Group.

Baltzer & Hansen. (2011). *Agricultural input subsidies in Sub-Saharan Africa. Evaluation study*. DANIDA.

Banaeian, N., & Zangeneh, M. (2011). Assessment Of Productive Efficiency For Irrigated And Dryland Wheat Farming In Iran Using Data Envelopment Analysis. *Or Insight*, *24*(4), 215–230.

Banson, K. E., Asare, D., Heng, L., Cobbinah, J. F., & Adu-Sarkodieh, A. (2014). Impact Of Small Scale Irrigation Technologies On Poverty Alleviation Among Peri-Urban And Urban Farmers. *Journal of Life Science*, *8*(2), 142–151.

Barnidge, E., Baker, E. A., Motton, F., Fitzgerald, T., & Rose, F. (2011). Exploring community health through the Sustainable Livelihoods Framework. *Health Education & Behavior*, *38*(1), 80–90. doi:10.1177/1090198110376349 PMID:21169478

Bayu, W., Rethman, N. F. G., Hammes, P. S., & Alamu, G. (2006). Effects of Farmyard Manure and Inorganic Fertilizers on Sorghum Growth, Yield and Nitrogen use in a Semi- Arid Area of Ethiopia. *Journal of Plant Nutrition*, *29*(2), 391–407. doi:10.1080/01904160500320962

Bekunda, M., Sanginga, N., & Woomer, P. L. (2010). Restoring Soil Fertility in Sub-Sahara Africa. Advances, 108(10), 183–236.

Belahsen, R., & Rguibi, M. (2006). Population health and Mediterranean diet in southern Mediterranean countries. *Public Health Nutrition*, *9*(8A), 1130–1305. doi:10.1017/S1368980007668517 PMID:17378952

Belay, A., Classens, A. S., Wehner, F. C., & De Beer, J. M. (2006). Influence of Residual Manure on Selected Nutrient Elements and Microbial Composition of Soil under Long-Term Crop Rotation. *South African Journal of Plant and Soil*, *18*(1), 1–6. doi:10.1080/02571862.2001.10634392

Bello, W. (2003). *De-globalization:Ideas for a New World Economy*. Fernwood.

Berbel, J., Mesa-Jurado, M. A., & Pistón, J. M. (2011). Value Of Irrigation Water In Guadalquivir Basin (Spain) By Residual Value Method. *Water Resources Management*, *25*(6), 1565–1579. doi:10.1007/s11269-010-9761-2

Berdegué, J.A., Balsevich, F., Flores, L., & Reardon, T. (2005). Central American supermarkets' private standards of quality and safety in procurement of fresh fruits and vegetables. *Food Policy, 30*, 254-269.

Berne Declaration. (2013). *Agropolicy. A handful of corporations control world food production.* Available at http://www.econexus.info/sites/econexus/files/Agropoly_Econexus_BerneDeclaration.pdf

Beyan, A., Jema, H., & Adem, K. (2014). Effect Of Small-Scale Irrigation On The Farm Householdsincome Of Rural Farmers: The Case Of Girawa District, East Hararghe, Oromia, Ethiopia. *Asian Journal of Agriculture and Rural Development*, *4*(3), 257.

Birhane, M., & Geta, E. (2016). Farmers' Willingness To Pay For Irrigation Water Use: The Case Of Agarfa District, Bale Zone, Oromia National Regional State. *International Journal Of Agricultural Economics*, *1*(2), 35–39.

Birner, R., Davis, K., Pender, J., Nkonya, E., Anandajayasekeram, P., Ekboir, J., & Cohen, M. et al. (2009). From Best Practice to Best Fit: A Framework for Designing and Analyzing Pluralistic Agricultural Advisory Services Worldwide. *Journal of Agricultural Education and Extension*, *15*(4), 341–355. doi:10.1080/13892240903309595

Birt, C., Buzeti, T., Grosso, G., & Justesen, L. (2017). Healthy and Sustainable Diets for European Countries. European Public Health Association – EUPHA. https://biblio.ugent.be/publication/8521128/file/8521129.pdf

Biswas, D., & Venkatachalam, L. (2015). Farmers' Willingness To Pay For Improved Irrigation Water — A Case Study Of Malaprabha Irrigation Project In Karnataka, India. *Water Economics And Policy*, *01*(01), 1450004. doi:10.1142/S2382624X14500040

Bjornlund, H., Van Rooyen, A., & Stirzaker, R. (2016). Profitability And Productivity Barriers And Opportunities In Small-Scale Irrigation Schemes. *International Journal of Water Resources Development*, 1–15.

Bokhtiar, S. M., & Sakurai, K. (2005). Effects of Organic Manure and Chemical Fertilizer on Soil Fertility and Productivity of Plant and Ratoon Crops of Sugarcane. *Archives of Agronomy and Soil Science*, *51*(3), 325–334. doi:10.1080/03650340500098006

Bos, M. G., Burton, M. A. S., & Molden, D. J. (2005). *Irrigation And Drainage Performance Assessment: Practical Guidelines*. Wallingford, UK: Cabi Pub.

Brabben, T. E., Angood, C., Skutsch, J. C., & Smith, L. (2004). *Irrigation Can Sustain Rural Livehoods: Evidence From Bangladesh And Nepal*. Academic Press.

Brenkert, G. (1998). The Marketing Challenge: Towards Being Profitable and Socially Responsible. *Journal of Business Ethics*, *7*(7), 497–507.

British Dietetic Association. (2013). *Policy on sustainable food*. Author.

Brown, E., & Jacobson, M. F. (2005). *Cruel Oil: How palm oil Harms health, rainforest & wildlife*. Center of Science in the Public Interest.

Bryant, R., & Bailey, S. (1997). *Third world political ecology: An introduction*. London: Routledge.

Bryceson, D. F. (2004). Agrarian vista or vortex: African rural livelihood policies. *Review of African Political Economy*, *31*(102), 617–629. doi:10.1080/0305624042000327831

Bulte, E. H., & Soest, D. P. V. (2001). Environmental degradation in developing ountries: Households and the reverse environmental Kuznets Curve. *Journal of Development Economics*, *65*(1), 225–235. doi:10.1016/S0304-3878(01)00135-3

Buri, M. M., Issaka, R. N., Wakatsuki, T. & Otoo, E. (2004). Soil Organic Amendments and Mineral Fertilizers: Options for Sustainable Lowland Rice Production in the Forest Agro – Ecology of Ghana. *Agricultural and Food Science Journal of Ghana*, *3*, 237 – 248.

Burt, C. M., & Styles, S. W. (1998). Modern Water Control And Management Practices. In *Irrigation: Impact On Performance (Water Reports No. 16)*. Rome: The World Bank.

Byerlee, D., & Heisey, P. (1986). Past and Potential Impacts of Maize Research in Sub-Saharan Africa: A Critical Assessment. *Food Policy*, *21*(3), 255–277. doi:10.1016/0306-9192(95)00076-3

California Environmental Protection Agency (CEPA). (2010). First Draft Detailed of The California-Modified GREET Pathway for Sugar beet Ethanol. California Air Resources Board.

Capone, R., Bilali, H. E., Debs, P., Cardone, G. & Driouech, N. (2014). Food system sustainability and food Security: Connecting the Dots. *Journal of Food Security, 2*, 13-22.

CarbonTrust. (2013). Energy and carbon conversions. Carbon Trust.

Carlsson-Kanyama, A. (1997). Weighted average source points and distances for consumption origin- tools for environmental impact analysis? *Ecological Economics*, *23*(1), 15–23. doi:10.1016/S0921-8009(97)00566-1

Carlsson-Kanyama, A., & Gonzalez, A. D. (2009). Potential contributions of food consumption patterns to climate change. *The American Journal of Clinical Nutrition*, *89*(5), 1704S–1709S. doi:10.3945/ajcn.2009.26736AA PMID:19339402

Cassman, K. G., Dobermann, A., Walters, D. T., & Yang, H. S. (2003). Meeting cereal demand while protecting natural resources and improving environmental quality. *Annual Review of Environment and Resources*, *28*(1), 315–358. doi:10.1146/annurev.energy.28.040202.122858

Chamala, S. (1990). Establishing a group: A participative action model. In S. Chamala & P. D. Mortiss (Eds.), *Working together for Landcare: Group management skills and strategies* (pp. 13–38). Brisbane: Australian Academic Press.

Chandrasekaran, K., Devarajulu, S., & Kuppannan, P. (2009). Farmers' Willingness To Pay For Irrigation Water: A Case Of Tank Irrigation Systems In South India. *Water (Basel)*, *1*(1), 5–18. doi:10.3390/w1010005

Chand, S., Anwar, M., & Patra, D. D. (2006). Influence of Long-Term Application of Organic and Inorganic Fertilizer to Build up Soil Fertility and Nutrient Uptake in Mint-Mustard Cropping Sequence. *Communications in Soil Science and Plant Analysis*, *37*(1-2), 63–76. doi:10.1080/00103620500408704

Chang, H. (2003). Kicking Away the Ladder: Development Strategy in Historical Perspective. In Rethinking Development Economics. Anthem Press.

Chan, L., Receveur, O., Sharp, D., Schwartz, H., Ing, A., & Tikhonov, C. (2011). *First Nations Food, Nutrition and Environment Study (FNFNES): Results from British Columbia (2008/2009). Prince George*. University of Northern British Columbia.

Charles, H., Godfray, J., Crute, I. R., Haddad, L., Lawrence, D., Mui, J. F., & Whiteley, R. et al. (2010). The future of the global food system. *Philosphical Transactions of the Royal Socieity B*, *365*(1554), 2769–2777. doi:10.1098/rstb.2010.0180 PMID:20713383

Chase-Dunn, C. (1989). *Global Formation*. Cambridge, MA: Blackwell.

Chazovachii, B. (2012). The Impact Of Small Scale Irrigation Schemes On Rural Livelihoods: The Case Of Panganai Irrigation Scheme Bikita District Zimbabwe. *Journal of Sustainable Development in Africa*, *14*(4), 217–231.

Chen, S., & Ravallion, M. (2001). How did the World's Poor Fare in the 1990s? *Review of Income and Wealth, 4*(3).

Chew, S. (1995). Environmental Transformations: accumulation,Ecological Crisis and Social Movements. In D. Smith & J. Borocz (Eds.), *A New World Order?Global Transformations in the Late Twentieth Century*. Westport, CT: Praeger Publishers.

Chirwa, Dorward, & Matita. (2012). *Thinking 'Graduation' from the Farm Input Subsidy Programme in Malawi*. Policy brief number 3.

Chirwa, E. W. (2011). *Dorward, and Mirriam Matita*. Conceptualising Graduation.

Chivenge, P., Vanlauwe, B., & Six, J. (2011). Does the Combined Application of Organic and Mineral Nutrient Sources Influence Maize Productivity? A Meta-Analysis Plant Soil. Springer.

CIMMYT. (1993). *The Adoption of Agricultural Technology: A Guide for Survey Design*. Author.

Clapp, J. (2017). Food self-sufficiency: Making sense of it, and when it makes sense. *Food Policy, 66*, 88–96. doi:10.1016/j.foodpol.2016.12.001

Connell, D. J. (2010). Sustainable livelihoods and ecosystem health: Exploring methodological relations as a source of synergy. *EcoHealth, 7*(3), 351–360. doi:10.1007/s10393-010-0353-7 PMID:21104294

Conway, D., & Hynen, N. (2006). *Globalization's Contradictions: Geographies of Discipline,destruction and Transformation*. New York: Routledge.

Cullet, P. (2004). Intellectual property rights and food security in the South. *The Journal of World Intellectual Property, 7*(3), 261–286. doi:10.1111/j.1747-1796.2004.tb00209.x

Dao, G. T. N. (2013). *An Analysis Of Technical Efficiency Of Crop Farms In The Northern Region Of Vietnam* (Unpublished PhD thesis). University of Canberra, Canberra, Australia.

Davidson, D., & Hatt, K. (2005). *Consuming Sustainability Critical Social Analysis of Ecological Change*. Fernwood Publishing.

Davies, K. E. (2009). Agriculture and Climate Change: An Agenda for Negotiation in Copenhagen. The Important Role of Extension Systems. International Food Policy Research Institute. Focus 16. *Brief, 11*, 1–2.

Davis, K., Swanson, B., Amudavi, D., Mekonnen, D. A., Flohrs, A., Riese, J., . . . Zerfu, E. (2010). *In-Depth Assessment of the Public Agricultural Extension System of Ethiopia and Recommendations for Improvement*. IFPRI Discussion Paper 01041. Washington, DC: IFPRI.

Davis, K., & Heemskerk, W. (2012). *Investment in extension and advisory services as part of agricultural innovation systems. Module 3 of agricultural innovation systems: an investment sourcebook*. Washington, DC: The World Bank.

Dawda, T. D., & Amosah, J. (2015). An Analysis Of Small Scale Irrigation Farming And Poverty Reduction In North-Eastern Ghana. *International Journal Of Engineering Research & Technology (Ijert)*, *4*(10), 473–481.

De Haan, J., & Rotmans, R. (2011). Patterns in transitions: Understanding complex chains of change. *Technological Forecasting and Social Change*, *78*(1), 90–102. doi:10.1016/j.techfore.2010.10.008

De Silva, S., Johnston, R., & Sellamuttu, S. (2014). *Agriculture, Irrigation And Poverty Reduction In Cambodia: Policy Narratives And Ground Realities Compared*. Cgiar Research Program On Aquatic Agricultural Systems, Working Paper: Aas-2014-13.

Debon, H., Parrot, L., & Moustier, P. (2010). Sustainable urban agriculture in developing countries. A review. *Agronomy for Sustainable Development*, *30*(1), 21–32. doi:10.1051/agro:2008062

Denning, G., Kabambe, P., Sanchez, P., Malik, A., Flor, R., Harawa, R., & Sachs, J. et al. (2009). Input subsidies to improve smallholder maize productivity in Malawi: Toward an African Green Revolution. *PLoS Biology*, *7*(1), e1000023. doi:10.1371/journal.pbio.1000023 PMID:19175292

Department of Agricultural Forestry and Fishery (DAFF). (2012). *Abstract Agricultural Statistics 2012*. DAFF.

Department of Energy (DOE). (2012). *Draft of Integrated Resources Plan For Electricity-2010-2030*. DOE.

Department of Energy (DOE). (2013). *Draft 2012of Integrated Energy Planning*. Pretoria, South Africa: DOE.

Department of Environment (DEA). (2011). *National Climate Change Response*. White Paper. DEA.

Department of Environmental Affaire (DEA). (2011). *South Africa's Second National Communication under the United Nations Framework Convention on Climate Change*. DEA.

Department of Government Communication and Information Systems (GCIS). (2012). *Energy, South Africa Yearbook 2011/12*. GCIS.

Department of Minerals and Energy (DOE). (2007). Biofuels Industrial Strategy of the Republic of South Africa. DoE.

Department of National Treasury (DNT). (2013). *Reducing Greenhouse Gas Emissions and Facilitating the Transition to a Green Economy*. Policy Paper For Public Comment Carbon Tax Policy Paper. DNT.

Department of Science and Technology (DST). (2010). *National Survey of Research & Experimental Development Main Results 2009/10*. DST.

Department of Transport (DOT). (2010). *National Transport Master Plan*. Pretoria, South Africa: DOT.

Derbile, E. K. (2012). Water Users Associations And Indigenous Institutions In The Management Of Community-Based Irrigation Schemes In Northeastern Ghana. *European Scientific Journal. Esj*, *8*(26), 118–135.

Deribe, R. (2008). *Institutional Analysis Of Water Management On Communal Irrigation Systems In Ethiopia: The Case Of Atsbi Wemberta, Tigray Region And Ada'a Woreda, Oromiya Region* (Msc Thesis). Addis Ababa University, Ethiopia.

Dillon, A. (2008). *Access To Irrigation And The Escape From Poverty: Evidence From Northern Mali*. Ifpri Discussion Paper 00782.

Dinye, R. (2013). Irrigated Agriculture And Poverty Reduction In Kassena Nankana District In The Upper-East Region, Ghana. *Journal Of Science And Technology (Ghana)*, *33*(2), 59–72. doi:10.4314/just.v33i2.6

Dittoh, S., Akuriba, M. A., Issaka, B. Y., & Bhattarai, M. (2010). *Sustainable Micro-Irrigation Systems For Poverty Alleviation In The Sahel: A Case For "Micro" Public-Private Partnerships*. Poster Presentation Presented At The Joint 3rd African Association Of Agricultural Economists (Aaae) And 48th Agricultural Economists Association Of South Africa (Aeasa) Conference, Cape Town, South Africa.

Dobermann, A. & Nelson, R. (2013). *Opportunities and solutions for sustainable food production*. Academic Press.

Dorward, A., & Chirwa, E. W. (2011). The Malawi Agricultural Input Subsidy Programme: 2005-6 to 2008-9. *International Journal of Agricultural Sustainability*, *9*(1).

Dube, K. (2016). Implications Of Rural Irrigation Schemes On Household Economy. A Case Of Lower Gweru Irrigation Scheme, Zimbabwe. *South African Journal Of Agricultural Extension (Sajae)*, *44*(1), 75–90. doi:10.17159/2413-3221/2016/v44n1a371

Dutta, S., Pal, R., Chakeraborty, A., & Chakrabarti, K. (2003). Influence of Integrated Plant Nutrient Phosphorus and Sugarcane and Sugar Yields. *Field Crops Research*, *77*, 43–49.

Eastwood, L. (2006). Contesting the Economic Order: Resisting the Media Construction of Reality. In S. Best & A. J. Nocella (Eds.), *Igniting a Revolution: Voices in Defense of Mother Earth*. Edinburgh, UK: AK Press.

Eberle, U., Hayn, D., Rehaag, R., & Simshäuser, U. (Eds.). (2006). *Ernährungswende. Eine Herausforderung für Politik, Unternehmen und Gesellschaft* [Nutrition Change: A Challenge for Politics, Business, and Society]. Munich: Oekom. (in German)

EC/JRC. (2009). *Environmental impacts of diet changes in the EU. Technical Report, European Commission (EC), Joint Research Centre (DG JRC)*. Brussels: IPTS.

Ecker & Breisinger. (2012). *The Food Security System*. Washington, DC: International Food Policy Research Institute. Retrieved from http://www.ifpri.org/sites/default/files/publications/ifpridp01166.pdf

Economist Intelligence Unit. (2015). *Fixing: Towards A More Sustainable Food System*. Available online at http://foodsustainability.eiu.com/whitepaper/

Ekekwe, E. (1986). *Class and State in Nigeria*. London: Longman.

Ekpenyong, A. S. (2015). Urbanization: Its implication for sustainable food Security, health and nutritional nexus in developing economies - A case study of Nigeria. *Journal of Studies in Social Sciences*, *11*, 29–49.

Ericksen, P. (2008). Conceptualizing food systems for global environmental change research. *Global Environmental Change*, *18*(1), 234–245. doi:10.1016/j.gloenvcha.2007.09.002

Esmaeili, A., & Shahsavari, Z. (2011). Valuation Of Irrigation Water In South-Western Iran Using A Hedonic Pricing Model. *Applied Water Science*, *1*(3–4), 119–124. doi:10.1007/s13201-011-0015-0

Estavo, G., & Prakash, M. (1998). *Grassroots Post-Modernism: Remaking the soil of cultures*. Zed Books.

Estruch, R., Martínez-González, M. A., & Corella, D. (2013). Effects of a Mediterranean-style diet on cardiovascular risk factors: A randomized trial. *Annals of Internal Medicine*, *2006*(145), 1–11. PMID:16818923

Etissa, E., Dechassa, N., Alamirew, T., Alemayehu, Y., & Desalegne, L. (2014). Irrigation Water Management Practices In Smallholder Vegetable Crops Production: The Case Of The Central Rift Valley Of Ethiopia. *Science. Technology And Arts Research Journal*, *3*(1), 74–83. doi:10.4314/star.v3i1.13

European Commission. (2014). Impact assessment on measures addressing food waste. Review of EU Waste Management Target. EC.

European Commission. (2015). *Sustainable food consumption and production in a resource-constrained world*. https://ec.europa.eu/research/scar/pdf/scar_3rd-foresight_2011.pdf.

Fagbenro, J. A. (2001). Organic Fertilizer Raw Material Situation in Nigeria. *Proceedings of a Workshop on Organic Fertilizer Development*, 1-9.

Fairhurst, T. (Ed.). (2012). *Handbook for Integrated Soil Fertility Management*. Nairobi, Kenya: Africa Soil Health Consortium.

Fanadzo, M. (2012). Revitalisation Of Smallholder Irrigation Schemes For Poverty Alleviation And Household Food Security In South Africa: A Review. *African Journal of Agricultural Research*, *7*(13), 1956–1969. doi:10.5897/AJARX11.051

FAO and Biodiversity International. (2010). *Biodiversity and Sustainable Diets United against Hunger*. International Scientific Symposium, FAO/Biodiversity International.

FAO, IFAD, & WFP. (2015). The State of Food Insecurity in the World 2015. In *Meeting the 2015 international hunger targets: Taking stock of uneven progress*. Rome: FAO.

FAO. (1997). The food system and factors affecting household food security and nutrition. In Agriculture food and nutrition for Africa - A resource book for teachers of agriculture. Rome: FAO.

FAO. (2002). The State of Food Insecurity in the World 2001. Rome: FAO.

FAO. (2006). *The right to Food: The Human Right to Adequate Food.* Retrieved from http://www.fao.org/righttofood/about-right-to-food/human-right-principles-panther/en/

FAO. (2007). *Paying farmers for environmental services*. Rome: Food and Agriculture Organization of the United Nations.

FAO. (2008). *From Subsistence farming to sugar-cane monoculture: impacts on agrobiodiversity, local knowledge, and food security*. Rome: Food and Agriculture Organization.

FAO. (2009). *Declaration of the World Food Summit on Food Security*. Rome: Food and Agriculture Organization of the United Nations. Retrieved from http://www.fao.org/fileadmin/templates/wsfs/Summit/Docs/Final_Declaration/WSFS09_Declaration.pdf

FAO. (2009). *Linking people, places and products. A guide for promoting quality linked to geographical origin and sustainable geographical indications*. Rome: FAO/SINER-GI.

FAO. (2010). *Biodiversity for a World without Hunger*. Retrieved from http://www. fao.org/biodiversity/biodiversity-home/en/

FAO. (2010). *Food insecurity in the World: addressing food insecurity in protracted crises*. Rome: Food and Agriculture Organization of the United Nations.

FAO. (2010). *Report of the Technical Workshop on Biodiversity in Sustainable Diets*. Rome: FAO.

FAO. (2011). *Global food losses and food waste – Extent, causes and prevention*. Rome: FAO. Available at: http://www.fao.org/docrep/014/mb060e/mb060e00.htm

FAO. (2011). *The state of the world's land and water resources for food and agriculture. Managing systems at risk*. Rome: Food and Agriculture Organization of the United Nations.

FAO. (2012). *Food Outlook Global Market Analysis*. FAO.

FAO. (2012). *The state of food insecurity in the World*. Rome: Food and Agriculture Organisation.

FAO. (2013). *The State of Food and Agriculture. Food Systems for Better Nutrition*. Rome: FAO.

FAO. (2014). Agriculture, Forestry and Other Land Use emissions by sources and removals by sinks. Rome: FAO.

FAO. (2015). *The State of Food and Agriculture. Social protection and agriculture: Breaking the cycle of rural poverty*. FAO.

FAO. (2017). *Soils. Key to unlocking the potential of mitigating and adopting to a changing climate. SDG goal 13*. FAO.

FAO. (2017). *The State of Food Security and Nutrition in the World*. FAO.

FAOSTAT. (2016). *Nigeria: Maize, green, production quality (tons)*. Retrieved from factfishwww.factfish.com

Farmar-Bowers, Q. (2013). Framing the research needs for food security in Australia. In Q. Farmar-Bowers, V. Higgins, & J. Millar (Eds.), *Food security in Australia: Fixing Business Media*. doi:10.1007/978-1-4614-4484-8_15

Fasoyiro, S. B., & Taiwo, K. A. (2012). Strategies for Increasing Food Production and Food Security in Nigeria. *Journal of Agricultural & Food Information, 13*(4), 338–355. doi:10.1080/10496505.2012.715063

Faulkner, J. W. (2006). Water Use, Productivity, And Profitability Of Small Scale Irrigation Schemes. In *Ghana's Upper East Region*. Cornell University.

Feder, G., Anderson, J., Birner, R., & Deininger, K. (2010). *Promises and Realities of Community-based Agricultural Extension*. IFPRI Discussion Paper 00959. Washington, DC: IFPRI.

Feder, G., Richard, E. J., & David, Z. (1985). Adoption of Agricultural Innovations in Developing Countries: A Survey. *Economic Development and Cultural Change, 33*(2), 255–298. doi:10.1086/451461

FMARD. (2016). *The Agriculture Promotion Policy (2016-2020): Building on the successes of the ATA, Closing Key Gaps*. Policy and Strategy Document of the Federal Ministry of Agriculture and Rural Development.

Food and Agricultural Organization (FAO). (2016). *Rice Market Monitor*. FAO.

Food and Agriculture Organization (FAO). (1997). The food system and factors affecting household food security and nutrition. In *Agriculture, food and nutrition for Africa: A resource book for teachers of agriculture*. Rome: Agriculture and Consumer Protection Department. Retrieved 15 October 2013 from http://www.fao.org/docrep/W0078E/W0078E00.htm

Food and Agriculture Organization of the United Nations (FAO). (2013). EX-ACT Version 3.4.1. FAO.

Fraser, E.D.G., Mabee, W., & Figge, F. (2005). A framework for assessing the vulnerability of food systems to future shocks. *Futures, 37*, 465-479.

Freeman, H. A., & Coe, R. (2002). Smallholder Farmers' Use of Integrated Nutrient-management Strategies: Patterns and Possibilities in Machakos District of Eastern Kenya. In C. B. Barrett, F. Place, & A. Abdillahhi (Eds.), *Natural Resources Management in African Agriculture:Understanding and Improving Current Practices* (pp. 143–154). Wallingford, UK: CABI Publishing. doi:10.1079/9780851995847.0143

Freibauer, Mathijs, Brunori, Damianova, Faroult, Gomis, O'Brien, & Treyer. (2011). Sustainable food consumption and production in a resource-constrained world. Belgium: European Commission-Standing Committee on Agricultural Research (SCAR).

Friedl, B., Omann, I., & Pack, A. (2014). *Socio-economic drivers of (non-) sustainable food consumption. An analysis for Austria*. Academic Press.

Frison, E. (2009). Director General calls for investment in true food security. *Biodiversity News*. Retrieved from http://www.bioversityinternational.org/news_and_events/news/news/article

FRN. (1999). Constitution of the Federal Republic of Nigeria.

Fuchs, D., & Lorek, S. (2001). Sustainable Consumption Governance in a Globalizing World. *Global Environmental Politics*, *2*(1), 19–45. doi:10.1162/152638002317261454

Galal, O. M. (2002). The nutrition transition in Egypt: Obesity, Under-nutrition and the Food Consumption Context. *Public Health Nutrition*, *5*(1a), 141–148. doi:10.1079/PHN2001286 PMID:12027277

Garnett, T., Appleby, M.C., Balmford, A., Bateman, I.J., Benton, T.G., Bloomer, P., … Godfray, H.C.J. (2013). Sustainable intensification in agriculture: Premises and policies. *Science, 341*, 33-34.

Garnett, M. C., Appleby, A., Balmford, I. J., Bateman, T. G., Benton, P., Bloomer, B., & Vermeulen, H. C. J. et al. (2016a). Sustainable intensification in agriculture: Premises and policies. *Scinece*, *30*, 1–2.

Garnett, T. (2014). *What is a sustaiable healthy diet Discussion paper*. Food Climate Research Network.

Garnett, T., Appleby, M., Balmford, A., & Bateman, I. (2016b). Sustainable Intensification in Agriculture: Premises and Policies. *Science*, *30*, 1–2. PMID:23828927

Garrett, J., & Ruel, M. (1999). *Are Determinants of Rural and Urban Food Security and Nutritional Status Different? Some Insights from Mozambique*. Washington, DC: International Food Policy Research Institute. Retrieved 15 October 2013 from http://www.ifpri.org/sites/default/files/pubs/divs/fcnd/dp/papers/dp65.pdf

Gary, S., & Bahla, G. S. (2008). Phosphorus Availability to Maize as Influenced by Organic Manures and Fertilizers P Associated Phosphate Activity in Soils. *Bioresource Technology*, *99*(13), 5773–5777. doi:10.1016/j.biortech.2007.10.063 PMID:18325765

Gasparatos, A., Stromberg, P., & Takeuchi, K. (2013, April). Sustainability impacts of first-generation biofuels. *Animal frontiers*, *3*(2), 12–26. doi:10.2527/af.2013-0011

Gauker, C. (2010). *The impacts of sustainable and industrial agriculture on human health*. Academic Press.

Gautam, Y., & Andersen, P. (2016). Rural livelihood diversification and household well-being: Insights from Humla, Nepal. *Journal of Rural Studies*, *44*, 239–249. doi:10.1016/j.jrurstud.2016.02.001

Gebregziabher, G., Namara, R. E., & Holden, S., & Others. (2012). Technical Efficiency Of Irrigated And Rain-Fed Smallholder Agriculture In Tigray, Ethiopia: A Comparative Stochastic Frontier Production Function Analysis. *Zeitschrift für Ausländische Landwirtschaft*, *51*(3), 203–226.

Gebrehiwot Yihdego, A. (2015). The Impact Of Small-Scale Irrigation On Income Of Rural Farm Households: Evidence From Ahferom Woreda In Tigray, Ethiopia. *International Journal Of Business And Economics Research*, *4*(4), 217–228. doi:10.11648/j.ijber.20150404.14

Georges, M. (2001). Recent progress in livestock genomics and potential impact on breeding programs. *Theriogenology*, *55*(1), 15–21. doi:10.1016/S0093-691X(00)00442-8 PMID:11198079

Gepts, P. (2006). Plant genetic resources conservation and utilization: The accomplishments and future of a societal insurance policy. *Crop Science*, *46*(5), 2278–2292. doi:10.2135/cropsci2006.03.0169gas

Ghaderzadeh, H., & Rahimi, M. H. (2008). Estimation Of Technical Efficiency Of Wheat Farms: A Case Study In Kurdistan Province, Iran. *American-Eurasian J. Agric. & Environ.*, *4*(1), 104–109.

Giddens, A. (1990). *The consequences of Modernity*. Stanford, CA: Stanford University Press.

Global Nutrition Report. (2015). *Basic Global Nutrition Report*. Retrieved from https://www.globalnutritionreport.org/files/2015/10/Basic-GNR-2015-presentation_102815.pdf

Godfray, H. C. J. & Garnett, T. (2014). Food security and sustainable intensification. *Philosophical Transaction of the Royal Society B, 369*, 1-12.

Godfray, H. C. J., Crute, I. R., Haddad, L., Lawrence, D., Muir, J. F., Nisbett, N., … Whiteley, R. (2010). The future of the global food system. *Philosophical Transaction of the Royal Society B, 365*, 2769-2777.

Godfray, H. C. J., Beddington, J. R., Crute, I. R., Haddad, L., Lawrence, D., Muir, J. F., & Toulmin, C. et al. (2010). Food Security: The Challenge of Feeding 9 Billion People. *Science*, *327*(5967), 812–818. doi:10.1126/science.1185383 PMID:20110467

Gonzalez, C. (2004). Trade Liberalization, Food Security and the Environment: The Neoliberal Threat to Sustainable Rural Development. *Transnat'l L. & Contemp. Probs.*, *14*(4), 419.

Goodman, D. (1997). World-scale processes and agro-food systems: Critique and research needs. *Review of International Political Economy*, *4*(4), 663–687. doi:10.1080/09672299708565787

Gorantiwar, S. D., & Smout, I. K. (2005). Performance Assessment Of Irrigation Water Management Of Heterogeneous Irrigation Schemes: 1. A Framework For Evaluation. *Irrigation and Drainage Systems*, *19*(1), 1–36. doi:10.1007/s10795-005-2970-9

Greaves, K. R. (2007). *Quantifying And Benchmarking Irrigation Scheme Performance With Water Balances And Performance Indicators* (MSc thesis). University of Kwazulu-Natal, South Africa.

Gregory, P. J., Ingram, J. S. I., & Brklacich, M. (2005). Climate change and food security. *Philosophical Transactions of the Royal Society of London. Series B, Biological Sciences*, *360*(1463), 2139–2148. doi:10.1098/rstb.2005.1745 PMID:16433099

Griggs, D. (2013). *Sustainable development goals for people and planet*. Academic Press.

Grilliches, Z. vi. (1980). Hybrid Corn: An Exploration in the Economics of Technological Change. *Econometrica*, *24*, 501–522.

Grunert, K. G., Hieke, S., & Wills, J. (2014). Sustainability labels on food products: Consumer motivation, understanding and use. *Food Policy*, *44*, 177–189. doi:10.1016/j.foodpol.2013.12.001

Gujarati, D. N. (1988). Basic Econometrics (2nd ed.). New York: McGraw Hill Publishers.

Gullick, L., & Urwich, C. L. (Eds.). (1957). *Papers on the Science of Administration*. New York: Institute of Public Administration.

Gupta, P., & Tsuchiya, T. (1991). *Chromosome Engineering in Plants: Genetics, Breeding Evolution*. Amsterdam: Elsevier.

Gussow, J. D. (1995). Mediterranean diets: Are they environmentally responsible? *The American Journal of Clinical Nutrition*, *61*(suppl), 1383S–1389S. PMID:7754992

Gustavsson, J., Cederberg, C., Van Otterdijk, R., & Meybeck, A. (2011). *Global Food Losses and Food Waste*. Rome: Food and Agriculture Organization of the United Nations.

Gyimah-Brempong, K., Johnson, M., & Takeshima, H. (2016). *The Nigerian Rice Economy: Policy Options for Transforming Production, Marketing, and Trade*. University of Pennsylvania Press. doi:10.9783/9780812293753

Hagos, F., Jayasinghe, G., Awulachew, S. B., Loulseged, M., & Yilma, A. D. (2012). Agricultural Water Management And Poverty In Ethiopia. *Agricultural Economics*, *43*, 99–111. doi:10.1111/j.1574-0862.2012.00623.x

Haile, H. T. (2008). *Impact Of Irrigation Development On Poverty Reduction In Northern Ethiopia* (PhD thesis). National University of Ireland, Cork, Ireland.

Haji, J., Aman, M., & Hailu, T. (2013). Impact Analysis Of Mede Telila Small Scale Irrigation Scheme On House Poverty Alleviation: Case Of Gorogutu District In Eastern Haratghe Oromia National Regional State Ethiopia. *International Journal Of Development And Economic Sustainability*, *1*(1), 15–30.

Hall, A. J., & Richards, R. A. (2013). Prognosis for genetic improvement of yield potential and water-limited yield of major grain crops. *Field Crops Research*, *143*, 18–33. doi:10.1016/j.fcr.2012.05.014

Hallensleben, N. (2012). *Characterizing Poor Smallholder Farmers In Nicaragua: Factors Influencing The Success Or Failure Of Micro-Irrigation Users* (Masters' Thesis). University Of St. Gallen (Hsg).

Han, Y., & Hansen, H. (2012). Determinants of Sustainable Food Consumption: A Meta-Analysis Using a Traditional and a Structura Equation Modelling Approach. *International Journal of Psychological Studies*, *4*(1), 22–44. doi:10.5539/ijps.v4n1p22

Hanyani-Mlambo, B. T. (2002). *Strengthening the pluralistic agricultural extension system: a Zimbabwean case study. Integrated Support to Sustainable Development and Food Security Programme (IP)*. Rome, Italy: FAO.

Harold, M., & Tabo, R. (2015). *Feeding Africa: United Nations Economic Commission for Africa, An Action plan for African Agricultural Transformation*. Abdon Diouf International Conference Center Dakar, Senegal.

Hart, K., Allen, B., & Karsten, J. (2013). *EU policy options to encourage more sustainable food choices*. Academic Press.

Harvey, D. (2005). *Brief History of Neoliberalism*. New York: Oxford University Press.

Hassan, R. M., Muriithi, F., & Kamau, G. (1998). Determinants of Fertilizer Use and the Gap between Farmers' Maize Yield and Potential Yields in Kenya. In R. M. Hassan (Ed.), *Maize Technology Development and Transfer: A GIS Application for Research Planning in Kenya*. Wallingford, UK: CAB International.

Hassen, I. W., Dereje, M., Minten, B., & Hirvonen, K. (2016). *Diet transformation in Africa: the case of Ethiopia*. Ethiopia Strategy Support Program (ESSP). Working Paper no. 87, March. International Food Policy Research Institute.

Havlin, J. L., Tisdale, S. L., Nelson, W. L., & Beaton, J. D. (2005). *Soil Fertility and Fertilizers. An Introduction to Nutrients Management*. Upper Saddle River, NJ: Pearson Educational Inc.

Hawkes, C. (2006). The links between agriculture and health: An intersectoral opportunity to improve the health and livelihoods of the poor. *Bulletin of the World Health Organization*, *84*(12), 984–990. doi:10.2471/BLT.05.025650 PMID:17242835

Haws, K., Winterich, K., & Naylor, R. (2014). Seeing the world through GREEN-tinted glasses: Green consumption values and responses to environmentally friendly products. *Journal of Consumer Psychology*, *24*(3), 336–354. doi:10.1016/j.jcps.2013.11.002

Haywood, L., von Maltitz, G., Setzkorn, K., & Ngepah, N. (2008). *Biofuel production in South Africa, Mozambique, Malawi and Zambia: A status quo analysis of the social, economic and biophysical elements of the biofuel industry in Southern Africa*. Pretoria: Council for Scientific and Industrial Research.

Helen, H. J. (2002). Food Insecurity and the Food Stamp Programme. *American Journal of Agricultural Economics*, *84*(5), 1215–1218. doi:10.1111/1467-8276.00382

Henao, J., & Baanante, C. (2006). *Agricultural Production and Soil Nutrient Mining in Africa: Implications for Resource Conservation and Policy Development*. Muscle Shoals, AL: An International Center for Soil Fertility and Agricultural Development.

Henning, R. K. (2013). *Jatropha curcas in Africa - an Evaluation.* Weissenberg: Bagani. Available at http://www.underutilized-species.org/Documents/PUBLICATIONS/jatropha_curcas_africa.pdf

Herrera-Estrella, L. R., & Alvarez-Morales, A. (2000). Genetically Modified Crops and Developing Countries. *EMBO Reports*, *2*(4), 256–258. doi:10.1093/embo-reports/kve075 PMID:11306538

Herrero, M., & Thorntonb, P. K. 2010. Livestock and global change: Emerging issues for sustainable food systems. *Proceedings of the National Academy of Sciences of the United States of America,* 110.

Herrmann, R., Jumbe, C., Bruentrup, M., & Osabuohien, E. (2017). Competition between Biofuel Feedstock and Food Production: Empirical Evidence from Sugarcane Outgrower Settings in Malawi. *Biomass and Bioenergy*. doi:10.1016/j.biombioe.2017.09.002

Hertwich, E., & Katzmayr, M. (2003). *Examples of sustainable consumption.* Academic Press.

Heyman, Y., Vignon, X., Chesne, P., Le Bourhis, D., Marchal, J., & Renard, J. P. (1998). Cloning in cattle: From embryo splitting to somatic nuclear transfer. *Reproduction, Nutrition, Development*, *38*(6), 595–603. doi:10.1051/rnd:19980602 PMID:9932293

Hinrichsen, D., Salem, R., & Blackburn, R. (2002) Meeting the Urban Challenge. Population Reports, Series M, No. 16. Baltimore, MD: The Johns Hopkins Bloomberg School of Public Health, Population Information Program.

HLPE. (2014). *Food losses and waste in the context of sustainable food systems. A report by the High Level Panel of Experts on Food Security and Nutrition of the Committee on World Food Security*. Rome: HLPE.

Hobson, K. (2002). Competing discourses of sustainable consumption: Does the rationalisation of lifestyles' make sense? *Environmental Politics*, *11*(2), 95–120. doi:10.1080/714000601

Hobson, K. (2003). Consumption, Environmental Sustainability and Human Geography in Australia: A Missing Research Agenda? *Australian Geographical Studies*, *41*(2), 148–155. doi:10.1111/1467-8470.00201

Holden, S., & Lunduka, R. (2010). *Too poor to be Efficient? Impacts of the Targeted Fertilizer Subsidy Programme in Malawi on Farm Plot Level Input Use, Crop Choice and Land Productivity*. Noragric Report No. 55.

Hollinger, F., & Staatz, J. M. (2015). *Agricultural Growth in West Africa Market and Policy Drivers*. Rome: African Development Bank and Food and Agriculture Organisation.

Holsti, O. (1969). *Content Analysis for the Social Sciences and Humanities.* Addison-Wesley.

Horrigan, L., Lawrence, R. S., & Walker, P. (2002). How sustainable agriculture can address the environmental and human health harms of industrial agriculture. *Environmental Health Perspectives*, *110*(5), 445–456. doi:10.1289/ehp.02110445 PMID:12003747

Huang, Q., Rozelle, S., Lohmar, B., Huang, J., & Wang, J. (2006). Irrigation, Agricultural Performance And Poverty Reduction In China. *Food Policy*, *31*(1), 30–52. doi:10.1016/j.foodpol.2005.06.004

Hunger Task Force. (2003). *Halving Hunger by 2015: A Framework for Action.* Interim Report, Millennium Project, UNDP.

Hussain, I., Biltonen, E., & International Water Management Institute. (Eds.). (2001). Irrigation Against Rural Poverty: An Overview Of Issues And Pro-Poor Intervention Strategies In Irrigated Agriculture In Asia: Proceedings Of National Workshops On Pro-Poor Intervention Strategies In Irrigated Agriculture In Asia, Bangladesh, China, India, Indonesia, Pakistan, And Vietnam. Colombo, Sri Lanka: International Water Management Institute.

Hussain, I., Sial, M. H., Hussain, Z., & Akram, W. (2009). Economic Value Of Irrigation Water: Evidence From A Punjab Canal. *The Lahore Journal Of Economics*, *14*(1), 69.

Hussain, I., & Hanjra, M. A. (2004). Irrigation And Poverty Alleviation: Review Of The Empirical Evidence. *Irrigation and Drainage*, *53*(1), 1–15. doi:10.1002/ird.114

ICESCR. (1999). *The International Covenant on Economic, Social and Cultural Rights. General comments 12. Twentieth session, 12th May, 1999.* Retrieved from http://www.fao.org/fileadmin/templates/righttofood/documents/RTF_publications/EN/General_Comment_12_EN.pdf

Ingawa, D. N. (1988). *Water and land use rights under the National Fadama Development Project: A matter of Conservation.* Paper presented at the annual planning, Monitoring and evaluation Workshop in soil and water conservation Institute.

Ingram, J. (2011). A food systems approach to researching food security and its interactions with global environmental change. *Food Security, 3*, 417-431.

Ingram, J. A. (2011). food systems approach to researching food security and its interactions with global environmental change. *Food Security*, *3*(4), 417–431. doi:10.1007/s12571-011-0149-9

International Assessment of Agricultural Knowledge, Science and Technology for Development (IAAKSTD). (2008). *Bio-energy and Biofuel: Opportunity and Constraints. Issue in brief.* Island Press.

International Atomic Energy Agency (IAEA). (2009). International Energy Statistics Report 2009. Washington, DC: IAEA.

International Atomic Energy Agency (IAEA). (2011). Technology Roadmaps: Biofuel for Transport. OECD/IEA Report.

International Food Policy Research Institute (IFPRI). (2013). *South Africa Social Accounting Matrix 2005.* Washington, DC: IFPRI.

International Institute Of Social Studies. (2015). *Global governance/politics, climate justice & agrarian/social justice: linkages and challenges.* Author.

International Trade Centre (ITC). (2013). *Trade Map Statistics Database*. ITC.

IOM (Institute of Medicine). (2015). *A framework for assessing effects of the food system*. Washington, DC: The National Academies Press.

IPCC. (2014). *Mitigation of Climate Change*. Contribution of Working Group III to the Fifth Assessment Report of the Intergovernmental Panel on Climate Change.

Jacobsen, S., Sorensen, M., Soren, M., & Weiner, J. (2013). Feeding the world: Genetically modified crops versus agricultural biodiversity. *Agronomy for Sustainable Development*, *0138-9*. doi:10.1007/s13593-013

Jaghdani, T. J., Brümmer, B., & Barkmann, J. (2012). Comparison Of Methods For The Valuation Of Irrigation Water: Case Study From Qazvin, Iran: Comparison Of Methods For The Valuation Of Irrigation Water. *Irrigation and Drainage*, *61*(3), 375–385. doi:10.1002/ird.683

Jansson, J., Marell, A., & Nordlund, A. (2010). Green consumer behavior: Determinants of curtailment and eco-innovation adoption. *Journal of Consumer Marketing*, *27*(4), 358–370. doi:10.1108/07363761011052396

Janvry, L. (1981). *The Agrarian Question and Reformism in Latin America*. Johns Hopkins University Press.

Jayne, T. S., & Rashid, S. (2013). Input subsidy programs in sub-Saharan Africa: A synthesis of recent evidence. *Agricultural Economics*, *44*(6), 1–16. doi:10.1111/agec.12073

Jin, S., Jansen, H. G., Muraoka, R., & Yu, W. (2012). The Impact Of Irrigation On Agricultural Productivity: Evidence From India. International Association Of Agricultural Economists (Iaee).

Johnson, M., Takeshima, H., & Gyimah-Brempong, K. (2013). *Assessing the Potential and Policy Alternatives for Achieving Rice Competitiveness and Growth in Nigeria*. The International Food Policy Research Institute (IFPRI). IFPRI Discussion Paper, 01301.

Johnson, J. L., Fanzo, J., & Cogill, B. (2014). Understanding sustainable Diets: A descriptive analysis of the determinants and processes that influence diets and their impact on health, food security, and environmental sustainability. *Advances in Nutrition*, *5*(4), 418–429. doi:10.3945/an.113.005553 PMID:25022991

Johns, T., & Eyzaguirre, P. B. (2001). Nutrition for sustainable environments. *SCN News*, *21*, 24–29.

Jongen, W., & Meerdink, G. (1998). *Food Product Innovation: How to Link Sustainability and the Market*. Wageningen Agricultural University.

Jose, D., Shanmugavelu, K. G., & Thamburaj, S. (1988). Studies on the Efficiency of Organic versus Inorganic form of Nitrogen in Brinjal. *Indian Journal of Horticulture*, *45*, 100–103.

Juliano, B.O. (2016). Rice: Role in Diet. *Reference Module in Food Science Encyclopaedia of Food and Health,* 641–645.

Kabore, D. P. (2016). Technical Efficiency Of Irrigated Rice Production In Burkina Faso Water Control Infrastructures: A Stochastic Frontier Approach. *Asian Journal of Agriculture and Rural Development*, *6*(10), 188–200.

Kamau, M., Smale, M., & Mutua, M. (2013). *Farmer Demand for Soil Fertility Management Practices in Kenya's Grain Basket*. Paper Prepared for Presentation at the *Agricultural and Applied Economics Association (AAEA) and CAES joint meeting*. Washington, DC.

Kamwamba-Mtethiwa, J., Weatherhead, K., & Knox, J. (2016). Assessing Performance Of Small-Scale Pumped Irrigation Systems In Sub-Saharan Africa: Evidence From A Systematic Review: Performance Of Small Pumped Irrigation Systems In Sub Saharan Africa. *Irrigation and Drainage*, *65*(3), 308–318. doi:10.1002/ird.1950

Kangile, R. J. (2015). *Efficiency In Production By Smallholder Rice Farmers Under Cooperative Irrigation Schemes In Pwani And Morogoro Regions, Tanzania* (Msc Thesis). Sokoine University Of Agriculture, Tanzania.

Kanyamuka, J. (2017). *Adoption of Integrated Soil Fertility Management Technologies and Its Effect On Maize Productivity: A Case of the Legume Best Bets Project in Mkanakhoti Extension Planning Area of Kasungu District in Central Malawi* (MSc. Thesis). Lilongwe University of Agricultural and Natural Resources, Lilongwe, Malawi.

Kareem, O. I. (2016). European Union's Standards and Food Exports from Africa: Implications of the Comprehensive Africa Agricultural Development Programme for Coffee and Fish. *Journal of African Development*, *18*, 83–97.

Kasa, S. (2008). Globalizing unsustainable food consumption:Trade policies, producer lobbies, consumer preferences, and beef consumption in Northeast Asia. *Globalizations*, *5*(2), 151–163. doi:10.1080/14747730802057480

Kastner, T., Rivas, M. J. I., Koch, W., & Nonhebel, S. (2012). Global changes in diets and the conseguences for land requirements for food. *Proceedings of the National Academy of Sciences of the United States of America*, *109*(18), 6868–6872. doi:10.1073/pnas.1117054109 PMID:22509032

Kaswamila, A. L., & Masuruli, B. M. (2004). The Role Of Traditional Irrigation Systems. In *Poverty Alleviation In Semi-Arid Areas: The Case Of Chamanzi In Lushoto District, Tanzania*. Dar Es Salaam, Tanzania: Mkuki Na Nyota Publishers.

Kaur, K., Kapoor, K.K. & Gupta, A.P. (2005). Impact of Organic Manures with and without. *Mineral and Soil Science, 168*, 117-122.

Kayuki, C. K., John, B., Zimwanguyizza, W., Katwijukye, A. K., Mawanda, A., Kabuye, F. M., & Iceduna, C. et al. (2012). Maize response to application of inorganic fertilizers and manure in eastern and northern Uganda. In *12th Integrated Soil Fertility Management in Africa: From Microbes to Market. Conference Information, Program and Abstracts*. CIAT.

Kearney, J. (2010). Food consumption trends and drivers. *Philosophical Transaction of the Royal Society B*, *365*(1554), 2793–2807. doi:10.1098/rstb.2010.0149 PMID:20713385

Keeney, D. R. (1997). Sustainable Agriculture and the Environment. In *Proceedings of the 1997 International Symposium on Sustainable Agricultural Development Compatible with Environmental Conservation in Asia*. JIRCAS.

Kesavan, P. C. & Swaminathan, M. S. (2008). Strategies and models for agricultural sustainability in developing Asian countries. *Philosophical Transaction of the Royal Society B, 363*, 877-891.

Khalid, S., Khan, M. B., Regina, K., Jos, K., & Gerd, A. (2003). Five steps to conducting a systematic review. *Journal of the Royal Society of Medicine*, 96. PMID:12612111

Khan, Z. R., Midega, C. A. O., Pittchar, J. O., Murage, A. W., Birkett, M. A., Bruce, T. J. A., & Pickett, J. A. (2014). Achieving food security for one million sub-Saharan African poor through push-pull innovation by 2020. *Philosophical Transactions of the Royal Society of London. Series B, Biological Sciences*, 369. PMID:24535391

Kharas, H. (2010). *The Emerging Middle Class in Developing Countries*. Development Centre Working Paper No. 285. Organization for Economic Co-operation and Development.

Kickbusch, I. (2010). *The food system: A prism of present and future challenges for health promotion and sustainable development*. Triggering the debate – white paper. Bern: Promotion Santé Suisse, 2010.

Kimaro, A. A., Timmer, V. R., Chamshama, S. A. O., Ngaga, Y. N., & Kimaro, D. A. (2009). Competition between Maize and Pigeon-Pea in Semi-Arid Tanzania: Effect on Yields and Nutrition of Crops. *Agriculture, Ecosystems & Environment, 134*(1-2), 115–125. doi:10.1016/j.agee.2009.06.002

King, F. S., & Burgess, A. (1993). Nutrition for Developing Countries (2nd ed.). Oxford, UK: Oxford University Press. doi:10.1093/acprof:oso/9780192622334.001.0001

Kjærga, B., Land, B., & Pedersen, K. B. (2013). Health and sustainability. *Health Promotion International*, *29*, 10–15. PMID:23300191

Klak, T. (2013). World System theory core,semi-peripheral and peripheral regions. In The companion to Development Studies. Routledge.

Klein, N. (2008). *The Shock Doctrine The Rise of Disaster Capitalism*. Metropolitan Books Henry Holt and Company.

Korten, D. (1995). *When Corporations Rule the World*. West Hartford, CT: Kumarian Press Inc.

Kutigi, A. U. (2005). *State of Fadama development in Niger state*. Paper presented at the agricultural week, Minna.

Kuwornu, J. K. M. (2012). Irrigation Access And Per Capita Consumption Expenditure In Farm Households: Evidence From Ghana. *Journal of Development and Agricultural Economics*, *4*(3), 78–92.

Laha, A., & Kuri, P. K. (2013). Measurement Of Technical Efficiency Under Alternative Institutions Of Tenurial Contracts And Its Determinants: Evidence From Rural West Bengal, India. *Discourse Journal Of Agriculture And Food Sciences*, *1*(11), 160–170.

Lambrecht, I., Vanlauwe, B., & Maertens, M. (2014). *Integrated Soil Fertility Management: From Concept to Practice in Eastern DR Congo*. Bioeconomics Working Paper Series. KU Leuven.

Lang, T., Barling, D., & Caraher, M. (2009). *Food policy: integrating health, environment and society*. Oxford, UK: Oxford University Press. doi:10.1093/acprof:oso/9780198567882.001.0001

Lawrence, G., & Burch, D. (2007). Understanding supermarkets and agri-food supply chains. In D. Burch & G. Lawrence (Eds.), *Supermarkets and agri-food supply chains: transformations in the production and consumption of foods*. Cheltenham, UK: Edward Elgar.

Lawrence, G., Lyons, K., & Tabatha, W. (2010). *Food Security, Nutrition and Sustainability*. New York: Earthscan.

Leathers, H., Smale, M., & Heisey, P. (1995). Maize of the Ancestors and Modern Varieties: The Microeconomics of High-Yielding Variety Adoption in Malawi. *Economic Development and Cultural Change*, *43*(2), 351–368. doi:10.1086/452154

Lefin, A. L. (2010). *Food consumption and sustainable development: An introduction*. Institut pour un Développement Durable.

Lehota, A. L. (2004). National and international trend of food consumption and behaviour. *Marketing*, *1*, 7–14.

Lenis Saweda, O. (2017). Is increasing inorganic fertilizer use for maize production in SSA a profitable proposition? Evidence from Nigeria. Elsevier. *Food Policy*, *67*, 41–51. doi:10.1016/j.foodpol.2016.09.011 PMID:28413245

Leontief, W. (1970). Environmental Repercussions and the Economic Structure: An Input-Output Approach. *The Review of Economics and Statistics*, *52*(3), 262–271. doi:10.2307/1926294

Lipton, M., Blackman, R., De Zoysa, D., Qureshy, L., & Waddington, H. (2003). Preliminary Review Of The Impact Of Irrigation On Poverty With Special Emphasis On Asia. Rome: Academic Press.

Littell, J. H., Corcoran, J., & Pillai, V. (2008). *Systematic reviews and meta-analysis*. New York, NY: Oxford University Press. doi:10.1093/acprof:oso/9780195326543.001.0001

Liverpool-Tasie, L., Adjognon, A., & Reardon, T. (2016). *Transformation of the food system in Nigeria and female participation in the Non-Farm Economy (NFE)*. 2016 Annual Meeting, Boston, MA.

Ljungman, C. M. (2004). Applying a Rights-Based Approach to Development: Concepts and Principles. *Conference Paper: The Winners and Losers from Rights-Based Approaches to Development*, 6.

Lock, I., & Ikeda, S. (2005). *Clothes Encounters: Consumption, Culture, Ecology and Economy. In Consuming Sustainability Critical Social Analysis of Ecological Change.* Fernwood Publishing.

Lönnroth, M. (2010). The Organization of Environmental Policy in Sweden - A Historical Perspective. *The Swedish Environmental Protection Agency Report*, *6404*, 319.

Lorek, S., & Fuchs, D. (2013). Strong sustainable consumption governance – precondition for a degrowth path? *Journal of Cleaner Production*, *38*, 36–43. doi:10.1016/j.jclepro.2011.08.008

Lukaszewski, A. (2004). Chromosome manipulation and crop improvement. In *Encyclopedia of Plant and Crop Science*. New York: Marcel Dekker. doi:10.1081/E-EPCS-120005593

Lunduka, R., & Kelly, V. (2012). *Economic Benefits of Integrated Soil Fertility Management (ISFM)*. ASHC policy brief 1.

Lunduka, R., & Kelly, V. (2009). *The Financial Benefits of Integrated Soil Fertility Management Using Inorganic Fertilizer in Combination with Other ISFM Practices. A Post-doctoral Fellow (Policy) Research, CABI Africa, ICRAF Complex, United Nations Avenue*. Nairobi, Kenya: Gigiri.

Macharia, J., Mugwe, J., Mucheru-Muna, M., Mang'uriu, D., & Mugendi, D. (2012). Factors Influencing ISFM Knowledge in the Central Highlands of Kenya. *12th Integrated Soil Fertility Management in Africa: From Microbes to Market. Conference Information, Program and Abstracts*. CIAT.

Machethe, C. L. (2011). Smallholder Irrigation And Agricultural Development In The Olifants River Basin Of Limpopo Province: Management Transfer, Productivity, Profitability And Food Security Issues. Gezina, South Africa: Water Research Commission.

Mackay, R., & Wolbring, G. (2013a). *Sustainable Consumption of Healthcare: Linking Sustainable Consumption with Sustainable Healthcare and Health Consumer Discourses*. The 3rd World Sustainability Forum.

Mackay, R., & Wolbring, G. (2013b). *Sustainable Consumption of Healthcare: Linking Sustainable Consumption with Sustainable Healthcare and Health Consumer Discourses*. Academic Press.

Madukwe, M.C. (2008). Practices without Policy. In *The Nigeria Agricultural Extension Service 30th inaugural lecture of the University of Nigeria*. University of Nigeria Press Ltd.

Magingxa, L. L., Alemu, Z. G., & Van Schalkwyk, H. D. (2006). Factors Influencing The Success Potential In Smallholder Irrigation Projects Of South Africa: A Principal Component Regression. *Presented At The International Association Of Agricultural Economists Conference*, Gold Coast, Australia.

Mahdhi, N., Sghaier, M., & Bachta, M. S. (2011). Water Use And Technical Efficiencies In Private Irrigated Perimeters In South-Eastern Of Tunisia. *Emirates Journal of Food and Agriculture*, *23*(5), 440.

Makhokha, S., Kimani, S., Mwangi, W., Verkuijl, H., & Musembi, F. (2001). Determinants of Fertilizer and Manure Use for Maize Production in Kiambu District, Kenya. CIMMYT (International Maize and Wheat Improvement Centre) and Kenya Agricultural Research Institute (KARI)

Makombe, G. (2011). *A Comparative Analysis Of The Technical Efficiency Of Rain-Fed And Smallholder Irrigation In Ethiopia*. Working Paper 143. International Water Management Institute.

Makombe, G., Kelemework, D., & Aredo, D. (2007). A Comparative Analysis Of Rainfed And Irrigated Agricultural Production In Ethiopia. *Irrigation and Drainage Systems*, *21*(1), 35–44. doi:10.1007/s10795-007-9018-2

Makombe, G., & Sampath, R. K. (1998). An Economic Evaluation Of Smallholder Irrigation Systems In Zimbabwe. *International Journal of Water Resources Development*, *14*(1), 77–90. doi:10.1080/07900629849510

Makombe, G., & Sampath, R. K. (1999). A Benefit-Cost Analysis Of Smallholder Irrigated Farms In Zimbabwe. *Irrigation and Drainage Systems*, *13*(1), 1–12. doi:10.1023/A:1006133220120

Maliwichi, L. L., Oni, S. A., & Obadire, O. S. (2012). An Investigation Into The Factors Affecting Food Availability, Choices And Nutritional Adequacy Of Smallholder Farming Households Under Irrigation And Dryland Farming In Vhembe District Of Limpopo Province, South Africa. *African Journal of Agricultural Research*, *7*(25), 3653–3664. doi:10.5897/AJAR11.1062

Malone, T. W., & Crowston, K. (1994). Toward an Interdisciplinary Theory of Coordination. *Computer Survey*, *26*(1), 87–119. doi:10.1145/174666.174668

Maltsoglou, I., & Khwaja, Y. (2010). *Bioenergy and food security: The BEFS analysis for Tanzania*. Rome: Food and Agriculture Organization.

Mangham, I. L. (1986). *Power and Performance in Organization*. New York: Basil Blackwell Inc.

Manzungu, E. (1995). Engineering Or Domineering? The Politics Of Water Control In Mutambara Irrigation Scheme, Zimbabwe. *Zambezia*, *22*(2), 115–136.

Marenya, P. P., & Barrett, C. B. (2007). Household-level Determinants of Adoption of Improved Natural Resources Management Practices among Smallholder Farmers in Western Kenya. *Food Policy*, *32*(4), 515–536. doi:10.1016/j.foodpol.2006.10.002

Martin, A. (2008, May 18). One Country's Table Scraps, Another Country's Meal. *The New York Times*.

Maruyama, A., Haneishi, Y., Okello, S. E., Asea, G., Tsuboi, T., Takagaki, M., & Kikuchi, M. (2014). Rice Green Revolution And Climatic Change In East Africa: An Approach From The Technical Efficiency Of Rainfed Rice Farmers In Uganda. *Agricultural Sciences*, *05*(04), 330–341. doi:10.4236/as.2014.54035

Marx, K. (1978). *The Marx-Engels Reader*. New York: Norton.

Mason, N. M., Jayne, T. S., Burke, W. J., & Ariga, J. (2016). *Agricultural Input Subsidy Programs in Africa: An Assessment of Recent Evidence*. MSU International Development Working Paper. Department of Agricultural, Food, and Resource Economics and the Department of Economics, Michigan State University.

Mason, N. M., Jayne, T. S., & Mofya-Mukuka, R. (2013). Zambia's input subsidy programs. *Agricultural Economics*, *44*(6), 613–628. doi:10.1111/agec.12077

Mason, N. M., Wineman, A., Kirimi, L., & Mather, D. (2017). The Effects of Kenya's 'Smarter' Input Subsidy Programme on Smallholder Behaviour and Incomes: Do Different Quasi-experimental Approaches Lead to the Same Conclusions? *Journal of Agricultural Economics*, *68*(1), 45–69. doi:10.1111/1477-9552.12159

Masters, B. (2004). *Safety of Genetically Engineered Foods: Approaches to Assessing Unintended Health Effects*. Washington, DC: The National Academies of Sciences.

Mather, D., Waized, B., Ndyetabula, D., Anna, T., & Minde, I. (2016). *The Profitability of Inorganic Fertilizer Use in Smallholder Maize Production in Tanzania: Implications for Alternative Strategies to Improve Smallholder Maize Productivity*. Guiding I investments in Sustainable Agricultural Intensification in Africa/Tanzania Working Paper #4.

Mckenzie, F. C., & Williams, J. (2015). Sustainable food production: Constraints, challenges and choices by 2050. *Food Security*, *7*(2), 221–233. doi:10.1007/s12571-015-0441-1

Mcmichael, A. J., Powles, J. W., Butler, C. D., & Uauy, R. (2007). Food, livestock production, energy, climate change, and health. *Lancet*, *370*(9594), 1253–1263. doi:10.1016/S0140-6736(07)61256-2 PMID:17868818

Mdemu, M. V., Mziray, N., Bjornlund, H., & Kashaigili, J. J. (2016). Barriers To And Opportunities For Improving Productivity And Profitability Of The Kiwere And Magozi Irrigation Schemes In Tanzania. *International Journal of Water Resources Development*, 1–15.

Meliko, M. O., & Oni, A., S. (2011). Effect Of Irrigation On Poverty Among Small-Scale Farmers In Limpopo Province Of South Africa. *The Journal of Agricultural Science*, *3*(3), 190–195.

Memudu, I. J. (2015). *Economic Analysis of Integrated Soil Fertility Management for maize production in Kogi and Kwara States, Nigeria* (Unpublished Ph.D. Thesis). Department of Agricultural Economics, Faculty of Agriculture, University of Ilorin.

Mengistie, D., & Kidane, D. (2016). Assessment Of The Impact Of Small-Scale Irrigation On Household Livelihood Improvement At Gubalafto District, North Wollo, Ethiopia. *Agriculture*, *6*(3), 1–22. doi:10.3390/agriculture6030027

Merumba, M. S., Ndyetabula, I. L., Nkuba, J. M., William, M. M., Mukandala, L., Sayi, B., & Shimwela, M. et al. (2012). Exploring the Effects of Different Fertilizers on Yields and Benefits of Important Crops Grown in Kagera, Tanzania Integrated Soil Fertility Management in Africa Conference Information. *Program with Abstracts (Geological Association of Canada)*, 44.

Mexico, D. F. Deksissa, T., Wyche-Moore, G. S. & Hare, W. W. (2008). Occurrence, Fate and Transport of 17 beta – estradiol and testosterone in the environment. *Proceeding of 2007 AWRA Summer Specialty Conference.*

Mhango, W. G., Snapp, S. S., & Phiri, G. Y. K. (2012). Opportunities and Constraints to Legume Diversification for Sustainable Maize Production on Smallholder Farms in Malawi. *Renewable Agriculture and Food Systems*, *28*(3), 234–244. doi:10.1017/S1742170512000178

Milner, D. (2012). *Consumption and Its Consequences.* Cambridge Polity Press.

Ministry of Budget and National Planning. (2017). *Economic Recovery and Growth Plan 2017-2020.* Abuja: Ministry of Budget and National Planning.

Minten, B., Tamru, S., Engida, E., & Tadesse, K. (2013). *Ethiopia's value chains on the move: the case of teff.* Ethiopia Strategy Support Program (ESSP). Working Paper no. 52, April. International Food Policy Research Institute.

Mission, E. (2008). Smallholder Irrigation Schemes In Nyanga North As Strategies For Poverty Reduction And Sustainable Rural Livelihoods. *Journal of Sustainable Development in Africa*, *10*(2), 653–676.

Mohammadi, S., Kalbasi, M., & Shariatmadari, H. (2009). Cumulative and Residual Effects of Organic Fertilizer on Selected soil Properties, Water Soluble P, Olsen-P and P-Sorbtion Index. *Journal of Agricultural Science and Technology*, *11*(4), 487–497.

Molden, D. (1998). *Indicators For Comparing Performance Of Irrigated Agricultural Systems.* Colombo: Iimi.

Molden, D. J., & Gates, T. K. (1990). Performance Measures For Evaluation Of Irrigation-Water-Delivery Systems. *Journal of Irrigation and Drainage Engineering*, *116*(6), 804–823. doi:10.1061/(ASCE)0733-9437(1990)116:6(804)

Mont, O., & Plepys, A. (2005). Motivating sustainable consumption. *Sustainable Development Research Network*, *29*, 30.

Moomaw, W., Griffin, T., Kurczak, K., & Lomax, J. (2012). *The Critical Role of Global Food Consumption Patterns in Achieving Sustainable Food Systems and Food for All, A UNEP Discussion Paper.* United Nations Environment Programme, Division of Technology, Industry and Economics.

Moore, J. (2000). Environmetal Crisis and the Methabolic Rift in World –historical perspective. *Organization & Environment*, *13*(2), 2. doi:10.1177/1086026600132001

Moyin, J. E. I. (2003). Incorporation of Agricultural Biomass and their Effects on Growth and Nutrient Content of Four Successive Crops of Amaranthus. *Conference Proceeding*, 102.

Moyo, M., Van Rooyen, A., Moyo, M., Chivenge, P., & Bjornlund, H. (2016). Irrigation Development In Zimbabwe: Understanding Productivity Barriers And Opportunities At Mkoba And Silalatshani Irrigation Schemes. *International Journal of Water Resources Development*, 1–15. Retrieved from Https://Doi.Org/10.1080/07900627.2016.1175339

Mózner, Z. V. (2013). *Towards Sustainable food consumption? An Ecological footprint in Hungary* (PhD thesis). Corvinus University of Budapest.

Mtambanengwe, F., & Kosina, P. (2007). *Cereal Knowledge Bank*. Retrieved from www.knowledgebank.irri.org/.../organic-and-inorganic-fertilizers.html

Mucheru-Muna, M., Ngetich, F., Mugendi, D., Mugwe, J., Pypers, P., Vanlauwe, B., & Merckx, R. et al. (2012). Organic Inputs and Mineral Fertilizer Effects on Soil Chemical Properties, Soil Water Content and Maize Productivity in Mbeere District, Kenya Integrated Soil Fertility Management in Africa Conference Information. *Program with Abstracts (Geological Association of Canada)*, 80.

Mujwahuzi, M., Ndetewio, P. I., Mwakaje, A. G., & Ngana, J. (2013). Factors Influencing Willingness To Pay For Watershed Services. In *Lower Moshi*. Pangani Basin.

Mullens, B. A., Szijj, C. E., & Hincle, N. C. (2002). Oviposition and Development of Fannia spp. (Diptera:Muscidae) on Poultry Manure of Low Moisture Levels. *Environmental Entomology*, *31*(4), 588–593. doi:10.1603/0046-225X-31.4.588

Muller, J. (2013). Capitalism and Inequality, What the Right and the Left Get Wrong. *Foreign Affairs*, *92*(2), 30–51.

Mupaso, N., Manzungu, E., Mutambara, J., & Hanyani-Mlambo, B. (2014). The Impact Of Irrigation Technology On The Financial And Economic Performance Of Smallholder Irrigation In Zimbabwe. Impact Of Irrigation Technology On Smallholder Irrigation. *Irrigation and Drainage*, *63*(4), 430–439. doi:10.1002/ird.1820

Mutambara, S., Darkoh, M. B. K., & Atlhopheng, J. R. (2016). A Comparative Review Of Water Management Sustainability Challenges In Smallholder Irrigation Schemes In Africa And Asia. *Agricultural Water Management*, *171*, 63–72. doi:10.1016/j.agwat.2016.03.010

Mutambara, S., & Munodawafa, A. (2014). Production Challenges And Sustainability Of Smallholder Irrigation Schemes In Zimbabwe. *Journal Of Biology Agriculture And Healthcare*, *4*(15), 87–96.

Mutegi, J., & Zingore, S. (2015). *Closing Crop Yield Gaps in sub-Saharan Africa through Integrated Soil Fertility Management*. IPNI-Sub-Saharan Africa Program.

Mutiro, J., & Lautze, J. (2015). Irrigation In Southern Africa: Success Or Failure? *Irrigation and Drainage*, *64*(2), 180–192. doi:10.1002/ird.1892

Mwendera, E., & Chilonda, P. (2013). Conceptual Framework For Revitalisation Of Small-Scale Irrigation Schemes In Southern Africa. *Irrigation and Drainage*, *62*(2), 208–220. doi:10.1002/ird.1723

Nakhumwa, T. O. (2004). *Dynamic Costs of Soil Degradation and Determinants of Adoption of Soil Conservation Technologies by Smallholder Farmers in Malawi* (PhD Thesis). University of Pretoria, South Africa.

Nambiro, E., Okoth, P., Kinyanjui, S., Murua, E., Kibe, P., & Macharia, R. (2012). Which Households are Most Likely to Take Up ISFM Technology and Why? Case of Sidindi, Western Kenya Integrated Soil Fertility Management in Africa Conference Information. *Program with Abstracts (Geological Association of Canada)*, 86.

Namyenya, A., Sserunkuuma, D., & Bagamba, F. (2014). Willingness To Pay For Irrigation Water And Its Determinants Among Farmers At Doho Rice Irrigation Scheme (Dris) In Uganda. *Journal of Development and Agricultural Economics*, *6*(8), 345–355. doi:10.5897/JDAE2014.0580

National Institute Of Consumer Research. (2010). *Sustainable Consumption and some contemporary European issues*. Author.

National Universities Commission (NUC). (2016). *Universities in Nigeria*. Retrieved 20th May 2016, from http://nuc.edu.ng/

Ndanu, A., Esilaba, A., Kibunja, C., Kamoni, P., & Okoth, P. (2012). A Situational Analysis of Opportunities and Challenges of Using ISFM Information and Knowledge in Kenya Integrated Soil Fertility Management in Africa Conference Information. *Program with Abstracts (Geological Association of Canada)*, 86.

Nemecek, T., Jungbluth, N., Canals, L. M., & Schenck, R. (2016). Environmental impacts of food consumption and nutrition: where are we and what is next? *Int J Life Cycle Assess, 21*, 607-620.

Nesheim, M., Oria, M., & Yih, P. (Eds.). (2015). *A Framework for Assessing Effects of the Food System*. Washington, DC: The National Academies Press.

Ng, M., Fleming, T., Robinson, M., Thomson, B., Graetz, N., Margono, C., … Abera, S.F. (2014). Global, regional, and national prevalence of overweight and obesity in children and adults during 1980–2013: a systematic analysis for the Global Burden of Disease Study 2013. *The Lancet*.

Ng, R. (2015). *Making Sustainable Food Choices: Influencing Consumer Behaviour for Food Sustainability in Oxford*. OxPolicy.

Ngenoh, E. K., Kirui, L. K., Mutai, B. C., Maina, M., & Koech, W. (2015). Economic Determinants Of The Performance Of Public Irrigation Schemes In Kenya. *Journal of Development and Agricultural Economics*, *7*(10), 344–352. doi:10.5897/JDAE2015.0676

Ngwira, A. R., Kabambe, V. H., Kambauwa, G., Mhango, W. G., Mwale, C. D., Chimphero, L., & Mapfumo, P et al.. (2012). Scaling Out Best Fit Legume Technologies for Soil Fertility Enhancement among Smallholder Farmers in Malawi. *African Journal of Agricultural Research*, *7*(6), 918–928. doi:10.5897/AJAR11.760

Nhundu, K., Gwata, C., & Mushunje, A. (2010). Impacts Of Zimbabwe European Union Micro-Project Programme (Zim/Eu Mpp) In Funding Smallholder Irrigation Projects On Food Security And Income Levels: A Case Study Of Mopane Irrigation Scheme In Zvishavane, Midlands Province, Zimbabwe. *African Journal of Agricultural Research*, *5*(14), 1759–1771.

Niger State Agricultural Development Programme. (1999). *Annual Evaluation Report*. Author.

Njoko, S. L. (2014). *Smallholder Farmers' Willingness And Ability To Pay For Improved Irrigation: A Case Of Msinga Local Municipality, Kwazulu-Natal Province* (Master thesis). University Of Kwazulu-Natal, South Africa.

Nkambule, B. L., & Dlamini, C. S. (2013). Towards Sustainable Smallholder Irrigation Development Projects: A Case Study Of The Maplotini Irrigation Scheme, Swaziland. *Journal of Agricultural Extension and Rural Development*, *5*(10), 216–224.

Nkhata, R. (2014). *Does Irrigation Have An Impact On Food Security And Poverty: Evidence From Bwanje Valley Irrigation Scheme In Malawi*. Ifpri Working Paper 04.

NOAN. (2012). *Nigerian Organic Agricultural Network*. Retrieved from www.noan.com

North, D. C. (2005). *Understanding the Process of Economic Change*. Princeton University Press. doi:10.1515/9781400829484

Nowak, P. J. (1992). Why Farmers Adopt Production Technology. *Journal of Soil and Water Conservation*, *47*(1), 14–16.

NRC (National Research Council) & IOM (Institute of Medicine). (2015). A framework for assessing effects of the food system. Washington, DC: The National Academies Press.

Nugent, R. (2011). *Bringing agriculture to the table: How agriculture can play a role in preventing chronic disease*. The Chicago Council on Global Affairs.

Nwaiwu, I. U., Ohajianju, D. O., Lemchi, J. I., Ibekwe, U. C., Nwosu, F. O., Ben-Chendo, N. G., & Kadiri, F. A. et al. (2010). Economics of Organic Manure Use by Food Crop Farmers in Ecologically Vulnerable Area of Imo State, Nigeria. *Researcher*, *2*(11), 56–61.

Nwanze, K. F., Mohapatra, S., Kormawa, P., Keya, S., & Bruce-Oliver, S. (2006). Rice Development in Sub-Saharan Africa. *Journal of the Science of Food and Agriculture*, *86*(5), 675–677. doi:10.1002/jsfa.2415

Nyarko, Y. (2017). Transforming Rural Africa— Economics, Technology and Governance. *Journal of African Development*, *19*(1), 160–167.

O'Connor, J. (1998). *Natural Causes: Essays in Ecological Marxism*. Guiford Press.

Obayelu, E. A. (2015). Transformation from Subsistence to Commercial Agriculture in Nigeria: The Effects of Large-Scale Land Acquisition on Smallholder Farmers. In E. Osabuohien (Ed.), *Handbook of Research on In-Country Determinants and Implications of Foreign Land Acquisitions* (pp. 409–431). Hershey, PA: Business Science Reference. doi:10.4018/978-1-4666-7405-9.ch021

OECD. (2008). *Environmental Performance of Agriculture in OECD Countries Since 1990*. Paris: OECD Publishing.

OECD. (2011). *Evaluation of Agricultural Policy Reforms in the European Union*. Paris: OECD.

Oenema, O., Oudendag, D., & Velthof, G. L. (2007). Nutrient losses from manure management in the European Union. *Livestock Science*, *112*(3), 261–272. doi:10.1016/j.livsci.2007.09.007

Ogada, M. J., Muchai, D., Mwabu, G., & Mathenge, M. (2014). Technical Efficiency Of Kenya's Smallholder Food Crop Farmers: Do Environmental Factors Matter? *Environment, Development and Sustainability*, *16*(5), 1065–1076. doi:10.1007/s10668-014-9513-1

Ogun State Government (OSG). (2016). *Ogun State, Nigeria, Investors' Forum*. Retrieved 30th June 2016, from http://ogunstateinvestorsforum.ng/agriculture/

Ogundare, K., Agele, S., & Aiyelari, P. (2012). Organic Amendment of an Ultisol: Effects on Soil Properties, Growth, and Yield of Maize in Southern Guinea Savanna Zone of Nigeria. *International Journal of Recycling of Organic Waste in Agriculture*, *1*(1), 11. doi:10.1186/2251-7715-1-11

Ogunsumi, L. O., Ajayi, A., Amire, C. M., & Williams, S. (2013). Sustainability of Agricultural Transformation Agenda. The place of Rice Farmers in Ogun State Nigeria. *Research on Humanities and Social Sciences*, *3*(13), 66–78.

Okeh, C., Onwosi, C. O., & Fredrick, J. O. C. (2014). Biogas Production from Rice Husks Generated From Various Rice Mills in Ebonyi State, Nigeria. *Renewable Energy*, *62*, 204–208. doi:10.1016/j.renene.2013.07.006

Okodua, H. (2017). *Assessing the Impact of Rice Sector Policy Reforms on Income Mobility of Rural Households in Nigeria*. AGRODEP Working Paper 0039, February.

Okoruwa, V., Jabbar, M. A., & Akinwumi, J. A. (1996). Crop-Livestock Competition in the West African Derived Savanna: Application of a Multi-objective Programming mModel. *Agricultural Systems*, *52*(4), 439–453. doi:10.1016/S0308-521X(96)00005-4

Okunmadewa, F. (2003). Risk and Vulnerability in Agriculture: concept and Context. *Staff seminar paper in the Department of Agricultural Economics*. University of Ibadan.

Oladimeji, Y. U., & Abdulsalam, Z. (2014). An Economic Analysis Of Dry Season Irrigated Farming In Asa River, Kwara State, Nigeria: Implications For Poverty Reduction. *Journalof Sustainable Development In Africa*, *16*(7), 1–15.

Olayide, S. O., & Heady, E. O. (1982). *Introduction to agricultural production economics*. Ibadan, Nigeria: Ibadan University Press, University of Ibadan.

Olayinka, A., & Ailenubhi, V. (2001). Influence of Combined Application of Cow Dung and Inorganic Nitrogen on Microbial Respiration and Nitrogen Transformation in an Alfisol. *Nigerian Journal of Soil Research*, *2*, 15–20.

Olayiwola, K. W., Osabuohien, E. S., Okodua, H., & Ola-David, O. (2015). Economic Integration, Trade Facilitation and Agricultural Exports Performance in ECOWAS Sub-Region. In M. Ncube, I. Faye, & A. Verdier-Chouchane (Eds.), *Regional Integration and Trade in Africa* (pp. 31–46). New York: Palgrave Macmillan. doi:10.1057/9781137462053_3

Ollenburger, M. H. (2012). *Modelling Integrated Soil Fertility Management Options in Malawi under Variable Soil and Climate Conditions* (MSc Thesis). Michigan State University.

Omisore, J. K. (2013). Response of Maize (Zea mays l.) to Organic and Inorganic Nitrogen Sources in the Southern Guinea Savanna of Nigeria (A Ph.D. Final Project Report). Department of Agronomy, University of Ilorin.

Omondi, S. O., Mbogoh, S. G., & Munei, K. (2014). An Evaluation Of The Factors Influencing Farmers'willingness To Pay (Wtp) For Irrigation Water: The Case Of Ahero Irrigation Scheme In Kenya. *International Journal Of Science. Environmental Technology*, *3*(5), 1778–1789.

Onabanjo, A. A. (2011). *Lessons and Experiences of Participatory Policy Making Process: The Case Study of Ogun State Rice Policy. In Abeokuta: Planning, Research and Statistics*. Ogun State Ministry of Agriculture.

Oni, O. A. (2010). *Food Poverty and Livelihoods Issues in Rural Nigeria*. Academic Press.

Onwueme, I. C. (1979). *Cassell's Tropical Agriculture Series*. Southampton, UK: The Camelot Press ltd.

Orapin, S. (1996). People's participation in community development. *TDRI Quarterly Review*, *11*(3), 19–25.

Oriola, E. O. (2009). Breaking The Vicious Cycle In Irrigation Farming System For Sustainable Food Security In Nigeria. *African Research Review*, *1*(3), 234–245.

Osabohien, R., Osabuohien, E., & Urhie, E. (2017). Food Security, Institutional Framework and Technology: Examining the Nexus in Nigeria Using ARDL Approach. *Journal of Current Nutrition and Food Science*, *13*(4), 1–10.

Osabuohien, E. (2016, June 20). Economy without Buffer must Experience Shocks. *Daily Times*, p. B8.

Osabuohien, E. (2014). Large-scale Agricultural Land Investments and Local Institutions in Africa: The Nigerian Case. *Land Use Policy*, *39*, 155–165. doi:10.1016/j.landusepol.2014.02.019

Osabuohien, E., & Efobi, U. R. (2013). Africa's Money in Africa. *The South African Journal of Economics*, *81*(2), 292–306. doi:10.1111/saje.12012

Osabuohien, E., Efobi, U. R., Gitau, C. M., & Bruentrup, M. (2015). Agents and Implications of Foreign Land Deals in East African Community: The Case of Uganda. In E. Osabuohien (Ed.), *Handbook of Research on In-Country Determinants and Implications of Foreign Land Acquisitions* (pp. 263–286). Hershey, PA: Business Science Reference. doi:10.4018/978-1-4666-7405-9.ch013

Otaha I. J. (2013). Food Insecurity in Nigeria: Way Forward. *Journal of African Research Review*, *7*(4), 26-35.

Ozor, N., & Nwankwo, N. (2008). The role of local leaders in community development programmes in Ideato Local Government Area of Imo State: Implication for extension policy. *Journal of Agricultural Extension*, *12*(2), 2.

Paalberg, R. (2013). *Food Politics: What Everyone Needs to Know*. Oxford University Press.

Paarlberg, R. (2008). *Starved for Science: How Biotechnology is Being Kept Out of Africa*. Harvard Univ. Press. doi:10.4159/9780674041745

Paehlke, R. (2003). Egalitarian Perspectives on Sustainability. In UNESCO's Encyclopaedia of Life Support Systems (vol. 36). UNESCO.

Pal, C. A., Gachengo, C. N., Delve, R. J., Cadisch, G., & Giller, K. E. (2001). Organic Inputs for Soil Fertility Management in Tropical Agro-Ecosystems. *Agriculture, Ecosystems & Environment*, *83*, 27–42. doi:10.1016/S0167-8809(00)00267-X

Panitch, L., & Leys, C. (Eds.). (2007). Coming to Terms with Nature: Socialist Register 2007. Fernwood.

Parra, J. C., & Wodon, Q. (2009). *SimSIP-SAM: A Tools for the Analysis of Input-Output Tables and Social Accounting Matrices*. World Bank.

Pearson, D., Friel, S., & Lawrence, M. (2014). Building environmentally sustainable food systems on informed citizen choices: Evidence from Australia. *Biological Agriculture and Horticulture*, *30*(3), 183–197. doi:10.1080/01448765.2014.890542

Peet, E., & Watts, W. (1996). Liberation Ecology: Development, Sustainability and Environment in an Age of Market Triumphalism. In R. Peet & M. Watts (Eds.), *Liberation Ecologies:Environment,Development,Social movements* (pp. 1–45). London: Routledge. doi:10.4324/9780203286784

Pender, J., Nkonya, E., Jagger, P., Sserunkuuma, D., & Ssali, H. (2004). Strategies to Increase Agricultural Productivity and Reduce Land Degradation: Evidence from Uganda. *Agricultural Economics*, *31*(2-3), 181–195. doi:10.1111/j.1574-0862.2004.tb00256.x

Pereira, H., & Marques, R. C. (2017). An Analytical Review Of Irrigation Efficiency Measured Using Deterministic And Stochastic Models. *Agricultural Water Management*, *184*, 28–35. doi:10.1016/j.agwat.2016.12.019

Peter, G. (2011). The Impact Of Small-Scale Irrigation Schemes On Household Food Security In Swaziland. *Journal of Sustainable Development in Africa*, *13*(6), 102–117.

Pinstrup-Andersen, P. (2010). *The advantages of genetic engineering in agriculture include increased food production and reduced hunger - benefits for hungry and malnourished in developing countries outweigh disadvantages*. Retrieved from http://www.monsanto.com/biotechGM/asp/experts.asp?id=P

Place, F., Barrett, C. B., Freeman, H. A., Ramisch, J. J., & Vanlauwe, B. (2003). Prospects for Integrated Soil Fertility Management Using Organic and Inorganic Inputs: Evidence from Smallholder African Agricultural Systems. *Food Policy*, *28*(4), 365–378. doi:10.1016/j.foodpol.2003.08.009

Pretty, J. N., & Noble, D. (2006). Resource-Conserving Agriculture Increases Yields in Developing Countries. *Environment Science and Technology.*

Pretty, J. N., Morison, J. I., & Hine, R. E. (2003). Reducing food poverty by increasing agricultural sustainability in developing countries. *Agriculture, Ecosystems & Environment*, *95*(1), 217–234. doi:10.1016/S0167-8809(02)00087-7

Pretty, J., Noble, A., Bossio, D., Dixon, J., Hine, R. E., & Morison, I. L. (2013). Resource-conserving agriculture increases yields in Developing Countries. *Environmental Science & Technology*, *40*(4), 1114–1119. doi:10.1021/es051670d PMID:16572763

Qureshi, M. E., Ranjan, R., & Qureshi, S. E. (2010). An Empirical Assessment Of The Value Of Irrigation Water: The Case Study Of Murrumbidgee Catchment. *The Australian Journal of Agricultural and Resource Economics*, *54*(1), 99–118. doi:10.1111/j.1467-8489.2009.00476.x

Qureski, A. A. (1989). *Water management in Fadama areas*. Paper presented at the National workshop on Fadama and Irrigation Development in Nigeria, Bauchi, Nigeria.

Rahji, M. A. Y., Fakayode, S. B., & Sanni, G. K. (2008). A Multinomial Logit Analysis of Agricultural Credit Rationing by Deposit Money Banks in Nigeria: A Case Study of Oyo State, South West, Nigeria. *Economic and Financial Review Central Bank of Nigeria*, *46*(2), 55–82.

Rahman, M. M., Sarker, M. A. M., Hasina, A., & Kashem, M. A. (1998). Role of some Indigenous Substances as Organic Fertilizers on the Growth and Yield of Brinjal Plants. *Bangladesh Journal of Scientific and Industrial Research*, *33*(2), 275–281.

Randolph, T., Schelling, E., Grace, D., Nicholson, C. F., Leroy, J., Cole, D., & Ruel, M. et al. (2007). Role of livestock in human nutrition and health for poverty reduction in developing countries. *Journal of Animal Science*, *85*(11), 2788–2800. doi:10.2527/jas.2007-0467 PMID:17911229

Rani, R., & Singh, H. N. (2015). A Comparative Study Of Technical Efficiency Of Rice Production In Irrigated And Rainfed Environment Of Uttrakhand. *Indian Journal Of Hill Farming*, *28*(2), 102–106.

Rao, P. S. (1993). *Review Of Selected Literature On Indicators Of Irrigation Performance*. Colombo, Sri Lanka: International Irrigation Management Institute.

Ravallion, M. (2001). *On the urbanization of poverty*. Washington, DC: World Bank. doi:10.1596/1813-9450-2586

Reardon, T. (2015). The hidden middle: The quiet revolution in the midstream of agrifood value chains in developing countries. *Oxford Review of Economic Policy*, *31*(1), 45–63. doi:10.1093/oxrep/grv011

Rees, T. (1998). *Mainstreaming Equality in the European Union: Education, Training and Labour Market Policies*. Routledge.

Regassa, B. A. (2013). *Technical Efficiency Variation For Smallholder Irrigated Maize Producers: Stochastic Frontier Approach (A Case Study Of Tibila Surface Water Irrigation Scheme)* (Master thesis). Mekelle, Ethiopia.

Regmi, A., & Meade, B. (2013). Demand side drivers of global food security. *Global Food Security*, *2*(3), 166–171. doi:10.1016/j.gfs.2013.08.001

Reisch, L., Eberle U., & Lorek S. (2013). Sustainable food consumption: an overview of contemporary issues and policies. *Sustainability: Science, Practice, & Policy*, *9*(2), 7–25.

Reisch, L., Eberle, U., & Lorek, S. (2013). Sustainable food consumption: an overview of contemporary issues and policies. *Sustainability: Science, Practice, & Policy*, *9*(2), 7–25.

Riley, L., Kohlbacher, F., & Hofmeister, A. (2012). A cross-cultural analysis of pro-environmental consumer behavior among seniors. *Journal of Marketing Management*, *28*(3-4), 290–312. doi:10.1080/0267257X.2012.658841

Rischke, R., Kimenju, S. C., Klasen, S., & Qaim, M. (2015). Supermarkets and food consumption patterns: The case of small towns in Kenya. *Food Policy*, *52*, 9–21. doi:10.1016/j.foodpol.2015.02.001

Risku-Norja, H. (2011). *From environmental concerns towards sustainable food provisioning. Material flow and suitable food consumption scenario studies on sustainability of agri-food systems* (Doctoral dissertation). Faculty of agriculture and Forestry, Faculty of Agriculture and Forestry, University of Helsinki.

Robbins, P. (2004). *What is political ecology? In Political ecology: A critical introduction* (pp. 5–16). Malden, MA: Blackwell Publishing.

Roberts, J., & Grimes, P. (2002). World system Theory and the environment: Towards a new Synthesis. In Sociological Theory and the Environment: Classical Foundations, Contemporary Insights. New York: Rowman and Littlefield.

Rogers, E. M. (2003). *Diffusion of innovations* (5th ed.). New York: Free Press.

Rogers, K. (1997). *Ecological Security and Multinational Corporations*. Kluwer Academic Publishers.

Rola, A. C., & Quintana-Alejandrino Jocelyn, T. (1993). Technical Efficiency Of Philippine Rice Farmers In Irrigated, Rainfed Low-Land And Upland Environments: A Frontier Production Function Analysis. *Philippine Journal of Crop Science*, *18*(2), 59–69.

Rosin, C., Stock, P., & Campbell, H. (Eds.). (2011). Food Systems Failure The Global Food Crisis and the Future of Agriculture. London: Routledge.

Saha, S., Mina, B. L., Gopinath, K. A., Kundu, S., & Gupta, H. S. (2008). Organic Amendments Affect Biochemical Properties of a Sub-Temperate Soil of the Indian Himalayas. *Nutrient Cycling in Agroecosystems*, *80*(3), 233–242. doi:10.1007/s10705-007-9139-x

Saidu, A., Abayomi, Y. A. & Aduloju, M. O. (2012). Evaluation of Complementary Use of Organic and Inorganic Fertilizers on the Performance of Upland Rice (*Oryza sativa* l.). *IJABR, 2*(3), 487-491.

Schloster, H., Deboer, J., & Boersema, J. J. (2012). Can we cut on the meat of the dish? Constructing consumer-oriented pathway towards meat substitution. *Appetite*, *158*, 39–49.

Schnaiberg, D., & Gould, K. (1994). *Environment and Society The Enduring conflict*. New York: St Martin's.

Schor, J. (2005). Prices and quantities: Unsustainable consumption and the global economy. *Ecological Economics*, *55*(3), 309–320. doi:10.1016/j.ecolecon.2005.07.030

Schutjer, W. A. (1991). Rural Development and Extension. *Journal of Extension*, *29*(1), 1–2.

Sears, E. (1981). Transfer of alien genetic material to wheat. In L. T. Evans & W. J. Peacock (Eds.), *Wheat Science, Today and Tomorrow* (pp. 75–89). Cambridge, UK: Cambridge University Press.

Segun-Olasanmi, O. A., & Bamire, A. S. (2010). *Analysis of Costs and Returns to Maize-Cowpea Intercrop Production in Oyo state, Nigeria*. Poster presented at the Joint 3rd African Association of Agricultural Economists (AAAE) and 48th Agricultural Economists Association of South Africa (AEASA) Conference, Cape Town, South Africa.

Seitzinger, S. P., Mayorga, E., Bouwman, A. F., Kroeze, C., Beusen, A. H. W., Billen, G., & Harrison, J. A. et al. (2010). Global river nutrient export: A scenario analysis of past and future trends. *Global Biogeochemical Cycles*, *24*(4), n/a. doi:10.1029/2009GB003587

Seligman, H. K., Laraia, B. A., & Kushel, M. B. (2009). Food Insecurity Is Associated with Chronic Disease among Low-Income NHANES Participants. *Journal of Nutrition and Diseases, 5*, 304-310.

Senaratna Sellamuttu, S., Aida, T., Kasahara, R., Sawada, Y., & Wijerathna, D. (2014). How Access To Irrigation Influences Poverty And Livelihoods: A Case Study From Sri Lanka. *The Journal of Development Studies*, *50*(5), 748–768. doi:10.1080/00220388.2013.841887

Seyfang, G., & Smith, A. (2007). Grassroots innovations for sustainable development: Towards a new research and policy agenda. *Environmental Politics*, *16*(4), 584–603. doi:10.1080/09644010701419121

Shantha, A., Ali, B. G. H., & Bandara, R. A. G. (2012). Efficiency And Managerial Ability Of Paddy Farming Under Minor Irrigation Conditions: A Frontier Production Function Approach. *Journal of Agricultural Sciences*, *7*(3), 145–158. doi:10.4038/jas.v7i3.4878

Shapouri, H., Gallagher, P., Nefstead, W., Schwartz, R., Noe, R., & Conway, R. (2010). *Energy balance for the corn–ethanol industry. Agricultural Economic Report 846. Office of the Chief Economist*. Washington, DC: U.S. Department of Agriculture.

Sharm, K. R., Pradha, N. C., & Leung, P. (2001). Stochastic Frontier Approach To Measuring Irrigation Performance: An Application To Rice Production Under The Two Systems In The Tarai Of Nepal. *Water Resources Research*, *37*(7), 2009–2018. doi:10.1029/2000WR900407

Shetty P. (2003). Malnutrition and Undernutrition. Science. *Medicine, 31*(4), 18-22. doi:10.1383/medc.31.4.18.27958

Shopouri, S., Rosen, M., Peters, S., Tandon, F., Mancino, G. L., & Bai, J. (2011). *International Food Security Assessment, 2011-12. GFA-22*. United States Department of Agriculture-Economic Research Service. Retrieved from http://www.ersusda.gov/Publications/GFA22/GFA22.pdf

Sinkaye, T. (2005). Agricultural Extension Participatory Methodologies and approaches in Agricultural Extension in Nigeria. AESON.

Sinyolo, S., Mudhara, M., & Wale, E. (2014). The Impact Of Smallholder Irrigation On Household Welfare: The Case Of Tugela Ferry Irrigation Scheme In Kwazulu-Natal, South Africa. *Water S.A.*, *40*(1), 145–156. doi:10.4314/wsa.v40i1.18

Sisay, B., & Fekadu, B. (2013). Small-Scale Irrigation And Household Income Linkage: Evidence From Deder District, Ethiopia. *African Journal of Agricultural Research*, *8*(34), 4441–4451. doi:10.5897/AJAR12.1793

Sisay, D., Jema, H., Degye, G., & Abdi Khalil, E. (2015). Technical, Allocative, And Economic Efficiency Among Smallholder Maize Farmers In Southwestern Ethiopia: Parametric Approach. *Journal of Development and Agricultural Economics*, *7*(8), 282–291. doi:10.5897/JDAE2015.0652

Small, L. E., & Svendsen, M. (1990). A Framework For Assessing Irrigation Performance. *Irrigation and Drainage Systems*, *4*(4), 283–312. doi:10.1007/BF01103710

Smith, L. E. D. (2004). Assessment Of The Contribution Of Irrigation To Poverty Reduction And Sustainable Livelihoods. *International Journal of Water Resources Development*, *20*(2), 243–257. doi:10.1080/0790062042000206084

Snapp, S., Jayne, T. S., Mhango, W., Benson, T., & Ricker-Gilber, J. (2014). *Maize Yield Response to Nitrogen in Malawi's Smallholder Production Systems.* IFPRI Working Paper 09. Malawi Strategy Support Programme.

Solomon, S., & Ketema, M. (2015). Impact Of Irrigation Technologies On Rural Households' Poverty Status: The Case Of Fogera District, North-Western Ethiopia. *AGRIS On-Line Papers in Economics and Informatics*, *7*(1), 59–67.

Sommer, R., Bossio, D., Dester, L., Dimes, J., Kihara, J., Koala, S., & Winowiecki, L. et al. (2013). *Profitable and Sustainable Nutrient Management Systems for East and Southern African Smallholder Farming Systems-Challenges and Opportunities: A Synthesis of the Southern and Eastern African Situation in Terms of the Past Experiences, Present and Future Opportunities in Promoting Nutrient Use in Africa.* CYMMYT.

Sonigo, D. P., Bain, M. J., Kong, M. M. A., Fedrigo, M. D., Withana, M. S., Watkin, M. E., & Dresner, D. S. et al. (2012). *Policies to encourage sustainable consumption Full.* France Europian Commsion.

Son, T. T. N., Thu, V., Man, L. H., Kobayashi, H., & Yamada, R. (2004). Effect of Long Term Application of Organic and Bio-Fertilizer on Soil Fertility Under Rice-Soyabeans-Rice Cropping Systems. *Omonrice*, *12*, 45–51.

Spaargaren, G. (1997). *The ecological modernisation of production and consumption: in environmental. sociology*. Wageningen, The Netherlands: Landbouw University Wageningen.

Speelman, S., Frija, A., Perret, S., D'haese, M., Farolfi, S., & D'haese, L. (2011). Variability In Smallholders' Irrigation Water Values: Study In North-West Province, South Africa. *Irrigation and Drainage*, *60*(1), 11–19. doi:10.1002/ird.539

Statistics South Africa. (2012). *Energy Accounts for South Africa 2002/2009*. Statistics South Africa.

Statistics South Africa. (2012). *South Africa Quarterly Labour Force Survey*. Pretoria: Statistics South Africa.

Steger, M. (2002). *Globalism: The new Market ideology*. New York: Rowman and Littlefield Publishers.

Stein, H. (2003). Rethinking African Development. In Rethinking Development Economics. Anthem Press.

Stern, P. (1997). *Environmentally Significant Consumption*. National Academy Press.

Stewart, R., Korth, M., Langer, L., Rafferty, S., & Silva, N. R. D. (2013). What are the impacts of urban agriculture programs on food security in low and middle-income countries? *Environmental Evidence*, *2*(1), 7. doi:10.1186/2047-2382-2-7

Stiglitz, J. (2010). *Free Fall: America, Free Markets and the Sinking of the World Economy*. New York: W.W. Norton.

Straughan, R., & Roberts, J. (1999). Environmental segmentation alternatives: A look at green consumer behavior in the new millennium. *Journal of Consumer Marketing*, *16*(6), 558–575. doi:10.1108/07363769910297506

Sutton, M.A., Bleeker, A., & Howard, C.M. (2013). *Our nutrient world: the challenge to produce more food and energy with less pollution*. Edinburgh, UK: NERC/Centre for Ecology & Hydrology.

Svendsen, M., Ewing, M., & Msangi, S. (2009). *Measuring Irrigation Performance In Africa.* International Food Policy Research Institute (IFPRI).

Swanson, B., & Rajalahti, R. (2010). *Strengthening Agricultural Extension and Advisory Systems: Procedures for Assessing, Transforming, and Evaluating Extension Systems.* ARD Discussion Paper 20. Washington, DC: World Bank

Swedish Environmental Protection Agency (SEPA). (2010). *The Climate Impact of Swedish Consumption.* Author.

Syaukat, Y., Arifah, F. N., & Minha, F. (2014). Economic Value And Service Fee Of Irrigation Water In The Districts Borgor And Kudus: Indonesia. *J. Issaas*, *20*(2), 157–172.

Tang, Z., Nan, Z., & Liu, J. (2013). The Willingness To Pay For Irrigation Water: A Case Study In Northwest China. *Global NEST Journal*, *15*(1), 76–84.

Tchale, H., & Saur. (2007). Soil Fertility Management and Maize Productivity In Malawi: *Curvature Correct Efficiency Modeling And Simulation. AAAE Conference Proceedings*, 151-163.

Temme, E. H., Bakker, H. M., Seves, M., Verkaik-Kloosterman, J., Dekkers, A. L., Raaij, J. M. V., & Ocké, M. C. (2015). How may a shift towards a more sustainable food consumption pattern affect nutrient intakes of Dutch children? *Public Health Nutrition*, *18*(13), 2468–2478. doi:10.1017/S1368980015002426 PMID:26344035

Tendall, D. M., Joerin, J., Kopainsky, B., Edwards, P., Shreck, A., Le, Q. B., & Six, J. et al. (2015). Food system resilience: Defining the concept. *Global Food Security*, *6*, 17–23. doi:10.1016/j.gfs.2015.08.001

Thibbotuwawa, M., Mugera, A., & White, B. (2013). *Production Efficiency And Technology Gap In Irrigated And Rain-Fed Rice Farming Systems In Sri Lanka: Non Parametric Approach.* In *57th Aares Annual Conference*, Sydney, Australia.

Thøgersen, J. (2010). Country differences in sustainable consumption: The case of organic food. *Journal of Macromarketing*, *30*(2), 171–185. doi:10.1177/0276146710361926

Tittonell, P. M., Corbeels, M. T., van Wijk, M. T., Vanlauwe, B., & Giller, K. E. (2008). Combining Organic and Mineral Fertilizers for Intergrated Soil Fertility management in Smallholder Farming Systems of Kenya: Exploration Using the Crop-Soil Model FIELD. *Agronomy Journal*, *100*(5), 1511. doi:10.2134/agronj2007.0355

Tiu, J. (2017). *Nestlé, Unilever, P&G Among Worst Offenders for Plastic Pollution in Philippines Beach Audit.* Available at https://www.ecowatch.com/plastic-pollution-philippines-beach-audit-2488280848.html

Tiwari, D. N. (1998). *Detemining Economic Value Of Irrigation Water: Comparison Of Willingness To Pay And Indirect Valuation Approaches As A Measure Of Sustainable Resource Use.* Cserge Working Paper Gec 98-05, Centre For Social And Economic Research On The Global Environment University College London And University Of East Anglia.

Tschirley, D., Reardon, T., Dolislager, M., & Snyder, J. (2015a). The rise of a middle class in east and southern Africa: Implications for food system transformation. *Journal of International Development*, *27*(5), 628–646. doi:10.1002/jid.3107

Tschirley, D., Snyder, J., Dolislager, M., & Snyder, J. (2015b). Africa's unfolding diet transformation: Implications for agrifood system employment. *Journal of Agribusiness in Developing and Emerging Economies*, *5*(2), 1–48. doi:10.1108/JADEE-01-2015-0003

Tweeten, L. (1999). The Economics of Global Food Security. *Review of Agricultural Economics*, *21*(2), 473–488.

Uduma, B. U., Samson, O. A., & Mure, U. A. (2016). Irrigation Potentials and rice self-sufficiency in Nigeria: A review. *African Journal of Agricultural Research*, *11*(5), 298–309. doi:10.5897/AJAR2015.10284

Ullah, M. S., Islam, M. S., Islam, M. A., & Haque, T. (2008). Effects of Organic Manures and Chemical Fertilizers on the Yield of Brinjal and Soil Properties. *Journal Bangladesh Agricultural University*, *6*(2), 271–276.

Ulsido, M. D., & Alemu, E. (2014). Irrigation Water Management In Small Scale Irrigation Schemes: The Case Of The Ethiopian Rift Valley Lake Basin. *Environmental Research, Engineering and Management*, *67*(1), 5–15. doi:10.5755/j01.erem.67.1.6240

Umeh, E. D. N., Joshi, R. C., & Ukwungwu, M. N. (1992). Biology, Status and Management of Rice Insect Pests in Nigeria. *Crop Protection (Guildford, Surrey)*, *11*(5I5), 408–413. doi:10.1016/0261 2194(92)90022-W

UN. (2002). *Rights-Based Approach to Development Programming: Training Manual*. Retrieved from http://www.un.org.ph/publications/RBAManual.pdf

UN. (2015). *The Zero Hunger Challenge – Advisory notes for action*. The UN Secretary General's High level Task Force on Global Food and Nutrition Security. Retrieved from http://www.un.org/en/issues/food/taskforce/pdf/HLTF%20 %20ZHC%20Advisory%20Notes.pdf

UN. (2017). *What is a human rights-based approach? Human Rights-Based Approach Portal*. Retrieved from http://hrbaportal.org/faq/what-is-a-human-rights-based-approach

UNCTAD. (2011). Sustainable agriculture and food security in LDCs. *Least Developed Countries Series, 20*, 1-2.

UN-DESA. (2015). *World Population Prospects: The 2015 Revision, Key Findings and Advance Tables*. Available online at http://esa.un.org/unpd/wpp/publications/files/key_findings_wpp_2015.pdf

UNEP. (2012). *The critical role of global food consumption patterns in achieving Sustainable Consumption*. Nairobi, Kenya: UNEP.

UNEP. (2012). *The Critical Role Of Global Food Consumption Patterns In Achieving Sustainable Food Systems And Food For All*. A UNEP Discussion Paper.

UNEP. (2016). *Food Systems and Natural Resources.* A Report of the Working Group on Food Systems of the International Resource Panel.

UNEP/MAP. (2005). *Mediterranean strategy for sustainable development: a framework for environmental sustainability and shared prosperity. Tenth Meeting of the Mediterranean Commission on Sustainable Development (MCSD)*, Athens, Greece.

United Nations Commission for Environment and Development (UNCED). (1992). *Agenda 21.* Oxford Press.

United Nations Committee on Economic, Social, and Cultural Rights. (1999). *The right to adequate food.* Geneva: United Nations.

USDA (United States Department of Agriculture). (2014). *Maize production in Nigeria (2004-2014).* USDA.

USGS. (2013). *Mineral Commodity Summaries: (various elements).* Washington, DC: U.S. Geological Survey.

Van Den Berg, M., & Ruben, R. (2006). Small-Scale Irrigation And Income Distribution In Ethiopia. *The Journal of Development Studies*, *42*(5), 868–880. doi:10.1080/00220380600742142

Van Halsema, G. E., Keddi Lencha, B., Assefa, M., Hengsdijk, H., & Wesseler, J. (2011). Performance Assessment Of Smallholder Irrigation In The Central Rift Valley Of Ethiopia. *Irrigation and Drainage*, *60*(5), 622–634. doi:10.1002/ird.613

Vanlauwe, B., Chianu, J., Giller, K. E., Merckx, R., Mokwunye, U., Pypers, P., . . . Sanginga, N. (2010). *Integrated Soil Fertility Management: Operational Definition and Consequences for Implementation and Dissemination.* 19th World Congress of Soil Science, Soil Solutions for a Changing World, Brisbane, Australia.

Vanlauwe, B. (2015). *Sustainable Agricultural Resources Management: Unlocking Land Potential for Productivity and Resilience. Background Paper.* Nairobi, Kenya: United Nations Economic Commission for Africa.

Vanlauwe, B., Coyne, D., Gockowski, J., Hauser, S., Huising, J., Masso, C., & Asten, P. V. et al. (2014). Sustainable Intensification and the African Smallholder Farmer. *Current Opinion in Environmental Sustainability*, *8*, 15–22. doi:10.1016/j.cosust.2014.06.001

Vanlauwe, B., Diels, J., Aihou, K., Iwuafor, E. N. O., Lyasse, O., Sanginga, N., & Merckx, R. (2002). Direct Interactions between Nitrogen Fertilizer and Organic Matter: Evidence from Trials with 15 Nitrogen –Labelled Fertilizer. In B. Vanlauwe, J. Diels, N. Sanginga, & R. Merckx (Eds.), *Integrated Plant Nutrient Management in Sub-Saharan Africa* (pp. 173–184). Wallingford, UK: CAB International.

Vanlauwe, B., & Giller, K. E. (2006). Popular Myths around Soil Fertility Management in Sub-Saharan Africa. *Agriculture, Ecosystems & Environment*, *116*(1-2), 34–46. doi:10.1016/j.agee.2006.03.016

Vanlauwe, B., Wendt, J. W., & Diels, J. (2001). Combined Application of Organic Matter and Fertilizer. *Journal - American Society of Agronomy, 50*, 247.

Verburg, P.H., Mertz, O., Erb, K.H., Haberl, H., & Wu, W. (2013). Land system change and food security: towards multi-scale land system solutions. *Current Opinion in Environmental Sustainability, 5*, 494-502. DOI: .cosust.2013.07.00310.1016/j

Vermeir, I., & Verbeke, W. (2006). Sustainable food consumption: Exploring the consumer "attitude–behavioral intention" gap. *Journal of Agricultural & Environmental Ethics, 19*(2), 169–194. doi:10.1007/s10806-005-5485-3

Vermeulen, S.J., Campbell, B.M., & Ingram, J.S.I. (2012). Climate Change and Food Systems. *Annual Review of Environment and Resources, 37*, 195-222.

Viinamäki. (2004). A theory of coordination and its implication on EU structural policy: A comparative study of the challenges for coordination in structural funds in Finland, Ireland and Sweden. Acta Wasaensia 132. Administrative Science 9. Vaasa: University of Vaasa/Publication Unit.

Vlaemincka, P., Jiangb, T., & Vrankena, L. (2014). Food labeling and eco-friendly consumption: Experimental evidence from a Belgian supermarket. *Ecological Economics, 108*.

Vogit, M. (2014). *Meanings attached to food and sustainable food consumption* (Master's thesis). Uppsala University, Department of Earth Sciences.

Von Malititz, G. P., & Brent, A. (2008). Assessing the biofuel options for Southern Africa. In *CSIR (Council for Scientific and Industrial Research), Science real and relevant: 2nd CSIR Biennial Conference*. Pretoria, South Africa: CSIR.

von Maltitz, G. P., Nickless, A., & Blanchard, R. 2010. Maintaining biodiversity during biofuel development. In Projects in developing countries: A framework for policy evaluation. Newcastle upon Tyne, UK: Newcastle University.

von Maltitz, G., & Stafford, W. (2011). *Assessing opportunities and constraints for biofuel development in sub-Saharan Africa*. Bogor: Center for International Forestry Research.

Vonbraun, J. (2007). The world food situation: New riving forces and required actions. *IFPRI's Bi-Annual Overview of the World Food Situation presented to the CGIAR Annual General Meeting*.

Vonmeyer-Höfer, M., Juarez Tijerino, A. & Spiller, A. (2015). *Sustainable food consumption in China and India.* Global Food Discussion Papers, No. 60.

Wakili, A. M., & Isa, A. H. M. (2015). *Technical Efficiency Of Small Scale Rice Production In Adamawa State, Nigeria*. Presented At The Chemical, Food And Environment Engineering (Iccfee'15), Dubai, UAE.

Wallace, M. B., & Knausenberger, W. I. (1997). *Inorganic Fertilizer Use in Africa: Environmental and Economic Dimensions. Winrock International Institute for Agricultural Development and Division of Productive Sector Growth and the Environment Office of Sustainable Development Bureau for Africa U.S. Agency for International Development with the Environmental and Natural Resources Policy and Training (EPAT)* Project.

Wallerstein, I. (1976). World System Analysis:the second Phase'. *RE:view*, *13*(2).

Wang, S., Just, D., & Pinstrup-Andersen, P. (2006). *Tarnishing Silver Bullets: Bt technology adoption, bounded rationality and the outbreak of secondary pest infestations in China.* Paper presented at American Agricultural Economics Association annual meeting, Long Beach, CA.

Wapner, P. (1996). *Environmental Activism and World Civic Politics.* State University of New York Press.

Watto, M. A., & Mugera, A. W. (2014). Measuring Production And Irrigation Efficiencies Of Rice Farms: Evidence From The Punjab Province, Pakistan: Irrigation Efficiencies Of Rice Farms. *Asian Economic Journal*, *28*(3), 301–322. doi:10.1111/asej.12038

Weinstein, B. (2008). Developing Inequality. *The American Historical Review*, *113*(1), 1–18. doi:10.1086/ahr.113.1.1

Westhoek, H., Ingram, J. S. I., Van Berkum, S., Özay, L., & Hajer, M. (2016). *Food Systems and Natural Resources. A Report of the Working Group on Food Systems of the International Resource Panel.* UNEP Nairobi.

WFP. (2002). Vulnerability Analysis and Mapping: Standard Analytical Framework. Rome: Author.

WFP. (2013). *Food and Nutrition Security in Democratic People's Republic of Korea. Vam Food security Analysis*. Retrieved from https://www.wfp.org/sites/.../files/DPRK_draft_FINAL%20 1%20October%202013.pdf

WFP. (2017). *Global Nutrition Report*. Retrieved from https://www.wfp.org/stories/10-fact-about-food-and-nutrition-nigeria

WHO. (2013). *Undernutrition: The Basics. Generation Nutrition 6*. Retrieved from http://www.who.int/nutgrowthdb/summary_jme_2013.pdf

WHO. (2013). *UNICEF-WHO-The World Bank Joint Child Malnutrition Estimates 2013.* Available at: http://www.who.int/nutgrowthdb/summary_jme_2013.pdf?ua=1

WHO. (2015). *Global Health Observatory (GHO) data*. Available at: http://www.who.int/gho/ncd/risk_factors/obesity_text/en/

Wise, T. (2015). Two roads diverged in the food crisis: Global policy takes the one more travelled. *Canadian Food Studies*, *2*(2), 9–16.

Wolf, E. (1997). *Europe and the People Without History*. Berkeley, CA: University of California Press.

Woomer, P. L., & Swift, M. J. (1994). *The Biological Management of Tropical Soil Fertility*. Chichester, UK: John Wiley and Sons.

World Bank. (1991). *Urban policy and economic development: An agenda for the 1990s*. Washington, DC: World Bank. Retrieved from http://wwwds.worldbank.org/servlet/WDSContentServer/WDSP/IB/1999/0/10/000178830_9810 9014 2135/Rendered/PDF/multi_page.pdf

World Bank. (2001). *Poverty and Hunger: Issues and Options for Food Security in Developing Countries*. Washington, DC: World Bank.

World Bank. (2006). *The World Bank Annual Report*. Retrieved from http:// siteresources.worldbank.org/INTANNREP2K6/ Resources/2838485 1158333614345/AR06_final_LO_RES.pdf

World Bank. (2013). Nigeria: Fertilizer consumption (%fertilizer production). *Factfish*. Retrieved from www.factfish.com

World Bank. (2013). Turn Down The Heat: Climate Extremes, Regional Impacts, And The Case For Resilience (A Report For The World Bank By The Potsdam Institute For Climate Impact Research And Climate Analytics). Washington, DC: World Bank.

World Bank. (2013). World Databank Statistics. World Bank.

World Drop Engineering, Inc. (1993). *Final Report on the Study of Irrigation Potential of Shallow Aquifer in Fudama Areas*. Benue Agricultural and Rural Development Authority BNARDA.

World Food Summit. (1996). *Rome Declaration on World Food Security*. Author.

World Health Organisation. (2003a). Diet, nutrition and the prevention of chronic diseases. *World Health Organization Technical Report Series*, 916. PMID:12768890

World Health Organisation. (2003b). *Poverty and Health*. Paris: OECD.

World Population Prospects, (2016). *World Population Prospects: The 2015 Revision – Key Findings and Advance Tables*. United Nations Department of Economic and Social Affairs, Population Division.

Yan, A., Ngadi, M., & Kok, R. (2016). Physical properties of rice husk and bran briquettes under low pressure densification for rural applications. *Biomass and Bioenergy*, *84*(January), 22–30.

Yedra, H., Mesa-Jurado, M. A., López-Morales, C. A., & Castillo, M. M. (2016). Economic Valuation Of Irrigation Water In South-Eastern Mexico. *International Journal of Water Resources Development*, *32*(6), 931–943. doi:10.1080/07900627.2015.1133404

Yuya, B. T. (2014). Determinants Of Technical, Allocative And Economic Efficiencies Among Onion Producing Farmers In Kobo District, Amhara Region, Ethiopia. *Journal Of Agricultural Economics, Extension And Rural Development, 2*(5), 54–62.

Zakaria, H. (2014). Factors Affecting Farmers Willingness To Pay For Improved Irrigation Service. *International Journal Of Agricultural Economics And Extension, 2*(1), 68–76.

Zewdu, S. (2016). The Role Of Irrigation In Household Food Security In Upper Blue Nile Basin: The Case Of Jedeb Irrigation Scheme, Amhara Region, Ethiopia. *International Journal Of Agricultural Economics, 1*(4), 108–116.

Zeweld, W., Huylenbroeck, G. V., Hidgot, A., Chandrakanth, M. G., & Speelman, S. (2015). Adoption of Small-scale Irrigation and Its Livelihood Impacts in Northern Ethiopia. *Irrigation and Drainage*.

About the Contributors

Abiodun Obayelu started his educational career in University of Ilorin, Kwara State Nigeria where he obtained Bachelor Degree in Agriculture (B. Agric) in 1995. In year 2002, he obtained a Master's degree in Business Administration (M.B.A) and a Master of Science degree in Agricultural Economics from University of Ilorin and Ibadan, Nigeria respectively. Obayelu obtained his Ph.D degree in Agricultural Economics in August 2009 from University of Ibadan. Between 2005 and 2007, he was University of Ibadan Postgraduate School Teaching and Research Assistant and currently lectures in the Department of Agricultural Economics and Farm Management, Federal University of Agriculture Abeokuta (FUNAAB) Ogun State, Nigeria. His areas of interest include: Food and Consumer Welfare Economics, Development Economics, Impact Evaluation and Agricultural Innovations. He has published extensively in both international and local journals, and has to his credit over 60 publications. Dr. Obayelu has won a number of awards and international recognitions. These include PhD Doctoral Research Award by African Economic Research Consortium (AERC) and International Development Research Centre (IDRC) in August 2009, PhD Thesis Award by Postgraduate School University of Ibadan in 2009 and United States Association for Small Business and Entrepreneurship (USASBE) recognition for outstanding service in reviewing, evaluating, and providing scholarly insights for the 2012 national conference, New Orleans, Louisiana, January 2012. Dr Obayelu is a member of editorial board of several international journals among which are: African Journal of Economic and Management Studies (AJEMS), Emerald Emerging Markets Case Studies (EEMCS) and associate editor of the Journal of Agricultural Science published by the Canadian Centre for Science Education. He is also a member of both international and local associations including the African Association of Agricultural Economists (AAAE), Nigerian Association of Agricultural Economists (NAAE), Nigerian Association of Evaluators (NAE), African Growth and Development Policy (AGRODEP) Modeling Consortium Impact Evaluation Network (AIEN), Emerald Literati Network and African Economics Research Consortium Network.

* * *

Godwin A. Abu is Professor of Agricultural Economics and currently, the Director, Institute of Food Security, University of Agriculture Makurdi, Benue State, Nigeria. He is the Country Coordinator, FERTORAO Project: Sustainable Increase in Agricultural Productivity through Integrated Management of Organic Fertilizers in West Africa. A two-year World Bank funded project for ECOWAS involving Burkina Faso, Côte d'Ivoire, Ghana, Mali, Nigeria, and Senegal. Previously, he had consulted and developed the agro potential for Gombe state, conducted impact evaluation for the West Africa Agricultural Productivity Programme (WAAPP) in Nigeria. He conducted studies for British-Oxfam Abuja, USAID-MARKETS,DFID-PrOpCom, World Bank, African Development Bank (AfDB) and UNICEF. He is a Member, Steering Committee, Rice value Chain in Benue and Nasarawa States under the German Federal Ministry of Economic Cooperation and Development Special Initiative ONE WORLD-No hunger 'Global Programme on Green Innovation Centers for the Agriculture and Food sector'.

Kemisola Adenegan lectures in the Department of Agricultural Economics, University of Ibadan. Her area of specializations are Marketing, Gender Issues and Welfare Economics. She has published widely in both National and International Journals.

Clementina Ajayi is a Lecturer at the Federal University of Technology, Akure, Nigeria and a doctoral student of Agricultural Economics at the University of Ibadan, Nigeria. She is currently researching on Rural Households Vulnerability to Food Insecurity and Choice of Coping Mechanisms in Nigeria. Her areas of focus are Food, Welfare and Environmental Economics with a special interest in rural households, food security, poverty and gender. She is a chartered banker with experience in management, retail marketing and agricultural finance. A respecter of individuality, promoter of diversity and multiculturalism, she believes our cultural differences are strengths and not weaknesses. She has won awards, grants and commendations locally and internationally. Clementina is also member of many professional organizations including Agricultural and Applied Economics Association (AAEA), International Food and Agribussiness Management Association (IFAMA), Working Group of African Political Economy (WGAPE), Organization of Women in Science from Developing World (OWSD), Association for the Advancement of African Women's Economists (AAAWE) and Institute of Chartered Economists of Nigeria (ICEN).

Oluwole Matthew Akinnagbe is a lecturer in the Department of Agricultural Extension and Communication Technology, Federal University of Technology, Akure, Nigeria. He obtained his professional training at the University of Nigeria Nsukka, and holds Ph.D in Agricultural Extension with specialization in Agricultural Planning and Evaluation. He is member of many professional organizations including African Evaluation Association (AfrEA), Monitoring and Evaluation Nigeria group (MandENigeria), European Evaluation Society (EES), American Evaluation Association (AEA), Agricultural Policy Research Network (APRNet), Agricultural Extension Society of Nigeria (AESON), African Growth and Development Policy Modelling Consortium (AGRODEP) IFPRI, and International Society of Extension Education (INSEE). Dr. Akinnagbe has participated in many national and internationally funded research studies on impact evaluation, rural development, agricultural innovation systems, agro-technology generation and institutional reforms. He has good mastery of the computer and its applications to problem solving situations including data analysis using computer softwares. He possesses excellent communication skills and has many years of experience in development work, teaching and research at the University level. Dr. Akinnagbe has published widely in reputable national and international scholarly journals with over 40 publications to his credit. He won the Young Extension Professional award in recognition of the outstanding contribution in the field of extension teaching, research and field extension services in 2013 by International Society of Extension Education, India. Dr. Akinnagbe has 6 months' working experience as Senior Education Fellow with African Network for Agriculture, Agroforestry and Natural Resources Education (ANAFE) - an international organization, based in Nairobi Kenya, between June and December 2015.

Henry Alapiki is a Professor of Political science with specialization in Development studies, Political Economy and Comparative Politics. Professor Alapiki has published widely in National and International journals and served as a resource person and lead researcher for notable international organizations on contemporary Development challenges in Nigeria. He was Dean, Faculty of Social Sciences, University of PortHarcourt and currently Executive Director of University of PortHarcourt Foundation.

Luke Amadi is a Senior Research Fellow at the Educational Support and Development Initiatives for the Less Privileged (ESDIL). His areas of research interest include political ecology, sustainable development, security, political economy and development administration. Since 2009, he has engaged in building the capacities of civil society organizations (CSOs), lecturing and consulting assignments on democracy, good governance and development. He was State Media contact for the UNDP/Rivers State 7th Country Programme and has several publications in scholarly journals.

Pedro Gomes holds a MSc in Environmental Engineering from the Faculty of Sciences and Technology - New University of Lisbon. Currently working in the Department of Economic Statistics - Environment and Agriculture at Statistics Portugal where the main tasks consist in collecting information internally or by external sources in order to prepare and analyse statistics on environment and economics applied to companies in order to provide information with National and International responsibilities. His work is highlighted in the contributions for publications developed in Statistics Portugal regarding Environmental Statistics and Environmental Management and Protection in Industrial Enterprises. He is an element of the Working Group "2030 Agenda for Sustainable Development Goals". He is the author of four chapters for the SPRINGER book "BEACH MANAGEMENT TOOLS: Concepts, Methodologies and Case Studies" edited by Camilo M. Botero, Omar D. Cervantes and Charles W. Finkl.

Charles B. L. Jumbe is a Professor of Economics at the Centre for Agricultural Research and Development (CARD) and the former Director of Research and Outreach at the Lilongwe University of Agriculture and Natural Resources (LUANAR) since 2013. He holds a PhD in Economics from the Norwegian University of Life Sciences (Norway). Before joining the University in 1996, he worked as an Economist in several government departments including the Office of the President and Cabinet in the Department of Economic Planning and Development (1990-1993), Ministry of Energy and Mining (1993-1996) and Ministry of Agriculture and Food Security (1995- 1996). With more than 20 years of research experience, he has worked on a wide range of areas or topics such as renewable and non-renewable energy, biofuels, monitoring and evaluation of development projects, environment and climate change, agriculture, food and nutrition security, natural resource management, forestry, efficient cookstoves, water and sanitation, rural development, poverty and vulnerability assessment, baseline (situational) analysis, mid-term evaluation and impact evaluation (end of project) of development interventions.

Joseph Kanyamuka is a Research and Policy Analyst at the New Alliance Policy Acceleration Support Project (NAPAS:Malawi) based at the Ministry of Agriculture, Irrigation and Water Development. He holds a Master of Science Degree in Agricultural and Applied Economics from the Lilongwe University of Agriculture and Natural Resources. He is a Quantitative Economist. His recent research interests have focused on agricultural productivity and value chain analysis, technology adoption and profitability analysis.

Constant Adeniyi Labintan is from Benin Republic. He is currently consultant at Climate Analytics while supporting the implementation of Climate Change Science Based Impact Assessment project so called PAS-PNA in Republic of Benin. He is also a lecturer in Technical and Economics Modeling of Energy System at the Institute for Renewable Energy (ESMER). With multiple disciplinary backgrounds and international experience in research and in consulting constant has accumulated years of experiences in area of Climate Change impact assessment and Mitigation Pathways Analysis, in Integrated Policy Analysis and have worked across Africa and Asia with UNDP, UNU-INRA, Climate Analytics.

Mahmud Mallam hails from Niger State, Nigeria. He holds an MSc. He holds degree in Agricultural Economics from the University of Nigeria Nsukka. He is currently the Monitoring and Evaluation specialist at the African Development Bank assisted Agricultural Transformation Agenda support program. In recent past, he was the monitoring and evaluation specialist at the World Bank assured Community and Social Development program CSDP Niger State.

Josue Mbonigaba is an Associate Professor in the School of Accounting Economics and Finance at the University of KwaZulu-Natal, Durban, South Africa. His research interests include development economics, health economics, regional economic integration and resource economics. He published widely in health economics in the field of efficiency of public health services, cost-effectiveness, health financing and has acted as a referee in many journals in his areas of expertise. His lecturing experience spans a period of more than eight years.

Edgar Muhoyi was born in Zimbabwe, Africa. He studied BSc. Honors in Economics and His MSc in Economics with the University of Zimbabwe. Currently, Edgar is a PhD student with the University of Kwazulu-Natal. His research interests lie in Development Economics.

Ibukun Joyce Ogwu was born on February, 11th 1984 to the family of Memudu of Odoroko, Ejuku in Kogi State Nigeria. She had her nursery, primary and secondary school education at ECWA George Campion Academy located in Egbe, Kogi State. She completed her B.Agriculture degree in 2006 at the University of Ilorin where she obtained my masters in 2011 and doctoral degree in 2016 respectively. She has been opportuned to work with the World Bank sponsored FADAMA III project and currently, she lectures at the University of Abuja.

Uchechukwu E. Okorie is a Doctoral student in the Department of Economics and Development Sciences, Covenant University, Nigeria. A researcher for the past six years, has conducted and completed research projects, attended international conferences and published over 20 scholarly articles in both local and international journals and conference proceedings.

Steve Adogwu Okpachu hails from Otukpo in Otukpo Local Government Area of Benue State. He is a principal lecturer with Federal College of Education (Technical), Potiskum. Yobe state, Nigeria. He is currently pursuing a doctoral degree in Agricultural Economics at University of Abuja. He has attended several conferences and workshops both local and international. His previous research and publications focused on impact of HIV/AIDS on farm labour productivity and the impact of FDI on agricultural growth and development.

Olubunmi Abayomi Omotesho hails from Kogi State, Nigeria. He was born on 28th May, 1959. He obtained his B.Sc. in Agricultural Economics in 1979, M.Sc. in Agricultural Economics in 1981 and then his Ph.D. in Agricultural Economics in 1991 from the University of Ibadan. He works in the Department of Agricultural Economics and Farm Management, Faculty of Agriculture, University of Ilorin. He joined the Department in 1984 and rose to the post of Professor in the year 2000. He is currently serving as the Director of Academic Planning of the Institution. His research interests spans through areas of agricultural resource economics using irrigation development, food security, poverty issues, resource productivity measures, risk in agriculture among others. He is a member of both local and international professional bodies in the field of agricultural economics. He has supervised and graduated over 20 Ph.D. holders. He is a very hardworking, diligent and amiable researcher and instructor who teaches and guides his students with strong professionalism.

Romanus Anthony Osabohien is a Master degree holder from the Department of Economics and Development Studies, College of Business and Social Science, Covenant University, Nigeria. He has currently concluded his Masters Thesis that centres on Social Protection Policies and Agricultural Performance in Nigeria. The results have been presented in conferences and published in international journals.

Evans Osabuohien is presently the Head, Department (Economics & Development Studies, Covenant University, Ota, Nigeria) where he started lecturing and researching in 2004. He was a research fellow at German Development Institute, Bonn with fellowship from Alexander von Humboldt Foundation (October 2013 –December 2014). He held same fellowship at German Institute for Global and

Area Studies (GIGA), Hamburg (January –September 2013). His other international awards: Swedish Institute's Guest Doctoral Fellowship (as guest Ph.D Candidate at Lund University 2009/2010); Council for the Development of Social Science Research in Africa (CODESRIA)'s grant for Ph.D Thesis in 2009; First Prize for FLACSO-WTO Chair Award in 2012. He currently coordinates the collaboration between Covenant University, Nigeria and German Development Institute, Bonn as well as the cooperation between Covenant University, Nigeria and United Nations Conference on Trade & Development Virtual Institute (UNCTAD Vi), Geneva. He has executed research projects for international organizations such as: UNCTAD Vi; The Centre for Economic and Policy Research (CEPR); African Economic Research Consortium (AERC); Economic Community for West African States (ECOWAS); Trade Policy Training Centre in Africa (TRAPCA), among others. He has research interest in Institutional, International, Agricultural and Land Economics where he has published over 65 scholarly articles and presented research findings in international conferences/workshops in five continents of the world. His edited book 'Handbook of Research on In-country Determinants and Implications of Foreign Land Acquisitions' was published 2015. He serves as reviewer to many international journals including: World Development; African Development Review; Land Use Policy; South African Journal of Economics; Journal of Environmental Policy and planning; Journal of Environmental Planning and Management; The Developing Economies; The Journal of Developing Areas; among others. He is a member of learned societies, namely: Royal Economic Society; Nigerian Economic Society; International Society of New Institutional Economics; African Growth and Development Policy Modelling Consortium (AGRODEP), CODESRIA; AERC; Africa Finance and Economic Association (AFEA); Young African Researchers in Agriculture (YARA), UNCTAD Virtual Institute; among others.

Jacob Ricker-Gilbert holds a Ph.D. in Agricultural Economics from Michigan State University. His M.S. is from Virginia Tech and his B.S. is from University of Vermont. Jacob's research and teaching activities are primarily related to economic growth and poverty reduction in developing countries. Jacob has a special interest in how technology, market access, and public policy affect peoples' decisions and well-being. He has international experience in Latin America, Asia, and Africa. Jacob's dissertation research, evaluating the impact of fertilizer subsidies on small farmers in Malawi, won the 2009 T.W. Schultz award for best paper at the Triennial Meeting of the International Association of Agricultural Economics. His dissertation work was also recognized by the African Association of Agricultural Economists in 2010. Prior to his doctoral studies, Jacob worked as an agricultural economist at the Economic Research Service in the U.S. Department of Agriculture from 2005 to 2007.

Harald Winkler is Director of the Energy Research Centre (ERC) at the University of Cape Town (UCT and has been a member of the SA delegation to the negotiations under the UN Framework Convention on Climate Change and been a lead author for the Intergovernmental Panel on Climate Change's Working Group III on mitigation. Harald's research interests are focused around climate policy, at international and national level. His current focus the future of the climate regime and implementation of the Paris Agreement, Equity and climate change, including work on 'equitable access to sustainable development', equity in nationally determined contributions and 'zero poverty zero emissions' pathways, Sustainable energy paths for South Africa, particularly in energy use and supply, Comparative analysis of mitigation actions in cities and in developing countries.

Index

G

H

I

K

L

M

N

O

P

R

S

U

V

Encyclopedia of
Information Science and Technology
Fourth Edition

www.ingramcontent.com/pod-product-compliance
Lightning Source LLC
LaVergne TN
LVHW080340011026
849161LV00009B/2139
9781522536314